AF682891

METAMORPHIC ROCKS: A CLASSIFICATION AND GLOSSARY OF TERMS

Recommendations of the International Union of Geological Sciences Subcommission on the Systematics of Metamorphic Rocks

Many common terms in metamorphic petrology vary in their usage and meaning between countries, and there is a range of specialized rock names that have been applied locally. The International Union of Geological Sciences (IUGS) Subcommission on the Systematics of Metamorphic Rocks (SCMR) has aimed to resolve this position and to present systematic terminology and rock definitions that can be used worldwide. This book is the result of discussion and consultation lasting 20 years and involving hundreds of geoscientists worldwide.

This volume presents a complete nomenclature of metamorphic rocks based on the recommendations of the IUGS Subcommission. Twelve multi-authored sections explain how to derive the correct names for metamorphic rocks and processes, and discuss the rationale and background behind the more important terms. These sections deal with rocks from high- to low- and very low-grade, including amphibolites, granulites and high-pressure rocks, as well as structural terminology, migmatites, metasomatism, contact metamorphism, metacarbonate rocks and impactites. It also gives a comprehensive glossary of definitions, sources and etymology of over 1100 terms, both those recommended by the Subcommission and those considered redundant, and a list of mineral abbreviations. Less common types of metamorphism such as lightning and combustion metamorphism are also covered.

A companion to *Igneous Rocks: A Classification and Glossary of Terms* (Second Edition, Cambridge University Press, 2002), this book will form a key reference and international standard for all geoscientists studying metamorphic rocks.

DOUGLAS FETTES is an Honorary Research Associate at the British Geological Survey, and Chairman of the SCMR. He was the compiler of the UK section of the Metamorphic Map of Europe and has over 40 years' involvement with structural and metamorphic studies.

JACQUELINE DESMONS, a researcher at the French 'Centre National de la Recherche Scientifique', has dedicated most of her career to the Alps and their metamorphism, including the high *P*/*T* events, metamorphic maps and reviews.

METAMORPHIC ROCKS

A Classification and Glossary of Terms

Recommendations of the International Union of Geological Sciences Subcommission on the Systematics of Metamorphic Rocks

Editors: D. FETTES and J. DESMONS

Contributing authors: P. Árkai, K. Brodie, I. Bryhni, E. Callegari, J. Coutinho, E. Davis, J. Desmons, D. Fettes, R. Grieve, B. Harte, H. Kräutner, N. Pertsev, O. Rosen, V. Rusinov, F. Sassi, R. Schmid, S. Sen, J. Siivola, W. Smulikowski, D. Stöffler, W. Wimmenauer, V. Zharikov

IUGS

CAMBRIDGE
UNIVERSITY PRESS

CAMBRIDGE UNIVERSITY PRESS
Cambridge, New York, Melbourne, Madrid, Cape Town, Singapore, São Paulo, Delhi

Cambridge University Press
The Edinburgh Building, Cambridge CB2 8RU, UK

Published in the United States of America by Cambridge University Press, New York

www.cambridge.org
Information on this title: www.cambridge.org/9780521868105

© Cambridge University Press 2007

This publication is in copyright. Subject to statutory exception
and to the provisions of relevant collective licensing agreements,
no reproduction of any part may take place without
the written permission of Cambridge University Press.

First published 2007
Reprinted 2008

Printed in the United Kingdom at the University Press, Cambridge

A catalogue record for this publication is available from the British Library

ISBN 978-0-521-86810-5 hardback

Cambridge University Press has no responsibility for the persistence or
accuracy of URLs for external or third-party internet websites referred to
in this publication, and does not guarantee that any content on such
websites is, or will remain, accurate or appropriate.

Contents

Figures

Tables

Contributing authors

P. Árkai, Institute for Geochemical Research, Hungarian Academy of Sciences, Budapest, Hungary

K. Brodie, Department of Earth Sciences, University of Manchester, UK

I. Bryhni, Natural History Museum, University of Oslo, Norway

E. Callegari, Department of Science, Mineralogy and Petrology, University of Torino, Italy

J. Coutinho, University of São Paulo, São Paulo, Brazil

E. Davis, Athens University, Greece

J. Desmons, CNRS, Nancy, France

D. Fettes, British Geological Survey, Edinburgh, UK

R. Grieve, Natural Resources Canada, Ottawa, Canada

B. Harte, School of Geosciences, University of Edinburgh, UK

H. Kräutner, LM University, Munich, Germany

N. Pertsev, IGEM, Moscow, Russia

O. Rosen, Geological Institute, Moscow, Russia

V. Rusinov, IGEM, Moscow, Russia

F. Sassi, Department of Mineralogy and Petrography, University of Padova, Italy

R. Schmid, ETH-Centre, Zurich, Switzerland

S. Sen, Indian Institute of Technology, Kharagpur, India

J. Siivola, Department of Geology, University of Helsinki, Finland

W. Smulikowski, Institute of Geological Sciences, Polish Academy of Sciences, Warsaw, Poland

D. Stöffler, Humboldt-University, Berlin, Germany

W. Wimmenauer, Freiburg University, Germany

V. Zharikov, Moscow State University, Russia

Preface

The Subcommission on the Systematics of Metamorphic Rocks (SCMR) is a branch of the IUGS Commission on the Systematics in Petrology (CSP). It started operating in 1985, with Rolf Schmid as chairman, succeeded by Douglas Fettes in 2001. The Subcommission consisted initially of 33 members, distributed in 11 Study Groups devoted to special topics, and a Working Group of more than 100 earth scientists spread worldwide. The Study Groups in addition to Subcommission members also drew membership from appropriate specialists worldwide. The main consultative work of the Subcommission was done initially by correspondence and during annual working meetings. Also, questionnaires were sent to members of the Working Group to improve prepared definitions and test international acceptance. The provisional recommendations were published on the SCMR website and critical comment encouraged. The final results were then drawn up and are now presented in printed form. The Subcommission's work was conducted in English and all its recommendations and definitions are designed only for English language usage. Transposition into other languages may follow.

The SCMR has dealt with all metamorphic rocks. This was taken to include rocks which are quenched melts produced by, or closely associated with metamorphic processes and which are not defined by the Igneous Subcommission. In addition the SCMR has defined a number of structural terms and processes closely associated with metamorphic systems.

The SCMR also includes the systematics of impactites. Although many impactite products and processes are not strictly metamorphic it was considered expedient to deal with the group as an entity, especially as no part of the subject was being considered elsewhere by the CSP.

Although the SCMR has taken every precaution to present a comprehensive and accurate set of recommendations it is inevitable in such a complex subject that omissions and mistakes will exist. Readers are encouraged to make these known to the editors (c/o Cambridge University Press). Constructive comment on how aspects of the nomenclature scheme might be improved are also welcome. See the SCMR website for updates (www.bgs.ac.uk/SCMR).

Lastly it would be wrong to think that these recommendations, even if they were perfect, would represent the final statement on metamorphic terminology. The science continues to evolve as new discoveries are made and new understandings develop: the terminology has to develop in parallel with these changes and it is hoped that the Subcommission will continue its work in this regard.

Acknowledgments

The Subcommission thanks the Chairmen (Peter Sabine, Jörg Keller, Giuliano Bellieni) and members of the Commission for Systematics in Petrology, and the IUGS for support of this work. Special thanks are due to the members of the Working Group (Appendix A.3) whose help and responses to questionnaires provided a view across the international community. Members of the various Study Groups played a critical role in providing expert opinion across the various subjects.

In addition the SCMR would like to thank the following individuals who helped in many different ways:

N. V. Aksamentova, K. Balogh, R. J. Bevins, K. Blanc, D. A. Carswell, H. Day, D. D. Eberl, W. von Engelhardt, V. I. Feldman, N. J. Fortey, B. French, A. I. Grabezhev, E. N. Gramenitskii, I. Herter, F. Hörz, A. Iijima, W. Johannes, M. Kanazirsky, K. Keil, H. J. Kisch, B. Kübler, A. Kunov, D. Laduron, L. Leoni, D. Lieger, F. Lippmann, J. Martini, O. Matejovska, K. R. Mehnert, R. J. Merriman, S. Morad, V. Z. Negrutsa, H. J. Nier, F. Nieto, R. Offler, B. I. Omel'yanenko, W. U. Reimold, B. Roberts, V. T. Safronov, P. Schiffman, K. A. Shurkin, V. Suchy, J. Touret, R. J. Tracy, I. A. Velinov, J. Verkaeren, D. Visconà, S. Vlad, M. Vuagnat, C. E. Weaver and V. L. Zlobin.

The members of the SCMR acknowledge the generous support given by their various home universities and organisations. The work of P. Árkai was supported by the Hungarian National Science Foundation (OTKA, Budapest), project nos. T007211/1993–1996, T022773/1997–2000 and T035050/2001–2004. F. Sassi acknowledges support from the Italian CNR (Instituto di Geoscienze e Georisorse) and MIUR. O. Rosen acknowledges support from the Petrography Committee of the Russian Academy of Sciences and financial support from the Russian Foundation for Basic Research, grants 99-05-65154 and 02-05-64397.

Lastly the editors would like to thank all those colleagues and in particular Arnošt Dudek, who helped in reviewing and checking these texts, and those who, along with many unnamed library staff and researchers, individually sought out and checked well over 1000 references and definitions for the glossary.

1 Introduction

Inzwischen herrscht in den mineralogischen Schriften eine erstaunliche Verwirrung in der Bestimmung dieser Gesteinsarten, und in dem Gebrauche der Benennungen.

Werner, 1786

(Meanwhile there prevails in mineralogical literature an astonishing confusion in the determination of rock types and in the nomenclature.)

This position described by Werner still has a certain resonance today. Despite many excellent works proposing classification schemes for metamorphic rocks, much of the terminology remains ill-defined and ambiguous. Practice differs across the international geological community and even such core words as 'gneiss' and 'schist' vary in their usage and meaning. To some extent this position reflects the fact that metamorphic rocks, and here we include metasomatic and diagenetic rocks, represent a vast range of lithologies and processes, overlapping with the nomenclature of sedimentary rocks on the one hand and with that of igneous rocks on the other. It also involves a range of processes from sedimentation and burial through deep-seated orogenesis to planetary collision.

Against this background the aim of the Subcommission on the Systematics of Metamorphic Rocks (SCMR) is to present a systematic nomenclature and to present an agreed definition of all related terms. The SCMR took its remit to cover all metamorphic rocks including quenched rocks not normally considered as igneous rocks, such as fulgurite, tektite and pseudotachylite. It also covered structural terms including fault structures and the systematics of impactites. The prime objective was to provide a scheme for naming and describing metamorphic rocks; no attempt has been made to cover the terminology relating to the detail or theory behind metamorphic processes, mineral chemistry, graphical presentations, etc.

It was agreed that rock names should, as far as possible, be applicable at the hand-specimen scale, that they should be based on non-genetic criteria and that these criteria should be measurable in the field or under the microscope. Nomenclature based on criteria such as rock chemistry or metamorphic grade was obviously unsuitable and a systematic scheme was devised based on compound names with structural root terms and mineral qualifiers (see Schmid *et al.*, Section 2.1). This scheme allows a systematic name to be given to any rock. However, it was accepted that there were many well-established specific names, such as *marble*, *amphibolite* and *eclogite*, that would have to remain and could potentially be used as alternative names. In addition many metamorphic rocks can be named by reference to their protolith. Thus, metamorphic rocks may potentially have up to three acceptable names, that is a systematic name, a specific name and a protolith-based name, for example *carbonate granofels – marble – metalimestone* or *hornblende-plagioclase schist – amphibolite – metabasalt*. The nomenclature scheme has to take account of this situation.

This book is accordingly structured in two main parts, the first dealing with the classification and nomenclature scheme, the second consisting of a glossary containing comprehensive definitions of all terms related to metamorphic terminology. The first part contains 12 sections: the first deals with the basis of the systematic scheme and how it should be used, the remainder deal with various specialist areas and discuss, as appropriate, the evolution of the terminology in these areas or the rationale that lies behind approved definitions. The second part is the glossary, which consists of around 1100 entries. It gives comprehensive definitions, etymology and source references, and categorizes the terms into three classes, namely: *recommended use, restricted use* and *obsolete* or *unnecessary*.

All the terms were fully researched and the source and significant references individually checked. In addition, all the principal terms were considered by the Study Groups. In some cases this involved intense debate and wide

Metamorphic Rocks: A Classification and Glossary of Terms. Recommendations of the International Union of Geological Sciences, eds. Douglas Fettes and Jacqueline Desmons. Published by Cambridge University Press. © Cambridge University Press 2007.

consultation with the international community before a consensus was reached. It is hoped therefore that it will be accepted that these definitions are designed for international use and may differ in detail from established usage in some areas.

1.1 Guidelines on how to use the book

1. The **systematic scheme for naming a rock** is given in section 2.1. The basis of the scheme is given in subsections 2.1.1 to 2.1.7. The procedure for naming a rock is given in subsections 2.1.8 onwards. This includes a flow chart with accompanying guidelines and a table outlining the use of qualifiers.
2. The **specialist sections** (2.2–2.12) should be consulted for extra information. A list of approved terms (recommended and restricted) arranged by specialist subject is given in section 3.3 of the glossary, and this may be used to find out the **terminology related to a specialist subject**.
3. The **glossary** should be used for the definition, source references etc. of all terms. It contains summary lists of all approved terms arranged alphabetically (section 3.2) and by subject (section 3.3).
4. All **references** cited in the book are in a consolidated list at the end.

2 Classification and nomenclature scheme

2.1 How to name a metamorphic rock

ROLF SCHMID, DOUGLAS FETTES, BEN HARTE, ELEUTHERIA DAVIS and JACQUELINE DESMONS

2.1.1 Introduction

The usage of some common terms in metamorphic petrology has developed differently in different countries and a range of specialized rock names have been applied locally. The SCMR aims to provide systematic schemes for terminology and rock definitions that are widely acceptable and suitable for international use. This first section explains the basic classification scheme for common metamorphic rocks proposed by the SCMR, and lays out the general principles which were used by the SCMR when defining terms for metamorphic rocks, their features, conditions of formation and processes. Subsequent sections discuss and present more detailed terminology for particular metamorphic rock groups and processes.

The SCMR recognizes the very wide usage of some rock names (for example, amphibolite, marble, hornfels) and the existence of many name sets related to specific types of metamorphism (for example, high P/T rocks, migmatites, impactites). These names and name sets clearly must be retained but they have not developed on the basis of systematic classification. Another set of metamorphic rock names, which are commonly formed by combining mineral names with structural terms (for example, quartz-mica schist, plagioclase-pyroxene granofels) is capable of being used in a systematic way. The SCMR recommends that such compound names are systematically applied using three root names (schist, gneiss and granofels), which are defined solely by structural criteria. Such systematic names are considered particularly appropriate when specific names are unknown or uncertain. A flow chart on 'How to name a metamorphic rock' enables any earth scientist to assign a name to a metamorphic rock, following this scheme. The section further gives guidelines on the appropriate use of these systematic names and on the use of possible alternatives based on the protolith and other specific names.

PRINCIPLES OF NOMENCLATURE

A nomenclature scheme consists of defined terms and the rules governing their use. In erecting a nomenclature scheme the SCMR was guided by the following underlying principles.

(a) The scheme must provide a consistent set of names to cover the spectrum of rock types and their characteristics without any terminology gaps.
(b) The scheme must ensure that all users can apply the same criteria to give any rock or its characteristic features the same name. These names should be understood uniquely and without ambiguity.

In any system of nomenclature a number of characteristic features or parameters are used to divide rocks into groups or sets, and the criteria for such divisions or subdivisions are fundamental to the terminology. The SCMR decided (see Schmid & Sassi, 1986) that the above principles would only be fulfilled if the criteria for any specific division/subdivision were defined using only one type of characteristic feature. For example, the criterion for a specific division/subdivision might be a particular feature of mineral content or structure, but it should not be both mineralogical and structural. In a series of divisions/subdivisions in a classification scheme, structure and mineral content may be applied at different stages, but they should not be applied simultaneously.

At a given stage of division/subdivision a set of rock groups may be recognized in a classification scheme, and these will be given group names (or **root names** in the case of major divisions).

Metamorphic Rocks: A Classification and Glossary of Terms. Recommendations of the International Union of Geological Sciences, eds. Douglas Fettes and Jacqueline Desmons. Published by Cambridge University Press. © Cambridge University Press 2007.

Such names form a fundamental element of the classification. The development of a nomenclature scheme in this way follows that used for the classification of igneous rocks (Le Maitre, 1989, 2002).

One of the main purposes of this section is to propose that a simple but comprehensive terminology for common metamorphic rocks may be based on their division into three major groups on the basis of their structure (as seen in hand specimen). These three groups are given the structural root names: **schist**, **gneiss** and **granofels**. In conjunction with the recognition of a systematic terminology of this type, the SCMR has also recognized a number of non-systematic names or *specific names*, which may be used as alternatives to the systematic names or to impart additional information. A flowchart and guidelines for the use of the nomenclature scheme are presented below.

2.1.2 Potential bases for the classification of metamorphic rocks

Ignoring characteristics like magnetic or electrical properties or age, which can rarely be determined or even inferred without special equipment, the major features of metamorphic rocks that can be widely used for classification are:

(a) the minerals present
(b) the structure of the rock
(c) the nature of the rock prior to metamorphism
(d) the genetic conditions of metamorphism (usually in terms of pressure and temperature, with or without deformation).
(e) the chemical composition of the rock.

Of the above, (a) and (b) form the most obvious major parameters for rock classification or nomenclature, and would also often be involved indirectly in classifications based on (c) and (d). The variety of minerals present would necessarily also provide much basic information for (e) if this were not to depend on the use of specialized techniques for chemical analysis.

Examination of metamorphic rocks shows great mineralogical, structural and chemical diversity. Their chemical and mineralogical diversity results in a large part from the fact that they may be formed from any pre-existing igneous or sedimentary rock. Added to this diversity of rock types subjected to metamorphism, there are wide variations in the conditions (temperature, pressure, deformation) of metamorphism itself; and as a consequence the metamorphic rocks derived from only one igneous or sedimentary precursor may show an extensive range of mineral assemblages and structures.

In contrast to igneous rocks, the large range in mineral content and chemistry for even common metamorphic rocks means that schemes of classification cannot be devised using a small number of parameters. Thus there are no simple metamorphic equivalents to classification plots based on SiO_2 vs. $Na_2O + K_2O$, or quartz, feldspar, feldspathoid ratios, as used by igneous petrologists (e.g. Le Maitre, 1989, 2002). The only way of reducing the number of mineralogical variables in metamorphic rocks to a small number of defining parameters is by inferring conditions of genesis (usually pressure–temperature conditions of formation). The metamorphic facies classification is very useful in this context, but assignment of facies to specific genetic conditions (e.g. pressure and temperature) rests on a number of assumptions and is susceptible to changes in knowledge and understanding. It also essentially ignores the structure of the rocks concerned. Furthermore, although facies terms are based on mineralogical changes, they do not imply that rocks of all chemical compositions have different mineral assemblages in each facies; nor do they imply that rocks of a particular chemical composition must have constant mineral content within a particular facies. Thus a facies terminology does not match one-to-one with the actual mineral assemblages seen in all rock compositions.

Following the precedents set by most other rock classification schemes, the SCMR decided, therefore, that the most comprehensive and applicable nomenclature scheme should be based on the following two principles:

1. Metamorphic rocks should be named, in the first instance, on **directly observable features, preferably at the mesoscopic scale, but where necessary at the microscopic scale**. (Thus the

definitions of rock terms recommended by the SCMR refer, as far as possible, to features observable in hand specimen, making allowance for the possible need for microscopic examination in some cases.)

2. **Genetic terms should not be the basis of primary definition of rock types**. (Genetic terms are clearly useful in genetic discussions, but should only be applied to a rock if the genetic process concerned is clearly defined and criteria for its recognition are clearly stated.)

The directly observable features of all rocks are their mineral content and structure. These have been the basis for common rock names in the past and, following the principles given above, are the primary basis for the metamorphic rock names recommended by the SCMR. (In some instances this allows for the use of a protolith term in describing and defining metamorphic rocks; see below.)

2.1.3 Previous terminology largely based on mineralogical and structural characteristics

COMPOUND NAMES

Metamorphic petrologists have traditionally coped with the variety and complexity of mineral content and structure, as outlined above, by using a series of compound hyphenated names (e.g. quartz-mica schist, lawsonite-glaucophane schist) in describing metamorphic rocks. The final or root word in such names may be based on structural, mineralogical or protolith characteristics (e.g. garnet-mica-quartz *schist*, garnet-biotite *amphibolite*, garnet-pyroxene *metabasic rock*, respectively), and the mineralogical prefixes provide further information on the mineral content of the rock being described. These compound terms have provided for an immense flexibility of description and naming of metamorphic rocks, and the SCMR has seen them as having considerable merit. However, their widespread usage has not usually been systematic, and the SCMR strongly recommends that they should now be used in a systematic way (see below) to provide a wide-ranging system of nomenclature for metamorphic rocks in general.

SPECIFIC ROCK NAMES AND NAME SETS

The existing terminology for metamorphism and metamorphic rocks includes many names based on specific mineralogical and/or structural and/or other criteria. These have been called *specific names* by the SCMR. Such names usually have very precise connotations, but have not been developed in a systematic way to embrace the whole range of metamorphic rocks: the exception being the metamorphic facies classification that, as discussed above, is not appropriate for a descriptive rock nomenclature.

Some of these specific names have become extremely widely used for common rock types. Examples of such terms are: amphibolite (for rocks largely made of amphibole and plagioclase); quartzite (in which quartz is by far the major constituent); marble (in which carbonate minerals predominate); slate (for a fine-grained rock with a well-developed regular fissility or schistosity). Amphibolite and slate illustrate names based on mineral content and structure respectively. The terms quartzite and marble are essentially mineralogical, as indicated, but it has also often been assumed that such rocks have equigranular or granofelsic structures.

Most of the specific terms including some of those just mentioned may be subdivided into groups associated with individual types of metamorphism (high P/T metamorphism, impactites, fault and shear rocks, migmatites, carbonate rocks, etc.). These groups have been called *specific name sets* by the SCMR. Many of the names making up these sets have a connotation for the context or genesis of the rock (ultramylonite, anatexite, skarn, etc.) and may provide important detail or additional information on these features.

As such, these specific terms are a fundamental part of metamorphic nomenclature. However, from the viewpoint of the development of an ordered system of classification, and the guiding principles outlined above, specific terms present a major problem. Specific names have not been developed into a general systematic framework that embraces the whole range of metamorphic rocks, even though some name sets, related to types of metamorphism, may possess a systematic structure – for example mylonites, which may be subdivided into protomylonite, mesomylonite, ultramylonite, etc. Despite this lack of systematization, it has to be recognized that the specific terms are an integral part of metamorphic terminology and that allowance for their use has to be made in any scheme of common nomenclature. This fact has been recognized by the SCMR which has attempted to produce a definitive list of specific names and has set up guidelines on their use (see below).

PROTOLITH NAMES

Metamorphic rock names based on protoliths (the lithological compositions of rocks before metamorphism) are very useful for two reasons:

(a) Determination of the original nature of the rock is often a fundamental consideration in establishing geological history.
(b) In weakly metamorphosed rocks and particularly those subjected to little deformation, the structural and mineralogical features of the protoliths may still be the principal observable features.

In the second instance, use of a protolith-based name may be a more appropriate name for a rock than one emphasizing metamorphic characteristics. Metaconglomerate (for a metamorphosed conglomerate) is an obvious example where the structure of the protolith is usually still the most obvious characteristic of the rock (and in any case the metamorphic mineral content of such a rock will change with the bulk chemical composition of each pre-existing clast).

The use of protolith names in the nomenclature of metamorphic rocks is very straightforward, and largely consists of prefixing the name of the protolith with 'meta' or 'meta-' (e.g. metagranite, metabasalt, meta-arkose). As we have seen, protolith terms may be used in compound names and carry mineralogical prefixes (e.g. biotite metasandstone, garnet metabasalt) or structural prefixes or qualifiers (e.g. schistose garnet metabasalt).

Protolith-based names are clearly useful in cases where the characteristics of the metamorphic rock largely reflect those of the protolith and the nature of the protolith can be fully determined.[1]

However, this is usually only the case in rocks of low metamorphic grade and/or those that have been only weakly deformed. In most metamorphic rocks applying the protolith name is not a matter of direct observation but is a matter of inference after its mineral content and microstructure have been taken into account, with the mineral content serving as a guide to bulk chemical composition when a chemical analysis is not available. Thus in many cases protolith names do not reflect the principal minerals and structural features of the rocks under observation.

It follows that although protolith terminology for metamorphic rocks is clearly very useful and straightforward, and the SCMR recommends its continued usage (see below), it provides a poor basis for a comprehensive and mainly descriptive terminology.

2.1.4 Systematic classification scheme using root names

The sets of names referred to above clearly provide a means for naming metamorphic rocks and allow for flexibility in nomenclature which is necessary given the diverse structural, mineralogical and protolith (chemical) nature of metamorphic rocks as a whole. **However, none of them in their present form provides for a systematic classification of common metamorphic rocks using a simple set of criteria**.

To tackle this problem, the SCMR suggests the adoption of a standard procedure for applying compound hyphenated names. As discussed above, this type of name allows for considerable flexibility, but the final or root term may be based on diverse criteria. Standardization on the basis of mineral content is impossible without a huge array of root names, but standardization

[1] It is important in using names such as metabasalt and metagabbro that the grain-size criteria of the protoliths can be fully established.

on the basis of structural terms using a single criterion can be achieved quite simply.[2]

USE OF THE TERMS SCHIST, GNEISS AND GRANOFELS

Following widespread usage in the English language, three terms essentially cover the principal varieties of structure found in metamorphic rocks, particularly as seen in hand specimen (and therefore easily applicable). These three terms are *schist, gneiss* and *granofels*. The SCMR proposes that these terms are used as the fundamental *root terms* in the adoption of a systematic terminology. It is proposed that these terms have **only a structural connotation, with no mineralogical or compositional implication**.[3] Essentially the terms reflect the degree of fissility or schistosity shown by the rock. Their definitions (see also glossary and Brodie *et al.*, Section 2.3) derive from the recommended SCMR definition of 'schistosity', which is as follows.

Schistosity: A preferred orientation of inequant mineral grains or grain aggregates produced by metamorphic processes. A schistosity is said to be *well developed* if inequant mineral grains or grain aggregates are present in a large amount and show a high degree of preferred orientation, either throughout the rock or in narrowly spaced repetitive zones, such that the rock will split on a scale of less than one centimetre. A schistosity is said to be *poorly developed* if inequant mineral grains or grain aggregates are present only in small amounts or show a low degree of preferred orientation or, if well developed, occur in broadly spaced zones such that the rock will split on a scale of more than one centimetre.

Thus, according to the SCMR scheme, if the schistosity in a metamorphic rock is well developed, the rock has a *schistose structure* and is termed a *schist*. If it is poorly developed, the rock has a *gneissose structure* and is termed a *gneiss*, and if schistosity is effectively absent the rock has a *granofelsic* structure and is termed a *granofels*.

It should be noted that each of these structural root terms will cover a number of specific rock names. Thus, the term 'schist' encompasses a number of names for rocks that possess a well-developed schistosity (as defined), for example, slate and phyllite. Similarly, the term granofels encompasses subsidiary names for rocks in which schistosity is essentially absent, for example hornfels.

Note: see below for the relative use of systematic names and specific names.

GENERAL PROCEDURE FOR NAMING A ROCK USING STRUCTURAL ROOT TERMS

In the system advocated by the SCMR the above fundamental or root terms (based on structure alone) are placed at the end of compound hyphenated names of the type described previously. The considerable diversity of mineralogical names found in metamorphic rocks can then be conveyed by the use of mineral names as prefixes to the root structural term (for example, staurolite-mica-quartz schist, plagioclase-pyroxene granofels, garnet-hornblende-plagioclase gneiss), the

[2] For the purposes of this discussion the term 'structure' refers to mesostructure or the structure of a rock at hand-specimen scale.

[3] The SCMR recognized that the use of the term 'gneiss' in a purely structural sense might prove a difficult concept to some geologists. For example, although the name has evolved in English language usage to imply a type of structure, for many non-English users the name also has mineralogical implications, in particular the presence of feldspar ± quartz. However, the suitability of the name as a structural root term to denote a poor fissility was very attractive and the SCMR decided to accept the English language meaning. This decision was taken after inquiry among the Working Group members, partly on the basis that the SCMR's recommendations were being made for English language use only, and also, critically, it was noted that all rocks currently considered as 'gneisses' would still be defined as such. A complementary concern was that a purely structurally based definition should not include rocks that in established practice could never be considered as gneisses, for example finely banded metasandstones and metamudstones at low/medium metamorphic grade. Although the SCMR accepted this difficulty, it was felt that in practice an adequate guideline could be provided to encourage the use of protolith-based terms to cover these limited cases. These points and the evolution of the terms 'schist' and 'gneiss' are further discussed in the paper on structural terms (Section 2.3), to which the interested reader is referred.

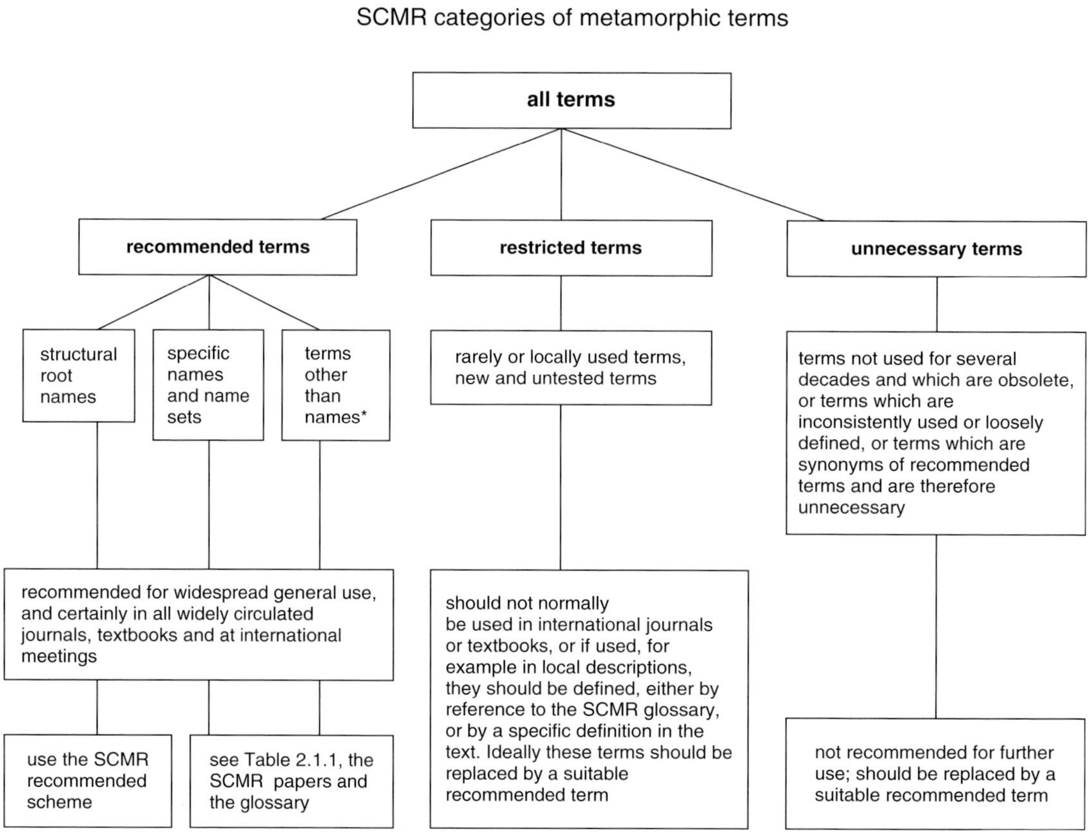

Fig. 2.1.1 Categories of rock terms as defined by the SCMR. The classification of a term as 'recommended', 'restricted' or 'unnecessary' is given in the SCMR glossary. * 'Terms' comprise the vocabulary of metamorphic nomenclature; 'names' are those terms used for the types of rock; 'terms other than names' therefore comprise all adjectives, process terms, etc.

mineral names being arranged in order of increasing modal abundance (see below).

Thus any metamorphic rock may be named by using one of the three terms to convey the basic structure, whereas the mineralogical features are given by prefixing the structural term with the names of the appropriate mineral constituents.

This nomenclature scheme for metamorphic rocks is set out in the lower part of the flowchart in Fig. 2.1.2. A compound hyphenated name of the type recommended may always be applied and allows a systematic set of names for petrographic descriptions. The only complexity to this simple scheme is the need to allow for the use of the specific names and name sets, described above, which have widespread usage.

2.1.5 Categories of rock terms: existing and proposed

In parallel with adopting the structural root name system as a comprehensive nomenclature, the SCMR has examined and categorized all the rock terms used in metamorphic nomenclature. Three classes of terms are recognized, namely 'recommended terms', 'restricted terms' and 'unnecessary terms' (Fig. 2.1.1). The **recommended rock names** are the basis of the SCMR nomenclature scheme.

They comprise the systematic **root names** and a comprehensive range of **specific names**.

THE RECOMMENDED SPECIFIC NAMES

In selecting the recommended specific names and name sets for use in the nomenclature scheme the SCMR relied on the work of its various Study Groups who established and defined the specific names and sets of names for their respective subjects. The conclusions of the Study Groups are contained in the following sections, which form part of the products of the SCMR. These sections are an essential element of the nomenclature scheme: they contain a range of terms related to their area of study (e.g. the specific name sets), background information on the terms, and figures and subsidiary flowcharts.

The recommended specific names range from particularly well-established terms for common rock types (e.g. amphibolite, marble, eclogite) to terms that describe relatively uncommon rock types or features of rocks (e.g. arterite, mesocataclasite). The latter are most likely to be used to give information when the context of the rock is known, whereas the former names may provide concise and widely acceptable alternatives to the structural root names (e.g. marble in place of calcite granofels).

Examples of well-established specific names, which may commonly be given preference over the equivalent structural root names, are listed in Table 2.1.1. The list is presented for information only; it is not intended to be exhaustive.

Table 2.1.1 *Examples of some of the most common specific names (definitions and full list of names by subject are given in the glossary).*

Such names would commonly be given precedence over the equivalent structural root names: see text for discussion.

Amphibolite	Greenschist	Phyllite
Calc-silicate rock	Hornfels	Quartzite
Cataclasite	Marble	Serpentinite
Eclogite	Migmatite	Skarn
Granulite	Mylonite	Slate

As discussed above, the specific names may also be grouped into *specific name sets* (e.g. migmatites, fault rocks) linked to individual types of metamorphism (see the other sections in this volume).

2.1.6 The SCMR glossary

During the course of its work the SCMR has sought to compile a comprehensive glossary of all metamorphic rock terms, structural terms and a few process-related terms, which will hopefully be of international usefulness. The list contains about 1100 entries. Each entry gives the approved SCMR definition, the first usage wherever possible, the etymology and the categorization of the terms as 'recommended', 'restricted' or 'unnecessary'. The basis of the categorization of the terms is given in Fig. 2.1.1. That is, 'recommended terms' are those that are required for an internationally applicable nomenclature; 'restricted terms' are those that are only used locally or rarely and require further definition if used; and 'unnecessary terms' are those that are no longer required.

2.1.7 Recommended guidelines for naming a rock

The procedure for giving a systematic name to any metamorphic rock, based on structural root terms, is given above and in the lower part of Fig. 2.1.2 starting with step 3. It is understood that this process does not encompass the use of specific rock names, which form an important aspect of the overall nomenclature scheme and which is outlined in the upper part of the flowchart, starting at step 1. Specific names may commonly provide a more concise, refined and detailed terminology than is available with the systematic structural root terms. In addition, it is recognized that under particular circumstances a protolith name may be the most descriptive name for a metamorphic rock.

It follows from these points that a single metamorphic rock may have up to three correct names, that is, a protolith, non-systematic/specific and systematic/structural root name

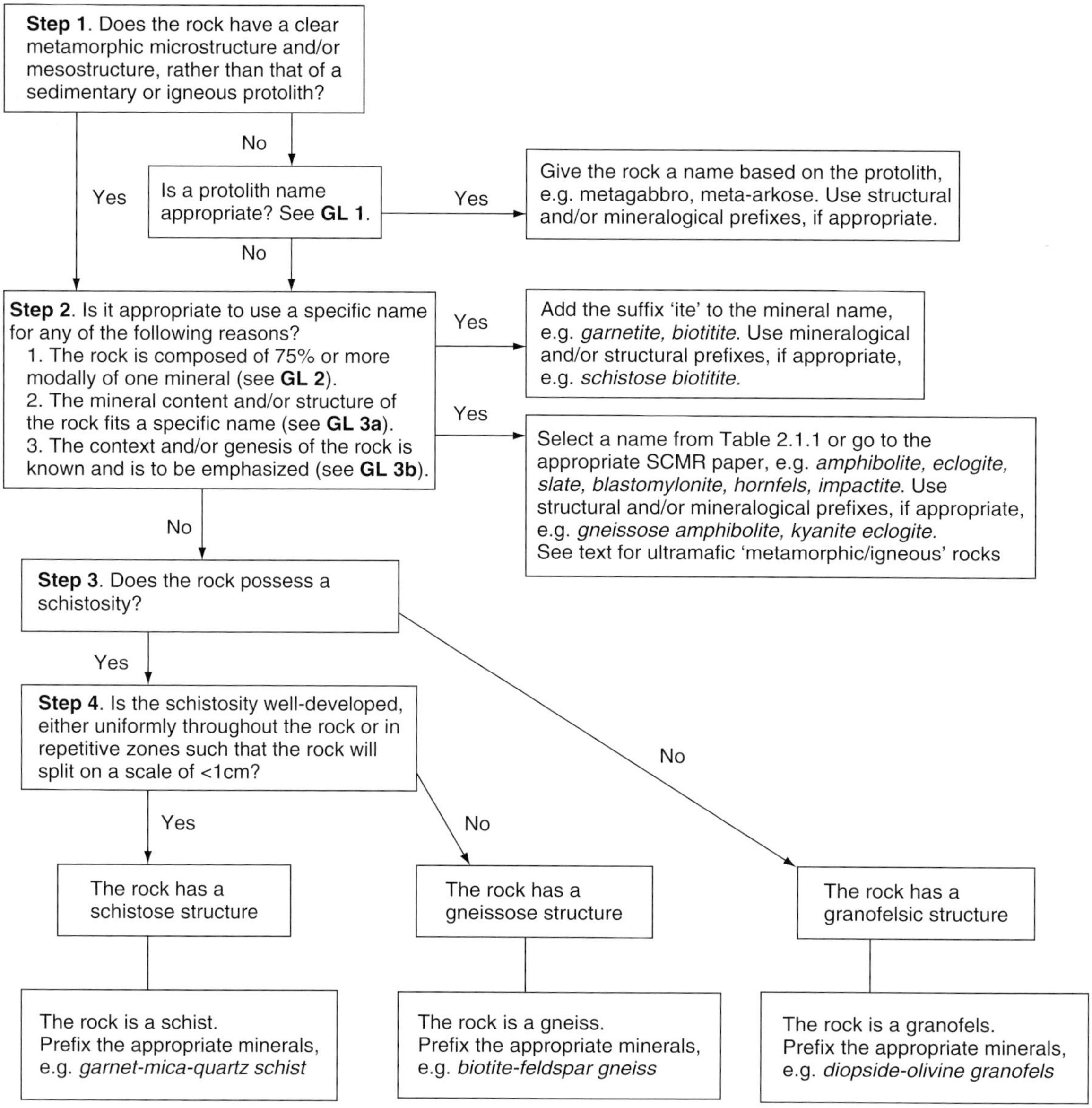

Fig. 2.1.2 How to name a metamorphic rock.
The upper part of the chart (steps 1 and 2) outlines the procedure for deriving a non-systematic name; the lower part (steps 3 and 4) outlines the procedure for deriving a systematic structural root name (use only the lower part to go directly to a structural root name).

(e.g. metabasalt, amphibolite, hornblende-plagioclase gneiss or metalimestone, marble, calcite granofels). The choice of which name to use depends partly on the information available to the user and partly on which aspect the user may wish to emphasize.

It is an underlying principle of the SCMR recommendations that the systematic nomenclature scheme exists in parallel with the use of non-systematic names (specific names, protolith-based names, etc.). The systematic nomenclature scheme is intended to complement, not replace, the non-systematic names. Well-founded non-systematic names will generally take precedence over systematic alternatives: systematic names will generally be appropriate when there is no suitable non-systematic name or the user is unsure or unaware about the applicability of a non-systematic name. Guidance on when to use a systematic or non-systematic name is given in the following subsection.

2.1.8 General procedure for naming a rock

The procedure for deriving an appropriate name is presented in Fig. 2.1.2. At several points the user has a choice. The first two steps relate to the use of non-systematic names (specific names, protolith-based names and names using the suffix 'ite'), the next two steps relate to the choice of the correct systematic name. At each of the first two steps the default position will direct the user towards the systematic structural root name, so that if the user is uncertain about any choice, they will always end up with a structural root name. Alternatively, of course, the user may proceed directly to the procedure for deriving the structural root name. In making the choices with the comprehensive flowchart (Fig. 2.1.2) some simple guidelines may be followed.

GL 1. If the rock features are dominated by those of the protolith or the **protolith** may be determined by the context of the rock,[4] then a protolith name may be applied. Protolith-based names are particularly recommended for weakly metamorphosed rocks, especially where the use of a structural root name would be considered contrary to established practice for example, with a metamorphosed sandstone the name 'biotite-quartz-feldspar metasandstone' should take precedence over 'biotite-quartz-feldspar gneiss (or granofels)'.

GL 2. If the rock contains ≥75% modally of one mineral then it may be named by adding the **suffix 'ite'** to the dominant mineral (for example, biotitite, epidotite, glaucophanite).

Note: there are several exceptions to this guideline, namely: *amphibolite*, this name refers to an amphibole + plagioclase rock (see glossary); *hornblendite* and *pyroxenite*, these names have been defined by Le Maitre (1989, 2002) as igneous rocks with ≥90% modal content of hornblende and pyroxene respectively, the SCMR recognizes that such rocks may occur as a result of metamorphic processes, however to avoid confusion the SCMR has adopted the same definition and modal values as Le Maitre (1989, 2002); *olivinite* and *plagioclasite* are reserved for igneous rocks (see Le Maitre, 1989, 2002) and should not be applied to metamorphic rocks. The suffix 'ite' should not be applied to *calcite, aragonite* or *dolomite*, because a rock consisting of 75% of any of these minerals is a marble and should be named as such (e.g. calcite marble).

GL 3a. If the rock fits the definition of one of the well-known and commonly used **specific names** then it is generally appropriate to use that specific term (e.g. amphibolite, eclogite, marble, slate, calc-silicate rock; see also Table 2.1.1 and glossary lists). There is no absolute rule on when to use or not to use a specific name. However a specific name will generally take preference over the equivalent systematic/structural root name if the specific name is well established or understood or if it is more concise or gives greater detail than the systematic alternative (e.g. marble rather than calcite granofels, amphibolite rather than hornblende-plagioclase granofels, slate and phyllite as types of schist). Conversely, a systematic name is more appropriate where there is no specific name or a possible specific name is little used, ambiguous or poorly defined.

GL 3b. If the context or genesis (that is, the metamorphic processes forming the rock) of the rock is known and particularly if it is desirable to emphasize this or give additional or detailed information about the context or genesis of the rock then the appropriate **specific name** should be used (e.g. nebulite, blastomylonite, tektite, hornfels). In this

[4] It is generally inappropriate to apply a protolith-based name to a rock at the hand-specimen scale if the rock does not clearly exhibit definitive features of the protolith.

case the names should conform to those in the glossary.

The specific and protolith names may carry mineralogical prefixes as outlined below and/or structural prefixes or qualifiers (see also Table 2.1.2) (e.g. garnet amphibolite, schistose marble, pyroxene-biotite amphibolite with gneissose structure).

USE OF 'METAMORPHIC/IGNEOUS' ULTRAMAFIC TERMS

Ultramafic rocks containing olivine, and/or pyroxene and/or hornblende such as peridotite, harzburgite, lherzolite, wehrlite, websterite, pyroxenite and hornblendite may be formed by either metamorphic or by igneous processes and therefore fall in the common ground between metamorphic and igneous terminology. The SCMR recommends that for these rocks the definitions, based on mineral content, as given by Le Maitre (1989, 2002) should be used. These definitions are adopted without any implication to the rock genesis. When garnet or other major or minor minerals are present they should be indicated by the appropriate prefix. If it is desirable to emphasize the metamorphic nature of one of these ultramafic rocks then this should be specifically stated. Alternatively, a structural root name may be given (e.g. pyroxene-olivine gneiss, ultramafic garnet-pyroxene granofels).

Note: under the SCMR rules, terms such as metaperidotite imply that the protolith was a peridotite, and do not make any statement about the present mineral content or structure of the rock.

USE OF PROTOLITH NAMES

When protolith names are used they should be prefixed with 'meta'. Care should be taken to ensure that the protolith name conforms to internationally applicable standards. For igneous rocks such standards are defined by Le Maitre (1989, 2002). To date the IUGS has not published recommendations on the nomenclature of sedimentary rocks; users should therefore consider giving the source reference for any sedimentary protolith name. The prefix 'meta' should never be used for former metamorphic rocks (e.g. meta-eclogite is not an acceptable term, see Table 2.1.2).

USE OF THE TERMS PELITE, PSAMMITE AND PSEPHITE

The terms pelite, psammite and psephite are generally considered to be sedimentary terms indicating increasing grain size and synonymous with lutite, arenite and rudite respectively, and broadly equivalent to mudstone, sandstone and conglomerate (e.g. Tomkeieff, 1983; Bates & Jackson, 1987). However, Tyrrell (1921) proposed that the terms pelite, psammite and psephite should be used for the metamorphosed equivalents of the sedimentary rocks. Although psephite is now obsolete the terms pelite and psammite still persist locally, at least in English literature, as terms for mica-rich and quartz-feldspar-rich metamorphic rocks respectively (e.g. Robertson, 1999). In addition, the term (meta)pelite has developed as a broad compositional term for an alumina-rich metasedimentary rock (e.g. Bowes, 1989a; Barker, 1990).[5] Thus the terms pelite and psammite have a grain-size connotation for sedimentary rocks and a more localized mineralogical and chemical (in the case of pelite) connotation for metamorphic rocks. Given this ambiguous usage, at least in English language literature, the SCMR recommends that the use of the terms pelite and psammite is restricted to sedimentary rocks and when an indicator of mineralogical or chemical composition in metamorphic rocks is intended the names metapelite and metapsammite may be used. However, whenever practical, the terms metamudstone and metasandstone should be given in preference.

USE OF MINERAL PREFIXES

The following rules have been set up by the SCMR.

[5] Note: the sedimentary term pelite may be taken to comprise all fine-grained sediments. As such it covers a broad range of rock chemistry including, for example, calcilutites. However, when used to imply a chemical composition the metamorphic term (meta)pelite has developed a more restricted meaning for rocks characteristically derived from alumina-rich sediments.

All the *major mineral constituents* (see Table 2.1.2) that are present in a rock should be prefixed. The prefixes should be hyphenated and placed in order of increasing abundance. For example biotite-quartz-plagioclase gneiss contains more plagioclase than quartz and more quartz than biotite. However mineral constituents whose presence is inherent in the definition of the rock, that is '*essential constituents*' (see Table 2.1.2), should not be added to the name (cf. garnet amphibolite and hornblende-plagioclase-garnet granofels).[6]

If *minor constituents* (see Table 2.1.2) are named, the form 'mineral'-bearing should be used and placed at the beginning of the name (e.g. rutile-bearing biotite-quartz-plagioclase gneiss). If more than one minor constituent is named, the names should be arranged in order of increasing modal abundance, for example, rutile-ilmenite-bearing quartz-plagioclase gneiss, where rutile is less abundant than ilmenite.

Prefixing of minor constituents is optional but is recommended for *critical mineral constituents* (see Table 2.1.2), conveying particular information on conditions of metamorphism.

If reference is made to a rock body in which some mineral constituents are not present throughout, the form ± may be used and placed at the end of the prefixes or after the rock name; in these cases no relative modal abundances are implied (for example quartz-biotite-plagioclase ± muscovite schist, amphibolite ± garnet).

Because the structural root terms used in the SCMR nomenclature scheme have no mineralogical implication the list of major constituents in a rock may be extensive (e.g. quartz-feldspar-staurolite-kyanite-biotite gneiss). The SCMR has therefore established a list of abbreviations of mineral names (Siivola & Schmid, Section 2.12) for the purpose of shorter prefixing of such names. It is recommended that these abbreviations are also used for other purposes, for example when writing chemical reactions and for inserting mineral names into figures, diagrams and tables.

[6] It may be appropriate to name an essential constituent if the mineral name is more specific than that given in the definition. For example, andesine amphibolite is an acceptable name even though plagioclase is an essential constituent of amphibolite.

USE OF OTHER QUALIFIERS, PREFIXES AND SUFFIXES

The more general qualifiers, prefixes and suffixes recommended by the SCMR are given in Table 2.1.2. More qualifiers and descriptive terms are given in the various specialist papers.

USE OF ACID, BASIC, METABASIC, MAFIC, ULTRAMAFIC, ETC.

The terms acid, basic, ultrabasic refer to the chemical composition of a rock (Table 2.1.2), as defined, for example, by Le Maitre (1989, 2002). Normally in metamorphic rock nomenclature the terms are only used to indicate the chemical composition of the protolith. Under the SCMR rules terms such as metabasic rock or metabasite describe a rock of basic chemistry that has been metamorphosed. Such terms give no indication of the current structure or mineral content of the rock.

The terms felsic, mafic, ultramafic refer to the relative content of felsic and mafic minerals in a rock (Table 2.1.2). If it is desirable to indicate that the rock is metamorphic then constructions such as mafic metamorphic rock, ultramafic metamorphic rock should be used. Terms such as metamafitite, meta-ultramafite, etc., which are ill defined, should be avoided. The adjectival terms may be used with compound or specific names, for example, mafic quartz-feldspar-biotite schist, ultramafic garnet-pyroxene granofels.

RETROGRADE OR RELICT MINERALS

For practical purposes, it is accepted that rock names that are defined on the basis of a characteristic diagnostic mineral assemblage may also be used for rocks that contain small amounts of retrograde or relict minerals not fitting into the definition. On the other hand the presence of small amounts of retrograde or relict minerals should not be reflected by the main name even if they are critical (see Table 2.1.2).

Table 2.1.2 *Qualifiers, prefixes and suffixes used and recommended by SCMR.*

Term	Usage
grain-size terms	
phaneritic	Individual grains visible with the unaided eye (*c*. >0.1 mm)
aphanitic	Individual grains not visible with the unaided eye (*c*. <0.1 mm)
	The SCMR decided, following extensive discussions, not to recommend absolute grain-size values for the expressions 'coarse-grained', 'fine-grained', etc. This decision reflected the feeling in the earth science community that there was currently no common standard for grain-size classification covering igneous, metamorphic and sedimentary rocks and that if fixed values were recommended the methods of grain-size measurement would also have to be defined. If absolute values are required then the most favoured values are: >16 mm: very coarse-grained; 16–4 mm: coarse-grained; 4–1 mm: medium-grained; 1–0.1 mm: fine-grained; 0.1–0.01 mm: very fine-grained; <0.01 mm: ultra-fine-grained. However, if this scale is used the fact should be specifically stated.
micro, micro-; meso, meso-; mega, mega-	Prefixes indicating that a feature is visible only at thin section (microscopic), hand specimen (mesoscopic), or outcrop or larger (megascopic) scale respectively.
	When used in conjunction with a rock name, mineral type, etc., the prefix implies that the object is unusually large or small compared with the standard for such objects (e.g. megacryst = crystal of much larger size than the other crystals in a rock, microtektite = tektite with a smaller grain size than most other tektites).
acid, intermediate, basic, ultrabasic	Terms defining the chemical composition of rocks based on SiO_2 wt%. The terms have been defined for igneous rocks by Le Maitre (1989, 2002) as acid $>63\%$, intermediate 52–63%, basic 45–52%, and ultrabasic $<45\%$ (all SiO_2 wt%).
colour terms for minerals and rocks	Because of the greater variety of mineral colours present in metamorphic rocks compared with igneous rocks, the SCMR recommends that the terms leucocratic, mesocratic and melanocratic are **not** used to indicate the colour of metamorphic rocks (cf. Le Maitre, 1989, 2002). For metamorphic rocks the SCMR recommends that simple terms such as light-, intermediate-, dark-coloured are used. However, the SCMR recommends the use of the following colour prefixes (following Le Maitre, 1989, 2002):
leuco-	prefix indicating that a rock contains considerably less coloured minerals than would be regarded as normal for that rock type.
mela-	prefix indicating that a rock contains considerably more coloured minerals than would be regarded as normal for that rock type.
%	percent by volume (if not otherwise specified).
$\pm$, +/−	Symbols to indicate that minerals are either present in variable, undefined quantities, or absent. For example, muscovite-biotite-quartz-plagioclase $\pm$ kyanite $\pm$ garnet schist (gneiss) indicates a schist (gneiss) that may contain kyanite and/or garnet.
'. . . ite'	Suffix added to a mineral name to generate a rock name where the rock contains $\geq 75\%$ modally of that mineral (for example garnetite, epidotite). The suffix should not be added to dolomite, calcite or aragonite. The following rocks are defined differently: amphibolite, hornblendite, pyroxenite, olivinite, plagioclasite and carbonatite.

Table 2.1.2 (*cont.*)

Term	Usage
monomineralic, bimineralic, trimineralic . . .	Expressions indicating the number of major constituents forming 95% of the rock.
'mainly composed of'	Used where a mineral(s) form(s) more than 50% by volume of the rock.
'mainly composed of mineral A and mineral B'	Used where both minerals are present at least as major constituents (see below) and together form more than 50% of the rock.
'mainly composed of mineral A ± mineral B'	Used where mineral A is present at least as a major constituent and mineral B may be present in an undefined quantity or absent, both minerals together forming more than 50% of the rock.
main constituent	Constituent (mineral) present in modal content ≥50%.
major constituent	Constituent (mineral) present in modal content ≥5%.
minor constituent	Constituent (mineral) present in modal content <5%.
essential constituent	Constituent (mineral) that must be present in a rock in a certain minimum amount to satisfy the definition of a rock. The minimum amount is given in the definition of the rock term. May be present as major or minor constituent.
critical constituent, critical phase assemblage	Constituent (mineral) or phase assemblage indicating by its presence or absence distinctive conditions for the formation of a rock and/or a distinctive chemical composition of a rock. May be present as a major or minor constituent.
felsic minerals	Collective term for modal quartz, feldspar and feldspathoids.
mafic minerals	Collective term for modal ferromagnesian and other non-felsic minerals.
meta . . ., **meta-**	Prefix in front of an igneous or sedimentary rock name indicating that the rock is metamorphosed (e.g. metasandstone, meta-andesite). The use of the prefix does not have any implications for the present mineral content or structure of the rock, which may or may not have been substantially changed from that of the protolith. The prefix should, of course, only be applied to a protolith name when the protolith can be fully identified by some means. The prefix 'meta' should never be used for a former metamorphic rock (e.g. meta-eclogite is not an acceptable term).* If the protolith was a metamorphic rock it should be referred to in the form 'metamorphosed eclogite', or more specifically, 'amphibolitized eclogite' 'retrogressed eclogite', 'contact-metamorphosed eclogite', etc.
ortho . . ., **ortho-**	Prefix indicating, when in front of a metamorphic rock name, that the rock derived from an igneous rock (e.g. orthogneiss).
para . . ., **para-**	Prefix indicating, when in front of a metamorphic rock name, that the rock derived from a sedimentary rock (e.g. paragneiss).
plagioclase	Feldspar of the series albite-anorthite (including albite). This usage conforms to the recommendations of the IMA but differs from Le Maitre (1989, 2002).

* Note: if the protolith is established as an ultramafic rock which may have formed as a result of metamorphic or igneous processes, it is acceptable to use the prefix 'meta', without any implication about the genesis of the protolith (for example, metaperidotite).

2.2 Types, grade and facies of metamorphism

Witold Smulikowski, Jacqueline Desmons, Douglas Fettes, Ben Harte, Francesco Sassi and Rolf Schmid

2.2.1 Types of metamorphism

Metamorphism: a process involving changes in the mineral content/composition and/or microstructure of a rock, dominantly in the solid state. This process is mainly due to an adjustment of the rock to physical conditions that differ from those under which the rock originally formed and from the physical conditions normally occurring at the surface of the Earth and in the zone of diagenesis. The process may coexist with partial melting and may also involve changes in the bulk chemical composition of the rock.

Metamorphism can be variably classified on the basis of different criteria such as:

1. the extent over which metamorphism took place, that is, regional metamorphism (m.) and local m.;
2. its geological setting, for example orogenic m., burial m., ocean-floor m., dislocation m., contact m. and hot-slab m.;
3. the particular cause of a specific metamorphism, for example impact m., hydrothermal m., combustion m., lightning m.; some of the terms listed under (2) also fall into this category, for example contact m. and hot-slab m.;
4. whether it resulted from a single or multiple event(s), that is, monometamorphism and polymetamorphism;
5. whether it is accompanied by increasing or decreasing temperatures, that is, prograde m. and retrograde m.

The main classification of metamorphism from the viewpoints of extent, setting and cause is shown in Figure 2.2.1. It does not include all terms known from the literature. Many terms such as thermal metamorphism, dynamic metamorphism, dynamothermal metamorphism, deformation metamorphism, upside-down metamorphism, cataclastic metamorphism etc. are not used here because they overlap with the terms used in Figure 2.2.1 or have ambiguous usage.

Regional metamorphism is a type of metamorphism that occurs over an area of wide extent, that is, affecting a large rock volume, and is associated with large-scale tectonic processes, such as ocean-floor spreading, crustal thickening related to plate collision, or deep basin subsidence.

Local metamorphism is a type of metamorphism of limited areal (volume) extent in which the metamorphism may be directly attributed to a localized cause, such as a magmatic intrusion, faulting or meteorite impact.

If the metamorphism, even over a very wide extent, can be related to a particular source, for example heat of an intrusion, or is restricted to a certain zone, for example dislocation, it is considered as local.

Orogenic metamorphism is a type of metamorphism of regional extent related to the development of orogenic belts.

The metamorphism may be associated with various phases of orogenic development and involve both compressional and extensional regimes. Dynamic and thermal effects are combined in varying proportions and timescales and a wide range of P–T conditions may occur.

Burial metamorphism is a type of metamorphism, mostly of regional extent, that affects rocks deeply buried under a sedimentary-volcanic pile and is typically not associated with deformation or magmatism.

The resultant rocks are partially or completely recrystallized and generally lack schistosity. It commonly involves very low to medium metamorphic temperatures and low to medium P/T ratios.

Ocean-floor metamorphism is a type of metamorphism of regional or local extent related to the steep geothermal gradient occurring near spreading centres in oceanic environments.

The recrystallization, which is mostly incomplete, encompasses a wide range of temperatures.

Metamorphic Rocks: A Classification and Glossary of Terms. Recommendations of the International Union of Geological Sciences, eds. Douglas Fettes and Jacqueline Desmons. Published by Cambridge University Press. © Cambridge University Press 2007.

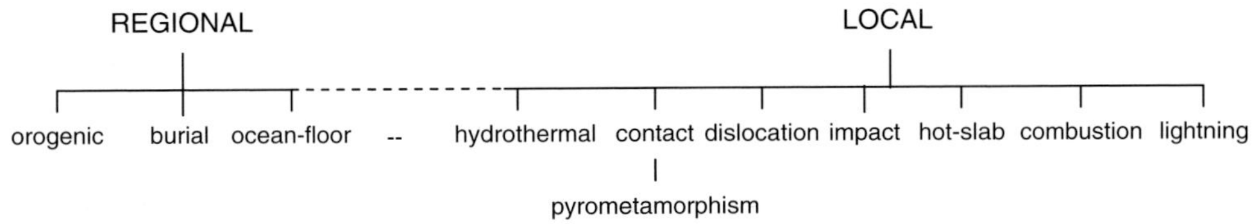

Fig. 2.2.1 Main types of metamorphism.

The metamorphism is associated with circulating hot aqueous fluids (with related metasomatism) and typically shows an increasing temperature of metamorphism with depth.

Dislocation metamorphism is a type of metamorphism of local extent, associated with fault zones or shear zones.

Grain size reduction typically occurs in the rocks, and a range of rocks commonly referred to as mylonites and cataclasites is formed.

Impact metamorphism is a type of metamorphism of local extent caused by the passage of a shock wave due to the impact of a planetary body (projectile or impactor) on a planetary surface (target).

It includes melting and vaporization of the target rock(s).

Contact metamorphism is a type of metamorphism of local extent that affects the country rocks around magma bodies emplaced in a variety of environments from volcanic to upper mantle depths, in both continental and oceanic settings.

It is essentially caused by the heat transfer from the intruded magma body into the country rocks. The range of metamorphic temperatures may be very wide. It may or may not be accompanied by significant deformation depending upon the dynamics of the intrusion.

Pyrometamorphism is a type of contact metamorphism characterized by very high temperatures, at very low pressures, generated by a volcanic or subvolcanic body.

It is most typically developed in xenoliths enclosed in such bodies. Pyrometamorphism may be accompanied by various degrees of partial melting (to form, for example, fritted rocks or buchites).

Hydrothermal metamorphism is a type of metamorphism of local extent caused by hot H_2O-rich fluids.

It is typically of local extent in that it may be related to a specific setting or cause (e.g. where an igneous intrusion mobilizes H_2O in the surrounding rocks). However, in a setting where igneous intrusion is repetitive (e.g. in ocean-floor spreading centres) the repetitive operation of circulating hot H_2O fluids may give rise to regional effects as in some cases of ocean-floor metamorphism. Metasomatism is commonly associated with this type of metamorphism.

Hot-slab metamorphism is a type of metamorphism of local extent occurring beneath an emplaced hot tectonic body.

The thermal gradient is inverted and usually steep.

Combustion metamorphism is a type of metamorphism of local extent produced by the spontaneous combustion of naturally occurring substances such as bituminous rocks, coal or oil.

Lightning metamorphism is a type of metamorphism of local extent that is due to a strike of lightning.

The resulting rock is commonly a fulgurite, an almost entirely glassy rock.

A rock or a rock complex may bear the effects of more than one metamorphic event (e.g. contact metamorphism following regional metamorphism), and thus the following types of metamorphism can be distinguished.

Monometamorphism is a metamorphism resulting from one metamorphic event (Fig. 2.2.2a and b).

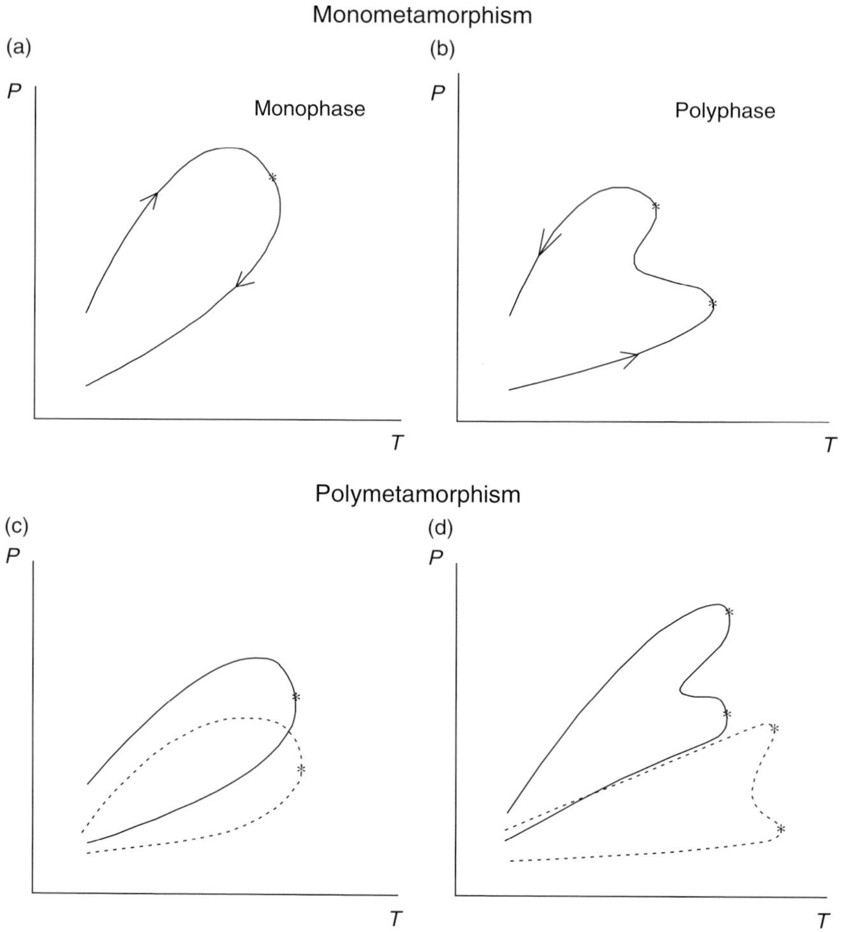

Fig. 2.2.2 Schematic *P–T–t* paths of monometamorphism (**a** and **b**) and polymetamorphism (**c** and **d**). Each line represents a metamorphic event: **a**, monophase with clockwise *P–T–t* path; **b**, polyphase with anticlockwise path; **c**, two monophase events; and **d**, two polyphase events. Asterisks represent thermal climaxes.

Polymetamorphism is a metamorphism resulting from more than one metamorphic event (Fig. 2.2.2c and d).

In these definitions a **metamorphic event** refers to a coherent sequence of metamorphic conditions (temperature, pressure, deformation) under which metamorphic reconstitution commences and continues until it eventually ceases. Typically a metamorphic event will involve a cycle of heating and cooling, which in orogenic metamorphism will be accompanied by pressure and deformation variations.

The series of metamorphic conditions in the metamorphic event may be represented on the pressure–temperature (*P–T*) diagram by a ***P–T–t* path**, where '*t*' refers to time. Thus in Figure 2.2.2 the continuous lines in 2.2.2a and 2.2.2b represent the sequence of *P–T* conditions which occurred in a given rock body over a period of time of a particular metamorphic event.

It is accepted that the changes in *P–T* conditions during a metamorphic event do not necessarily involve only one phase of heating and then cooling and/or one phase of increasing then decreasing pressure. Thus a metamorphic event may be **monophase** (e.g. with one thermal climax, see Fig. 2.2.2a) or **polyphase** (with two or more climaxes, Fig. 2.2.2b). Polymetamorphism is illustrated in Figure 2.2.2; this shows two monophase metamorphic events (2.2.2c) and two

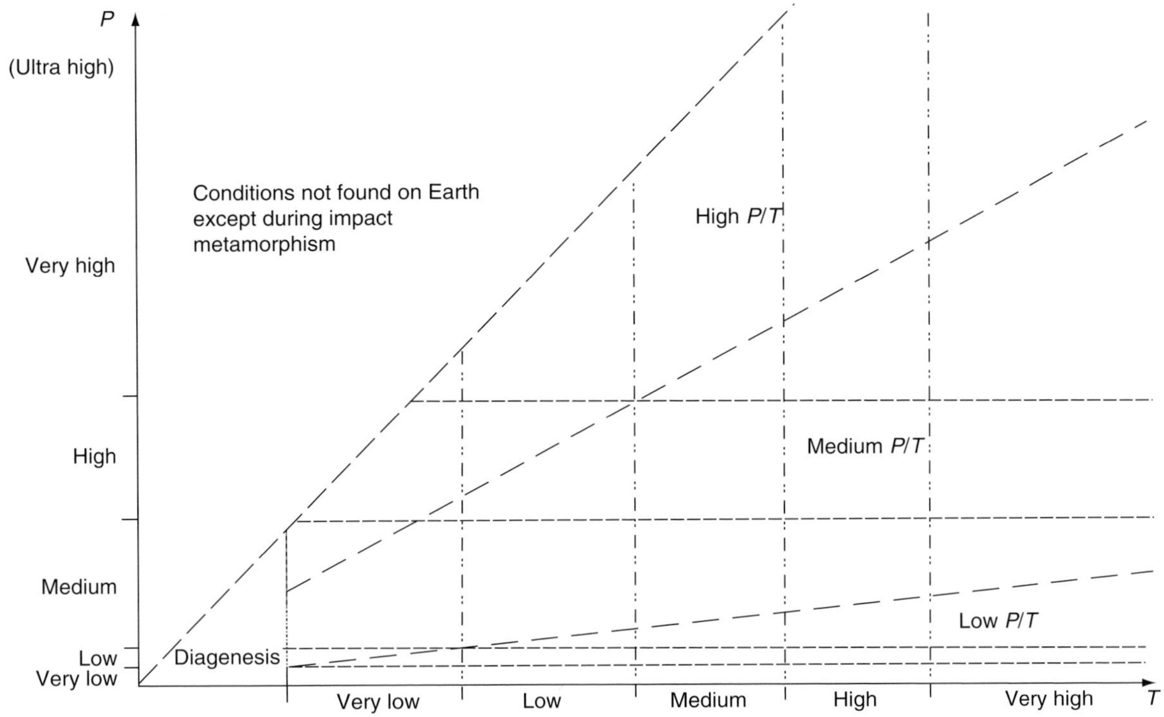

Fig. 2.2.3 Schematic representation in P–T space of the five isothermal, five isobaric bands and three P/T radial sectors.

polyphase events (2.2.2d) which have left their imprints on a rock body. Note that points on a P–T–t path may be labelled with specific ages and that, even in a monophase event, stages of metamorphism corresponding to restricted sections of the P–T–t path may be distinguished. The path may be clockwise (Fig. 2.2.2a) or anticlockwise (Fig. 2.2.2b) according to whether thermal climaxes are reached under conditions of decreasing or increasing pressure respectively. In practice it may be a difficult matter to differentiate polyphase monometamorphism from polymetamorphism.

The term 'plurifacial metamorphism', as defined by de Roever & Nijhuis (1963) and de Roever (1972), may correspond to either polyphase metamorphism or polymetamorphism and is not recommended for general use by SCMR.

2.2.2 Metamorphic temperature, pressure, grade, isograd

Relative terms such as high-temperature or low-pressure are often used to refer to the physical conditions of metamorphism but without precise designation of the temperatures and pressures involved. In order to maintain similarity of meaning it is proposed that the whole spectrum of temperature conditions encountered in metamorphism be divided into five parts, and the corresponding metamorphism may be designated as: **very low-**, **low-**, **medium-**, **high-**, **very high-temperature metamorphism**. Likewise the broad range of pressure conditions may be divided into five to give: **very low-**, **low-**, **medium-**, **high-**, **very high-pressure metamorphism**. In the highest part of the very high-pressure **ultra-high-pressure metamorphism** may be distinguished (Desmons & Smulikowski, this vol.). In a P–T grid the above divisions are represented by five isothermal and five isobaric bands respectively (Fig. 2.2.3). Circumstances of temperature and pressure may be combined together, for example medium-pressure/low-temperature metamorphism.

Related terms may be used to describe the ratio of pressure to temperature during metamorphism. The whole range of P/T ratios encountered may be divided into three fields (radial sectors in a *P–T* diagram) to give **low**, **medium** and **high** ***P*/*T* metamorphism** (Fig. 2.2.3), broadly reflecting the main divisions of facies series (see below). In addition, terms such as **very low** and **very high** ***P*/*T* metamorphism** may be used.

The term **metamorphic grade** is widely used to indicate relative conditions of metamorphism, but it is used variably. Within a given metamorphic area, the terms lower and higher grade have been used to indicate the relative intensity of metamorphism, as related to either increasing temperature or increasing pressure conditions of metamorphism or often both. Unfortunately this may give rise to ambiguity about whether grade refers to relative temperature or pressure, or some combination of temperature and pressure. To avoid this it is recommended that **metamorphic grade should refer only to temperature of metamorphism**, following Turner and Verhoogen (1951), Miyashiro (1973a) and Winkler (1974). If the whole range of temperature conditions is again divided into five (as in Fig. 2.2.3), then we may refer to **very low**, **low**, **medium**, **high**, **very high grade of metamorphism** in the same way as for 'very low, low, ...' etc. temperature of metamorphism and with the same meaning.

Depending on whether metamorphism is accompanied by increasing or decreasing temperature, two types can be distinguished.

Prograde (= progressive) metamorphism is a metamorphism giving rise to the formation of minerals which are typical of a higher grade (i.e. higher temperature) than the former phase assemblage.

Retrograde (= retrogressive) metamorphism is a metamorphism giving rise to the formation of minerals which are typical of a lower grade (i.e. lower temperature) than the former phase assemblage.

Isograd is a surface across the rock sequence, represented by a line on a map, defined by the appearance or disappearance of a mineral, a specific mineral composition or a mineral association, produced as a result of a specific reaction, for example, the 'staurolite-in' isograd defined by the reaction: garnet + chlorite + muscovite = staurolite + biotite + quartz + H_2O.

Isograds represent mineral reactions, not rock chemical composition. Hence, although the expressions like 'isoreactiongrad' (Winkler, 1974) and 'reaction isograd' (Bucher & Frey, 1994) may convey this meaning more accurately, they are unnecessary.

2.2.3 Metamorphic facies and facies series

Metamorphic facies is a fundamental notion in metamorphic petrology. The concept of metamorphic facies replaced that of depth zones, that is, epi-, meso- and catazone (Grubenmann & Niggli, 1924), when it became obvious that metamorphic grade is not necessarily correlated with depth.

The concept of metamorphic facies was first proposed by Eskola (1915) who later (Eskola, 1920) gave the following definition: a *metamorphic facies* is 'a group of rocks characterized by a definite set of minerals which, under the conditions obtaining during their formation, were at perfect equilibrium with each other. The quantitative and qualitative mineral composition in the rocks of a given facies varies gradually in correspondence with variation in the chemical bulk composition of the rocks.' In the same paper he also defined mineral facies as a more general term applicable to both igneous and metamorphic rocks. A *mineral facies* 'comprises all the rocks that have originated under temperature and pressure conditions so similar that a definite chemical composition has resulted in the same set of minerals ...' Subsequently Eskola (1939) wrote (translated from German by Fyfe *et al.*, 1958), 'In a definite facies are united rocks that for identical bulk composition exhibit an identical mineral composition, but whose mineral composition for varying bulk composition varies according to definite laws.'

The Subcommission proposes the following definition of facies, which follows Eskola's writings and the commentaries of other workers (in particular Turner, 1981).

A **metamorphic facies** is a set of metamorphic mineral assemblages, repeatedly associated in time and space and showing a regular relationship between mineral composition and bulk chemical composition, such that different metamorphic facies (sets of mineral assemblages) appear to be related to different metamorphic conditions, in particular temperature and pressure, although other variables, such as P_{H2O}, may also be important.

It is one of the strengths of the metamorphic facies classification that it identifies the regularities and consistencies in mineral assemblage development, which may be related to *P–T* conditions, but does not attempt to define actual pressures and temperatures.

In the broad sense, considering the exceptionally wide range of chemical compositions of rocks, and narrow ranges of *P–T* conditions over which mineral assemblages may change, it is theoretically possible to define a very large number of facies. In practice it has been found most convenient to define a reasonably small number of facies, which cover the broad range of crustal *P–T* conditions. These have been based principally on major changes in the mineral assemblages of rocks of basaltic composition, because such rock types are widespread and they show changes in mineral assemblages that are both distinct and reasonably limited in number, as realized by Eskola himself.

Within such major and broad facies, subunits or **subfacies** have been defined showing, for example, more detailed changes in pelitic assemblages. However, no widely used scheme of subfacies exists, and we make no attempt to define such here, since they may be defined for specific circumstances when necessary.

Eskola (1920, 1939) distinguished eight facies, namely **greenschist facies (f.)**, **epidote-amphibolite f.**, **amphibolite f.**, **pyroxene-hornfels f.**, **sanidinite f.**, **granulite f.**, **glaucophane-schist f. and eclogite facies**. Coombs *et al.* (1959), building on a suggestion of Eskola's, added a **zeolite facies** and a *prehnite-pumpellyite zone*, which Turner (1968) called *prehnite-pumpellyite metagreywacke facies*. Miyashiro (1973a) used the above ten facies, renaming the last one as the *prehnite-pumpellyite facies*. More recently various authors have recognized distinctions in the assemblages containing prehnite and pumpellyite, and erected three facies or subfacies based on the assemblages *prehnite-pumpellyite*, *prehnite-actinolite* and *pumpellyite-actinolite* (Árkai *et al.*, Section 2.5). These facies or subfacies, involving prehnite and pumpellyite, may be collectively referred to as the **subgreenschist facies** (e.g. Bucher & Frey, 1994, Merriman & Frey, 1999) and this term has accordingly been provisionally accepted by the SCMR as a general term covering a range of very low-grade metamorphism (Árkai *et al.*, Section 2.5, Fig. 2.5.1).

The merits of recognizing such a group of facies are evident from their extensive use over many years, and **the SCMR recommends that these ten facies be adopted as the major facies for general use**. Note, however, that blueschist facies is commonly used as a synonym for the glaucophane-schist facies and that the epidote-amphibolite facies is sometimes considered as part of the greenschist facies (on the basis of the coexistence of epidote with sodic plagioclase, i.e. anorthite content less than 17%).

The diagnostic minerals and mineral parageneses of the facies occurring in metamorphosed basaltic rocks for each of the facies are given in Table 2.2.1. It must be emphasized that diagnostic mineral assemblages for these facies may also be listed for other rock compositions. In the case of pelitic rocks there would be several assemblages in each of these facies because the phase assemblages of pelitic rocks are more sensitive to changes in the *P–T* conditions than those of basaltic rocks. Mineral parageneses for other rock compositions in these facies are given in many textbooks such as Turner (1968, 1981), Miyashiro (1973a, 1994), Bucher and Frey (1994), and Kretz (1994).

The relative positions of the ten facies in *P–T* space are shown in Fig. 2.2.4.

In many areas of regional metamorphism and contact metamorphism, sequences of mineral assemblages may be mapped that reflect increasing temperatures and increasing pressures of metamorphism. Such sequences typically reflect a variety of *P*/*T* values (Fig. 2.2.3) and show a sequence of isograds where mineral reactions define changes in the stability of mineral parageneses. In order to relate such prograde sequences

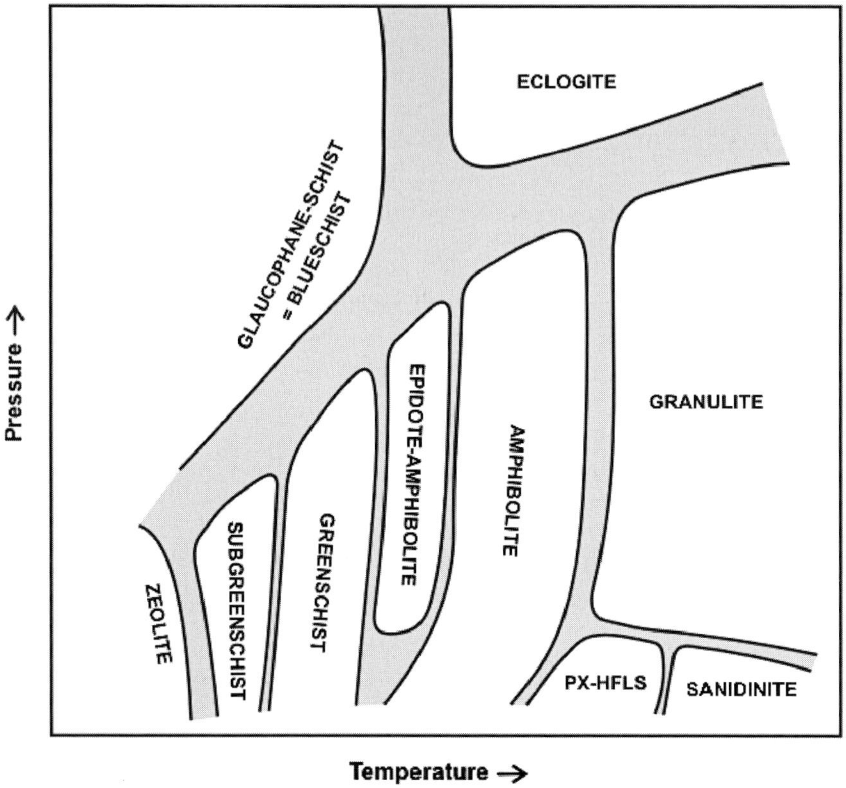

Fig. 2.2.4 Diagram showing the relative position of the ten facies (Table 2.2.1) in the *P–T* field. PX-HFLS = pyroxene-hornfels. Many similar diagrams exist, for example Turner (1968, 1981), Miyashiro (1973a, 1994), Winkler (1974), Yardley (1989) and Bucher and Frey (1994). The SCMR has not discussed the various presentations and makes no recommendation on the absolute *P* and *T* values, the precise fields of the facies or the nature of the areas of uncertainty between the fields.

of mineral parageneses to the broadly recognized metamorphic facies (Fig. 2.2.4), Miyashiro (1961) developed the concept of the ***metamorphic facies series***.

> **Metamorphic facies series** is a sequence of metamorphic facies developed under a particular range of *P*/*T*.

For regional metamorphism Miyashiro (1961) suggested three principal facies series and the existence of some intermediate facies series. Later (Miyashiro, 1973a), he referred to them as **baric types of metamorphism** because they broadly indicate different radial sectors in a *P–T* diagram such as Figure 2.2.3 (e.g. Spear, 1993, Fig. 2.3; Miyashiro, 1994, Fig. 8.1) and are distinguished by their range of *P*/*T* rather than their range of pressures or temperatures. The three principal baric types are:

1. a ***low-P/T type*** (also referred to as the *andalusite-sillimanite series* or *Abukuma type*) characterized by andalusite at lower grades and sillimanite at higher grades and typified by the sequence greenschist f. → amphibolite f. → granulite f.;
2. a ***medium-P/T type*** (also referred to as the *kyanite-sillimanite series* or *Barrovian type*) characterized by kyanite at lower grades and sillimanite at higher grades and typified by the sequence greenschist f. → epidote-amphibolite f. → amphibolite f. → granulite f.;

3. a ***high-P/T type*** (also referred to as *glaucophanic metamorphism*) characterized by the presence of glaucophane and typified by the sequence subgreenschist f. (prehnite-pumpellyite) → glaucophane-schist/blueschist f.

The three principal metamorphic facies series of Miyashiro have been generally adopted (e.g. Yardley, 1989; Spear, 1993; Kornprobst, 2002) although it is accepted that subdivisions, intermediates and variants exist (e.g. Harte & Hudson, 1979; Miyashiro, 1994).

Table 2.2.1 *Metamorphic facies and their characteristic minerals and mineral parageneses in metamorphosed rocks of basaltic chemical composition.*

Facies	Minerals and mineral parageneses
Zeolite facies	Zeolites such as laumontite and heulandite etc. (in place of other Ca-Al silicates such as prehnite, pumpellyite and epidote)
Subgreenschist facies	Prehnite-pumpellyite, pumpellyite-actinolite, prehnite-actinolite (prehnite and pumpellyite are the diagnostic Ca-Al silicates rather than minerals of the epidote or zeolite groups)
Greenschist facies	Actinolite-albite-epidote-chlorite (an epidote group mineral is the diagnostic Ca-Al silicate rather than prehnite or pumpellyite)
Epidote-amphibolite facies	Hornblende-albite-epidote(-chlorite)
Amphibolite facies	Hornblende-plagioclase (plagioclase more calcic than An_{17})
Pyroxene-hornfels facies	Clinopyroxene-orthopyroxene-plagioclase (olivine stable with plagioclase)
Sanidinite facies	Distinguished from the pyroxene-hornfels facies by the occurrence of especially high-temperature varieties and polymorphs of minerals (e.g. pigeonite, K-rich labradorite)
Glaucophane-schist or blueschist facies	Glaucophane-epidote(-garnet), glaucophane-lawsonite, glaucophane-lawsonite-jadeite
Eclogite facies	Omphacite-garnet-quartz (no plagioclase, olivine stable with garnet)
Granulite facies	Clinopyroxene-orthopyroxene-plagioclase (olivine not stable with plagioclase or with garnet)

2.3 Structural terms including fault rock terms

KATE BRODIE, DOUGLAS FETTES and BEN HARTE

2.3.1 Introduction

A Study Group (SG), under the leadership of K. Brodie, was set up to look at nomenclature relating to structural terms. At an early stage a questionnaire was sent to around 60 structural geologists throughout the world, with a series of initial definitions. The response did much to guide the work of the SG and the SCMR in finalizing its recommendations.

2.3.2 Background

Many of the definitions given below were adopted by the SCMR without difficulty; others gave rise to considerable debate. Problems arose for a variety of reasons: the usage of terms across the geological community (e.g. gneiss and schist) is variable; terms such as slate and cleavage proved difficult because there are no similar terms in many non-English speaking countries; equally, the difference between cleavage and schistosity and the use of texture and microstructure proved major sticking points. In other cases it was difficult to differentiate between rock types solely on features observable in the field at hand-specimen scale or in thin section examination as required by the SCMR scheme (Schmid *et al.*, Section 2.1). Also, although the SCMR scheme seeks to avoid terms based on processes wherever possible, this was not practical with many terms related to fault rocks (e.g. mylonite and cataclasite).

The following subsections discuss the main problem areas and the basis that the SCMR used in deciding on its recommended definitions.

2.3.3 The terms schist, gneiss and granofels

The SCMR decided to base its systematic rock names on structural root terms with mineral qualifiers (Schmid *et al.*, Section 2.1). One possibility would have been structural root terms based on grain size. However, although the sequence slate–phyllite–schist–gneiss could be used in this way, it was generally felt that these terms were too specific to be widely applicable and that there was an absence of other suitable names. The SCMR therefore decided that the most appropriate structural root terms were those reflecting the degree of fissility or development of schistosity in a rock. The terms selected for this purpose were schist, gneiss and granofels. These names are well entrenched in the literature and are generally acceptable as terms reflecting different degrees of fissility. However, the adoption of these names as purely structure-based terms gave rise to considerable debate within the SCMR. The following subsections discuss the current definitions of these names and consider the advantages and disadvantages of their use as structural root terms.

SCHIST

The term schist is derived from the Greek schistos, to split, and according to Tomkeieff (1983) was first used by Pliny. In its simplest form schist can be regarded as a rock possessing a schistosity. In the English language, however, many workers differentiate between a slate, which is a fine-grained rock possessing a well-developed schistosity, and a schist, which is a medium-grained rock with a good schistosity (e.g. Holmes, 1920; Spry, 1969; Barker, 1990).

The SCMR decided, however, that defining schist as a rock with a well-developed schistosity and thus making use of it as a structural root term was acceptable within general usage. In this way schist becomes the systematic root term covering all rocks with a well-developed schistosity including slates and phyllites. These latter names, however, are still retained within the SCMR nomenclature scheme as recommended specific names (Schmid *et al.*, Section 2.1).

GNEISS

Gneiss is thought to be a term originally used by miners in Bohemia for the host rock in which the metalliferous veins occurred. According to Tomkeieff (1983), it was first recorded by Agricola in 1556. The basis of modern usage was probably laid down by Werner (1786) who refined the list of components and defined gneiss

Metamorphic Rocks: A Classification and Glossary of Terms. Recommendations of the International Union of Geological Sciences, eds. Douglas Fettes and Jacqueline Desmons. Published by Cambridge University Press. © Cambridge University Press 2007.

as a feldspar-quartz-mica rock with a coarse schistosity (*dickschiefriges Gewebe*) or gneissosity.

In modern usage gneiss may be taken as a medium- to coarse-grained rock with a poorly developed schistosity, and feldspar and quartz as characteristic (to some, essential) components. The rock commonly has a banded structure reflecting compositional and/or structural variations. Gneiss is also generally presumed to be the product of medium- to high-grade metamorphism. Definitions of gneiss, however, vary greatly between authors, reflecting the emphasis they place on the various features. Two main areas of division may be considered.

(a) Mineral content and structure

Schist has a better-defined schistosity than gneiss. This may be taken to reflect the mineral content: that is, schist is richer in phyllosilicates and poorer in granular minerals, such as quartz and feldspar, than gneiss. The emphasis placed on these interdependent criteria has polarized the definition of gneiss.

Much of European literature has built on the definitions of the early workers and places emphasis on the mineral content. As such, related definitions of gneiss regard feldspar + quartz as essential components. As an extension of this requirement Fritsch *et al.* (1967) proposed a boundary between gneiss and schist at >20% modal feldspar (see also Lorenz, 1996).

Conversely, for other workers, particularly those writing in English, the definition of gneiss has evolved with an increasing emphasis on the structure (Winkler, 1974; Barth, 1978) and to some workers the mineral content is no longer regarded as an essential part of the definition (Harker, 1954; Bates & Jackson, 1987).

(b) The significance of banding

The significance of banding in the definition of gneiss also varies, for example, Yardley (1989) notes: 'English and North American usage emphasizes a tendency for different minerals to segregate into layers parallel to the schistosity, known as gneissic layering; typically quartz and feldspar-rich layers segregate out from more micaceous or mafic layers. European usage of gneiss is for coarse, mica-poor, high-grade rocks, irrespective of their fabric.'

Banding may be defined in a variety of ways; some workers place the emphasis on the alternation of schistose and granulose layers (Tyrrell, 1929; Tomkeieff, 1983; Bates & Jackson, 1987), while others regard mineral banding as the characteristic feature (Mason, 1978; Barker, 1990).

The SCMR debated these differences at length, particularly the requirement for feldspar + quartz in the definition of gneiss. However, the strong association in geological usage of the name gneiss with a rock possessing a poorer fissility than schist and the usefulness of gneiss as such a term proved decisive. Also, although this could be construed as showing a bias towards English-language usage, it was felt that this was acceptable because of the SCMR's decision to make its definitions in English.

The SCMR also noted that a structural-only definition did not exclude any rocks currently defined as gneiss. On the other hand a major concern for the SCMR was that a structure-only definition might include rocks that in current usage would never be considered as gneisses, for example metasandstone of low metamorphic grade. In these cases it was felt that a guideline encouraging the use of protolith-based names would provide an adequate safeguard (Schmid *et al.*, Section 2.1).

The SCMR chose the boundary between schist and gneiss based on the definition proposed by Wenk (as given by Winkler, 1974), namely: 'When hit with a hammer, rocks having a schistose fabric (schists) split perfectly parallel to 's' into plates, 1–10 mm in thickness, or parallel to the lineation into thin pencil-like columns.'

GRANOFELS

Granofels was introduced by Goldsmith (1959) as a term to describe rocks in which schistosity was absent or virtually absent. Previously, the term granulite had been used but as it also had grade and lithological connotations it was generally considered unsuitable.

2.3.4 Cleavage and schistosity

Early workers distinguished slate and (slaty) cleavage from schist and schistosity. Although it was subsequently recognized that there was no significant difference between the two, the four terms became well established in English literature and are now in common usage. In some languages, however, the distinction between cleavage and schistosity and slate and schist is not made. In recognition of this latter position the SCMR discussed recommending 'schistosity' as a term to cover both cleavage and schistosity. This, however, proved impractical and both terms have been retained and defined as recommended terms. The basis of their use and definition is given below.

Cleavage is the property of a rock to split on a set of regular parallel or subparallel planes.

Cleavage was the subject of extensive study and discussion among early structural geologists and its classification was largely based on the assumed mechanism of its formation (see review in Wilson, 1961). This led to a great confusion of terms. In order to address this problem, Powell (1979) proposed a systematic classification based on morphological rather than genetic criteria. Powell's scheme forms the basis of current definitions and is largely adopted here. Powell divided cleavage into continuous cleavage and spaced cleavage. Continuous cleavage is present throughout the rock at the grain-size scale and may be subdivided into fine continuous cleavage as found in fine-grained rocks, and coarse continuous cleavage as found in coarse-grained rocks. Spaced cleavage is subdivided into crenulation cleavage and disjunctive cleavage. The latter is developed independently of any pre-existing mineral orientation in the rock (e.g. fracture cleavage, pressure solution cleavage). (See below for full definitions of these terms.)

Schistosity is the preferred orientation of inequant minerals in a rock.

Schistosity, in some form, is present in most cleavage types. Only in certain disjunctive cleavages (e.g. fracture cleavage) is schistosity absent. Well-developed schistosity is characteristic of continuous cleavage and *is independent of grain size*. Thus, at the simplest level, all rocks with such a structure may be termed schists. However, it is common practice to refer to fine continuous cleavage as slaty cleavage and the associated rocks as slates.

2.3.5 Foliation

Foliated structure was used by Macculloch (1821) to denote a coarse mineral layering with a poor splitting: the equivalent, in modern terms, of gneissose structure. Darwin (1846) defined *foliation* and gave it the same meaning. This usage and meaning became established, particularly in British petrological literature (e.g. Harker, 1939; Fairbairn, 1949; see review in Wilson, 1961). In American literature, however, the term was generally taken to include schistosity and cleavage (e.g. Knopf & Ingerson, 1938; Turner & Weiss, 1963). This latter, wider meaning is now prevalent (e.g. Spry, 1969; Park, 1983; Tomkeieff, 1983; Barker, 1990; Davis & Reynolds, 1996) and is the one adopted by the SCMR. In this sense it is equivalent to *s-surface* (Turner & Weiss, 1963, p. 97).

2.3.6 Structure, fabric and texture

The use of the terms structure, fabric and texture may give rise to ambiguity. This is particularly true when the same words in other languages may have different meanings.

Structure is the arrangement of the parts of a rock mass irrespective of scale, including spatial relationships between the parts, their relative size and shape and the internal features of the parts.

The term *fabric* is a translation of Sander's (1930) term *Gefüge*. Fabric was defined by Knopf and Ingerson (1938) as 'the spatial data that govern the arrangement in space of the component elements that go to make up any sort of external form.' In current practice these elemental parts are only considered as contributing to a fabric if 'they occur over and over again in a reproducible manner from one sample of a rock to another' (Hobbs *et al.*, 1976). This means that although the fabric of a body may be considered at any scale the term is normally used at the crystallographic or mineral aggregate scale. Thus, for example, the preferred orientation of inequant mineral grains will produce planar or linear fabrics.

The term *texture* is used in two ways. The commonest way is as a term for the spatial arrangement and relative size of mineral grains and their internal features (Spry, 1969). In this sense texture is synonymous with *microstructure* or at least certain aspects of microstructure.

On the other hand, in material science and increasingly for some geologists (e.g. Barker, 1990; Vernon, 2004) texture means the presence of preferred orientation. In this sense texture is synonymous with *microfabric*.

Given this dual use the SCMR recommends that only (micro)structure and (micro)fabric are used. If the term texture is used, its meaning should be made quite clear.

2.3.7 Fault rocks

While some fault rocks might be considered to fall outside the remit of metamorphic nomenclature, many undergo chemical as well as structural changes, and the deformation occurs within the P–T range of metamorphism. The definition of fault rocks is problematic and many of the definitions involve processes. Different minerals deform in different ways depending on the temperature conditions, and this precludes definitions of fault rock terms that are mineralogically based. The definitions presented are systematic and general.

As might be expected, **mylonite** proved difficult to define. Since its original definition by Lapworth (1885) there have been many nomenclature schemes proposed for mylonites and related rocks (e.g. Quensel, 1916; Knopf, 1931; Spry, 1969; Higgins, 1971; Sibson, 1977). More recently a Penrose conference on mylonites (Tullis *et al.*, 1982) failed to arrive at an agreed definition of a mylonite, mainly because of the problem of knowing if, for example, plastic processes have been involved in the grain size reduction. Commonly, detailed microstructural analysis of thin sections is required which makes it difficult to apply these definitions on a hand-specimen scale. The definition given below is non-genetic from the point of view of the mechanism of deformation. The definition of mylonite also covers cohesive foliated cataclasites and this is in recognition of the fact that, in many cases, these are difficult to distinguish in the field. It is sometimes difficult to look at a very fine-grained fault rock in the field and know whether it is an ultramylonite, an ultracataclasite or indeed a pseudotachylite. Any division on the basis of percentage of crystal plasticity versus brittle deformation is not practical using thin sections let alone from field observations. In addition, grain-size sensitive flow is being recognized as an important deformation mechanism in many mylonites. Thin section observation may allow more specific terms or qualifiers to be applied. A review of the historic perspective of the nomenclature and classification of fault rocks is provided in Snoke *et al.* (1998).

Deformed **ultramafic rocks** have their own descriptive terminology (e.g. Bouillier & Nicolas, 1975; Harte, 1977) that grew from detailed petrographic studies of kimberlite xenoliths. Many of these terms can be replaced with more general terms such as mylonite and cataclasite, and have not been re-examined by the SCMR.

These examples serve to illustrate the considerable difficulties that have arisen within the SCMR in attempting to erect a practical and widely acceptable scheme. Some notes are included after particular definitions in order to explain the reasoning behind the recommended definition. In considering the definitions it is important to remember that they may be a compromise but one that is, it is hoped, workable.

All the terms given below fall into the SCMR category of 'recommended names' as defined by Schmid *et al.* (Section 2.1).

2.3.8 Definitions: 1. Main structural terms

A full list is given in the glossary.

Structure: the arrangement of the parts of a rock mass irrespective of scale, including spatial relationships between the parts, their relative size and shape and the internal features of the parts.

The terms micro-, meso- and mega- can be used as prefixes dependent on the scale of the feature.

Microstructure: structure on the thin section or smaller scale.

Mesostructure: structure on the hand-specimen scale.

Megastructure: structure on the outcrop or larger scale.

Texture: (a) the relative size, shape and spatial interrelationship between grains and internal features of grains in a rock;
(b) the presence of a preferred orientation on the microscopic scale.

Note: the use of 'texture' as defined in (a) above is common in geological literature (e.g. Spry, 1969), and as such is synonymous with *microstructure*. Because of this widespread usage the SCMR decided to accept 'texture' as a recommended term. However, in material science and in some languages 'texture' is used as defined in (b) above (e.g. Barker, 1990; Vernon, 2004) and as such is synonymous with *microfabric*. The SCMR encourages the use of (micro)structure and (micro)fabric to avoid ambiguity. If 'texture' is used then its meaning must be clear.

Fabric: the relative orientation of parts of a rock mass.

This is commonly used to refer to the crystallographic and/or shape orientation of mineral grains or groups of grains, but can also be used on a larger scale. Preferred linear orientation of the parts is termed *linear fabric*, preferred planar orientation *planar fabric*, and the lack of a preferred orientation is referred to as *random fabric*.

Foliation: any repetitively occurring or penetrative planar feature in a rock body.

Examples include:

- layering on a scale of a centimetre or less
- preferred planar orientation of inequant mineral grains
- preferred planar orientation of lenticular or elongate grain aggregates.

More than one kind of foliation with more than one orientation may be present in a rock. Foliations may become curved or distorted. The surfaces to which they are parallel are called *s-surfaces*. More precise terms should be used wherever possible.

Schistosity: a preferred orientation of inequant mineral grains or grain aggregates produced by metamorphic processes.

A schistosity is said to be well developed if inequant grains or grain aggregates are present in a large amount and show a high degree of preferred orientation. If the degree of preferred orientation is low or if the inequant grains or grain aggregates are only present in small amounts the schistosity is said to be poorly developed. See general comment above.

Schistose structure: a type of structure characterized by a schistosity that is well developed, either uniformly throughout the rock or in narrowly spaced repetitive zones such that the rock will split on a scale of one centimetre or less.

Gneissose structure: a type of structure characterized by a schistosity which is either poorly developed throughout the rock or, if well developed, occurs in broadly spaced zones, such that the rock will split on a scale of more than one centimetre.

Gneissosity: synonymous with and to be replaced by gneissose structure.

Granofelsic structure: a type of structure resulting from the absence of schistosity such that the mineral grains and aggregates of mineral grains are equant, or if inequant have a random orientation. Mineralogical or lithological layering may be present.

Cleavage: the property of a rock to split along a regular set of parallel or subparallel closely spaced surfaces.

More than one cleavage may be present in a rock. See general comment above.

Continuous cleavage: a type of cleavage characterized by the preferred orientation of all the inequant mineral constituents of a rock, and in which the cleavage planes are developed at the grain-size scale.

See general comment above.

Spaced cleavage: a type of cleavage in which the cleavage planes are spaced at regular intervals and separated by zones known as microlithons. The structure is visible to the unaided eye.

See general comment above.

Disjunctive cleavage: a type of spaced cleavage that is independent of any pre-existing mineral orientation in the rock.

See general comment above.

Slaty cleavage: a type of continuous cleavage in which the individual grains are too small to be seen by the unaided eye.

Spaced schistosity: a type of spaced cleavage characterized by regularly spaced zones with schistose structure that are structurally distinct from and separate rock layers called microlithons. The structure is visible to the unaided eye.

Fracture cleavage: a regular set of closely spaced parallel or subparallel fractures along which the rock will preferentially split.

Crenulation cleavage/schistosity: a type of spaced cleavage developed during crenulation of a pre-existing foliation, and orientated parallel to the axial plane of the crenulations.

Crenulation: a type of regular folding with a wavelength of one centimetre or less.

Lineation: any repetitively occurring or penetrative visible linear feature in a rock body.

It may be defined by:

- alignment of the long axes of elongate mineral grains (*mineral lineation*)
- alignment of elongate mineral aggregates
- parallelism of hinge lines or small-scale folds (*crenulation lineation*)
- intersection of two foliations (*intersection lineation*)
- slickenside striations or fibres.

More than one kind of lineation, with more than one orientation, may be present in a rock. Lineations may become curved or distorted. The lines to which they are parallel are called l-lines. Where possible the type of lineation should be indicated.

Fracture: a general term for any break in a rock mass, whether or not it causes displacement.

Fracture includes cracks, joints and faults.

Slate: an ultrafine- or very fine-grained metamorphic rock displaying slaty cleavage.

Slate is usually of very low metamorphic grade, although it may also occur under low-grade conditions.

Phyllite: a fine- to medium-grained metamorphic rock characterized by a lustrous sheen and a well-developed schistosity resulting from the parallel arrangement of phyllosilicates.

Phyllite is usually of low metamorphic grade.

Schist: a metamorphic rock displaying schistose structure.

For phyllosilicate-rich rocks the term schist is commonly used for medium- to coarse-grained varieties, whereas finer-grained rocks may be given the more specific names *slate* or *phyllite*. The term schist may also be applied to rocks displaying a linear fabric rather than a schistosity, and which will split on a scale of one centimetre or less. In this case the expression '*lineated schist*' is applied.

Gneiss: a metamorphic rock displaying a gneissose structure.

The term gneiss may also be applied to rocks displaying a dominant linear fabric rather than a gneissose structure, but which will split on a scale of more than one centimetre. In this case the term '*lineated gneiss*' is applied.

Granofels: a metamorphic rock displaying a granofelsic structure.

For granofels containing layers of different composition the term '*layered (*or *banded) granofels*' may be used.

2.3.9 Definitions: 2. Fault rock terms

Fault: (see Fig. 2.3.1) a fracture surface along which rocks have moved relative to each other.

Fault zone: a zone of sheared, crushed or foliated rock, in which numerous small dislocations have occurred, adding up to an appreciable total offset of the undeformed walls. All gradations may occur between multiple fault planes and single shear zones.

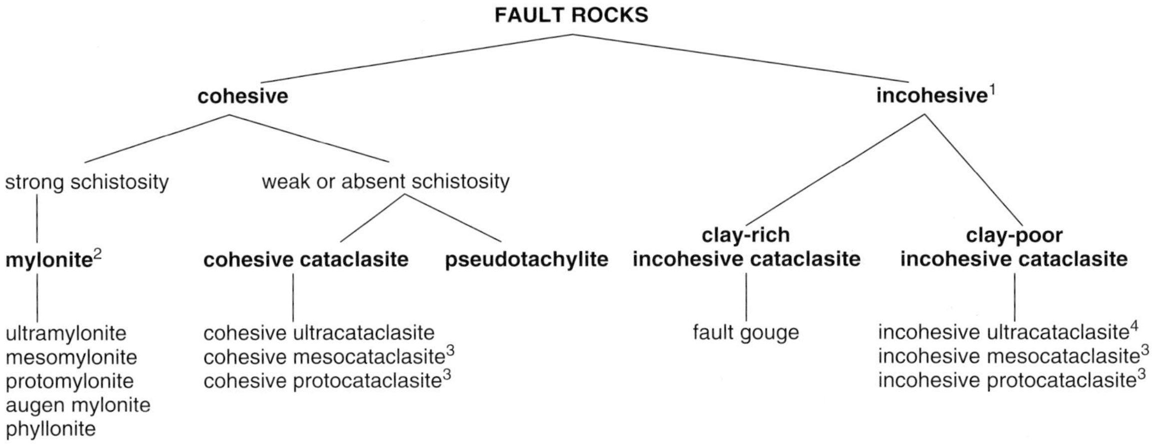

1. incohesive (incoherent, friable, uncemented): capable of being broken into component granules with fingers or with the aid of a pen knife.
2. cohesive and foliated cataclasite is indistinguishable in the field from a mylonite.
3. coarse-grained cataclasite may also be termed 'fault breccia' (with the prefix cohesive or incohesive).
4. incohesive very fine-grained ultracataclasite may also be termed 'fault gouge'.

Note: the incohesive rock names may be subdivided into foliated and non-foliated varieties.

Fig. 2.3.1 Organogram for fault rocks.

Fault rock: rock formed as a result of deformation in a fault zone.

Mylonite: a fault rock which is cohesive and characterized by a well-developed schistosity resulting from tectonic reduction of grain size, and commonly containing rounded porphyroclasts and lithic fragments of similar composition to minerals in the matrix.

Fine-scale layering and an associated mineral lineation or stretching lineation are commonly present. Brittle deformation of some minerals may be present, but deformation is commonly by crystal plasticity. Mylonites may be subdivided according to the relative proportion of finer-grained matrix into *protomylonite, mesomylonite* and *ultramylonite*.

When the protolith is known terms such as *mylonitized granite*, *granite mylonite* or *granite-derived mylonite* may be used.

Note: in the field it is often not possible to distinguish a foliated fault rock formed by brittle deformation (foliated cataclasite) from one formed by crystal plastic or grain boundary sliding processes, or a combination of different deformation mechanisms. Hence these are all defined in the field by the term mylonite.

Protomylonite: a mylonite in which less than 50% of the rock volume has undergone grain size reduction.

Mesomylonite: a mylonite in which more than 50% and less than 90% of the rock volume has undergone grain size reduction.

Note: as mylonite is the general term, to be consistent it is necessary to have a prefix for the more specific term intermediate between proto- and ultramylonite.

Ultramylonite: a mylonite in which more than 90% of the rock volume has undergone grain size reduction.

An ultramylonite need not be 'ultra' fine-grained.

Augen mylonite: a mylonite containing distinctive large crystals or lithic fragments around which the foliated fine-grained matrix is wrapped, often forming symmetric or asymmetric trails.

Blastomylonite: a mylonite that displays a significant degree of grain growth related to or following deformation.

Phyllonite: a phyllosilicate-rich mylonite that has the lustrous sheen of a phyllite.

Cataclasite: a fault rock which is cohesive with a poorly developed or absent schistosity, or which is incohesive, characterized by generally angular porphyroclasts and lithic fragments in a finer-grained matrix of similar composition.

Generally no preferred orientation of grains of individual fragments is present as a result of the deformation, but fractures may have a preferred orientation. A foliation is not generated unless the fragments are drawn out or new minerals grow during the deformation. Plastic deformation may be present but is always subordinate to some combination of fracturing, rotation and frictional sliding of particles.

Cataclasite may be subdivided according to the relative proportion of finer-grained matrix into *protocataclasite, mesocataclasite* and *ultracataclasite*.

Protocataclasite: a cataclasite in which the matrix forms less than 50% of the rock volume.

Mesocataclasite: a cataclasite in which the matrix forms more than 50% and less than 90% of the rock volume.

Note: as cataclasite is the general term, to be consistent it is necessary to have a prefix for the more specific term intermediate between proto- and ultracataclasite.

Ultracataclasite: a cataclasite in which the matrix forms more than 90% of the rock volume.

Fault breccia: a medium- to coarse-grained cataclasite containing > 30% visible fragments.

Fault gouge: an incohesive, clay-rich fine- to ultrafine-grained cataclasite, which may possess a schistosity and containing <30% visible fragments. Lithic clasts may be present.

Pseudotachylite: ultrafine-grained vitreous-looking material, usually black and flinty in appearance, occurring as thin planar veins, injection veins or as a matrix to pseudo-conglomerates or breccias, which infills dilation fractures in the host rock.

2.4 High *P*/*T* metamorphic rocks

Jacqueline Desmons and Witold Smulikowski

2.4.1 Introduction

A Study Group under the leadership of J. Desmons was set up to prepare the nomenclature of rocks that form at high *P*/*T*. Questionnaires were sent to a worldwide panel of specialists on high *P*/*T* metamorphism. Twenty-two agreed to cooperate. Provisional reports on these discussions were given by Desmons *et al.* (1997, 2001). This paper summarizes the final results of the discussions and presents the recommended definitions.

2.4.2 Eclogite

The term eclogite was created by Haüy (1822).

The SCMR definition of eclogite is:

Eclogite: plagioclase-free metamorphic rock composed of ≥75% vol. of omphacite and garnet, both of which are present as major constituents, the amount of neither of them being higher than 75% vol.

In this definition omphacite is taken as defined by the IMA Subcommittee on pyroxene nomenclature (Morimoto *et al.*, 1988; Fleischer & Mandarino, 1991; see also Carswell, 1990b), that is, based on its chemical composition, which, in the field, can be more or less inferred from its colour.

The wording of the above definition resulted from the many comments given by members of the Study Group and the Subcommission. Here follow a few explanatory comments.

1. 'Major constituent' means, according to the SCMR rules, present in an amount of ≥5% vol. This percentage may be considered to be too low, especially for omphacite. However, together with the ≥75% vol. amount of both omphacite and garnet it is intended to allow enough, and not too much, flexibility to the term.
2. Rocks in which either omphacite or garnet is present in an amount exceeding 75% vol. have to be called, according to the SCMR rules, garnet omphacitite or omphacite garnetite (see the use of the suffix '-ite' in Schmid *et al.*, Section 2.1).
3. K-feldspar can be present in an eclogite (hence plagioclase-free, not feldspar-free).
4. The basaltic composition (which in the field is only inferred from the mineral composition) or basic to ultrabasic composition does not need to be mentioned because it is implicitly given by the mode specified in the definition.
5. The colour term 'melanocratic' is not used in the definition of eclogite because:
   (i) The SCMR recommends that colour terms for metamorphic rocks are not used, owing to the greater variability of colours in such rocks compared with igneous rocks (Schmid *et al.*, Section 2.1).
   (ii) 'Melanocratic' implies a lower boundary of 65% of mafic and related minerals (Le Maitre, 1989, p. 6), which is too low an amount for eclogite.
6. Consistent with IMA recommendations (Fleischer & Mandarino, 1991; Mandarino & Back, 2004), the discredited Na-pyroxene chloromelanite has now been included under omphacite and its name does not therefore appear in the definition of eclogite.
7. The SCMR defines rocks, as far as possible, on their properties visible to the eye, that is, as they appear in the field. The amount of ≥75% has, therefore, to be taken as roughly three-quarters or more of the rock as seen in hand specimen. From a rapid literature survey it appears that a majority of eclogites have modal compositions consistent with the above definition. However, there is a tendency to put the stress on the presence of eclogite minerals by naming the rock eclogite even when the amount of the characteristic minerals is lower than required by the definition. The SCMR recommends that the term eclogite should not apply to such rocks but that another rock term preceded by the adjective eclogitoid should be used (as long as both omphacite and garnet are present; see the next subsection).
8. The additional presence of rutile may be mentioned as complementary information, as also

Metamorphic Rocks: A Classification and Glossary of Terms. Recommendations of the International Union of Geological Sciences, eds. Douglas Fettes and Jacqueline Desmons. Published by Cambridge University Press. © Cambridge University Press 2007.

may the names of other possible additional phases, for example paragonite, kyanite, etc. These minerals are not essential in the definition of eclogite. According to the SCMR rules (Schmid *et al.*, Section 2.1), the following style should be used: 'kyanite-paragonite eclogite' (if both minerals are major constituents, that is, present as $\geq$5%, and where paragonite is more abundant than kyanite), or 'rutile-bearing eclogite' (if rutile is a minor constituent, that is, present as $\leq$5%).

9. The SCMR defines rocks primarily according to their appearance in hand specimen. Therefore, classifications of eclogites based on garnet and/or whole rock chemical compositions, or with respect to their environment of formation (Smulikowski, 1964a, 1972, 1989; Coleman *et al.*, 1965; Banno, 1970), are beyond the scope of the SCMR.
10. Rocks consisting of pyrope-rich garnet and Cr-rich Ca-clinopyroxene, associated with ultrabasic rocks have been called eclogites in the past (Eskola, 1939). The correct name for these rocks is pyrope pyroxenite or pyroxene garnetite (see also griquaite below).

2.4.3 *Names for rocks related to eclogites*

There are two different groups of rocks related to eclogites. Terms referring to them should be clearly distinguished.

1. Various suffixes have been proposed, such as '-oid', '-oidic' and '-ic', for unaltered rocks that contain omphacite and garnet, but in a smaller amount than required by the definition for eclogite, owing to an unsuitable bulk composition.

   The Study Group members felt that the most appropriate use of the suffix '-ic' is to form an adjective the meaning of which is strictly related to the noun from which it derives. *Eclogitic* thus means: 'applying to eclogite'. The suffix '-oid', commonly used in igneous nomenclature (Le Maitre, 1989, 2002), rather loosely means, 'in some way related to the noun', including the noun itself (e.g. the use of granitoid, syenitoid, etc.). Accordingly, an *eclogitoid* would be a metamorphic rock containing both omphacite and garnet.

   It may be noted, for example, that the Eclogitic Micaschists ('Micaschisti eclogitici') of the Sesia-Lanzo zone in the Western Alps thus would remain as a formation name only. The omphacite-garnet rocks contained in this formation would encompass, on the above definitions, both 'eclogitoids' and 'eclogites'.
2. Eclogites that have experienced a later metamorphism and that now contain omphacite and garnet as abundant relics in an amphibolite groundmass have been given a long list of names. These include amphibolitized eclogite, post-eclogite amphibolite, retro-eclogite amphibolite, eclogitogene amphibolite, etc. Among the Study Group members, no one term emerged as favourite and acceptable to the SCMR. For the time being, it is left to each author to use their personally preferred term and to define it clearly.

2.4.4 *Eclogite facies, eclogite facies rocks*

Eclogite is also the name of a facies, which was established by Eskola (1915, 1920). As is the case with the other metamorphic facies, not all rocks metamorphosed in the eclogite facies are called eclogites (e.g. 'whiteschist', as defined below). For referring to all such rocks, the term '*eclogite-facies rocks*' or '*rocks in eclogite facies*' may be used. The definition of 'metamorphic facies' and a diagrammatic representation of metamorphic facies are given in Smulikowski *et al.* (Section 2.1).

2.4.5 *Glaucophane schist, glaucophane-schist facies and glaucophane-schist facies rocks*

According to the rules of the SCMR (Schmid *et al.*, Section 2.1) the term 'glaucophane schist' is used for a schist having only one major constituent, that is, glaucophane or ferroglaucophane ('major' meaning present in a modal proportion $\geq$5%). If additional minerals are present, as major or minor constituents, they have to be given as prefixes, for example, jadeite-bearing glaucophane-phengite schist, epidote-glaucophane schist, etc. Therefore the definition of 'glaucophane schist' is covered by the SCMR rules and no further special definition is required.

The term, however, has evolved to cover all rocks belonging to the glaucophane-schist facies (Eskola, 1915, 1920), commonly called by the synonym 'blueschist facies' (see below). This ambiguity and such a genetic meaning are things that the SCMR wishes to avoid. Accordingly, the SCMR recommends that the term 'glaucophane schist' as a rock name is used in the manner given above, and that rocks in the glaucophane-schist facies are explicitly called '*glaucophane-schist facies rocks*'. If a more specific name than 'rocks' is required, then one of the structural root terms (schist, gneiss or granofels) may be used (Schmid *et al.*, Section 2.1; Brodie *et al.*, section 2.1), and reference to the glaucophane-schist facies can be made by adding 'in glaucophane-schist facies'.

2.4.6 Glaucophanite

The term glaucophanite can be, or has been, used in three different ways.

1. As stated above, the SCMR proposes to use the suffix '-ite', appended to a mineral name, for metamorphic rocks that consist of $\geq 75\%$ of that mineral (Schmid *et al.*, Section 2.1). Following this proposal, glaucophanite is a general name for a rock consisting of $\geq 75\%$ of glaucophane or ferroglaucophane. Thus, the term does not need any further definition under the SCMR rules.
2. For Japanese and Russian geologists, among others, glaucophanite is a gneiss or granofels consisting of $\geq 50\%$ glaucophane.
3. The terms 'glaucophanite' and 'glaucophane schist' are used by some scientists as synonyms, in spite of the structural connotation of glaucophane schist.
   The SCMR, therefore, recommends the use as defined under (1) (above).

2.4.7 Blueschist, blueschist facies and blueschist facies rocks

The name 'blueschist' was established by Bailey (1962). At that time doubts had arisen whether glaucophane was stable only at elevated pressures or also under greenschist facies conditions. Jadeite and/or lawsonite were considered to be more reliable pressure indicators and the term 'blueschist' was introduced. It has now become clear that glaucophane requires a high P/T for its formation. Eskola's term 'glaucophane schist' has, therefore, regained its former significance, but in the meantime 'blueschist' has become well entrenched both as a rock and as a facies term.

As a rock name, blueschist is currently defined by the SCMR in a loose descriptive way with a recommendation to replace it by other terms if a more precise description of the rock is required. The proposed SCMR definition is as follows:

Blueschist: schist whose bluish colour is due to the presence of alkali amphibole.

The definition is complemented by:

1. More precise terms should be used wherever possible (e.g. jadeite-bearing glaucophane schist).

A schist containing blue amphibole is more explicitly called glaucophane schist, with the addition of the names of main or critical minerals, for example jadeite-bearing glaucophane-phengite schist, or epidote-glaucophane schist, etc.

2. This term is also used as a facies name and if used in this sense the facies context should be made clear (by saying 'blueschist-facies rock').

As stated above, the terms 'glaucophane schist facies' and 'blueschist facies' are regarded as synonyms (Smulikowski *et al.*, Section 2.2).

2.4.8 Whiteschist

Whiteschist (Schreyer, 1973) is defined by the SCMR in a loose descriptive manner, in a similar way to blueschist:

Whiteschist: light-coloured schist containing kyanite and talc. More precise rock terms should be used wherever possible (e.g. kyanite-talc-phengite schist),

complemented by:

Whiteschists represent Al-rich rocks metamorphosed under the conditions of the eclogite facies.

2.4.9 *Ultrahigh-pressure metamorphism*

The discovery of coesite and diamond in rocks of crustal origin showed that these rocks had been subject to pressures equivalent to those found in the mantle. This increased the range of metamorphic conditions considered to have operated on crustal rocks. The additional field of metamorphism was termed ultrahigh-pressure metamorphism or UHPM for short (Coleman & Wang, 1995; Carswell & Compagnoni, 2003). The SCMR defines UHPM as follows:

Ultrahigh-pressure metamorphism: that part of the metamorphic *P–T* field where pressures exceed the minimum necessary for the formation of coesite. Pressure may be sufficient for the formation of diamond.

2.4.10 *Ultramafic rock names*

As stated in Schmid *et al.* (Section 2.1), the SCMR definitions of the ultramafic rocks, such as harzburgite, lherzolite and websterite, follow those given by Le Maitre (1989, 2002), without implication for the genesis of the rock. If it is desirable to emphasize the metamorphic nature of the rock, mineral–structural root names should be used, for example pyroxene-olivine gneiss. The presence of garnet or other accessory minerals should be indicated by the appropriate prefix.

2.4.11 *Restricted and unnecessary rock names: griquaite, grospydite, mucronite, alkremite*

The SCMR regards the three terms griquaite, grospydite and alkremite as restricted terms and mucronite as an unnecessary term (sensu Schmid *et al.*, Section 2.1). Their further use in international journals is not recommended, although the restricted names may be given if qualified with a full, unambiguous definition. A widely accepted equivalent term is preferable (see below).

Griquaite (Beck, 1907): rock composed of pyrope-rich garnet, diopside ± orthopyroxene.

It may contain diamond. An equivalent term is, for example, orthopyroxene-bearing garnet-diopside gneiss.

Grospydite (Bobrievich *et al.*, 1960): plagioclase-free rock mainly composed of Ca-Al clinopyroxene and calcic garnet together with some kyanite.

An equivalent term is, for example, kyanite-bearing grossular-clinopyroxene gneiss.

Alkremite (Ponomarenko, 1975): ultramafic rock composed of spinel and pyrope-rich garnet, which together form about 75% of the rock and both of which are present in large proportion.

An equivalent term is, for example, spinel-pyrope granofels.

Mucronite (Reinsch, 1977): metamorphic rock mainly composed of jadeitic clinopyroxene, garnet, phengite, quartz ± K-feldspar.

An equivalent term is, for example, jadeite-garnet-phengite schist.

2.5 Very low- to low-grade metamorphic rocks

Péter Árkai, Francesco Sassi and Jacqueline Desmons

2.5.1 Introduction

A Study Group was set up in 1987 under the leadership of P. Árkai to study the nomenclature and systematics of rocks and processes related to the so-called very low- to low-grade metamorphism, that is, from the diagenetic field to the greenschist facies or epizone. The Study Group consisted of 30 scientists from five continents who were specialists in this area and who agreed to cooperate. This paper presents the definitions that were agreed, by repeated iterations, considering on one hand the various opinions within the Study Group (largely gathered through questionnaires), and by discussing these results within the SCMR on the other. Being the result of multiple discussions and compromises, the definitions presented here are recommended for international use.

2.5.2 Very low- and low-grade metamorphism

The SCMR agreed that for the purposes of its discussions and definitions the grade of metamorphism would be taken as equivalent to the temperature of metamorphism. The whole temperature/grade range of metamorphism has been divided into five parts: very low, low, medium, high and very high (Smulikowski *et al.*, Section 2.2) The present paper deals with the nomenclature of the very low- and low-grade rocks. It also covers the transition from diagenesis to metamorphism and from very low grade to low grade (the greenschist facies or epizone).

2.5.3 From diagenesis to metamorphism

DIAGENESIS

According to the SCMR (Smulikowski *et al.*, Section 2.2), metamorphism is: 'a process involving changes in the mineral content/composition and/or microstructure of a rock, dominantly in the solid state. This process is mainly due to an adjustment of the rock to physical conditions that differ from those under which the rock originally formed and that also differ from the physical conditions normally occurring at the surface of the Earth and in the zone of diagenesis. The process may coexist with partial melting and may also involve changes in the bulk chemical composition of the rock.' Consistent with this definition, the term diagenesis covers the lowest temperature part of the changes in the outer part of the Earth's crust, excluding weathering.

Diagenesis (sensu lato): all the chemical, mineralogical, physical and biological changes undergone by a sediment after its initial deposition, and during and after its lithification, exclusive of superficial alteration (weathering) and metamorphism.

The changes involved in diagenesis are the result of such processes as compaction, cementation, reworking, authigenesis, replacement, crystallization, leaching, hydration, dehydration, bacterial action and formation of concretions. These processes occur under conditions of pressure and temperature that are normal at the Earth's surface and in the outer part of the Earth's crust.

Diagenesis (sensu lato) may be subdivided into:

- *shallow diagenesis* (diagenesis sensu stricto): the chemical, mineralogical, physical and biological changes that take place in a sediment under physical conditions that do not differ significantly from those under which the sediment originated. It is characterized by the absence of alteration of detrital minerals.
- *deep diagenesis:* changes that are characterized by clay mineral reactions (such as the transformation of smectite to illite, kaolinite to dickite, etc., and the increase of the proportion of illite layers in interstratified clay minerals).

Instead of shallow and deep diagenesis the adjectives 'early' and 'late' are also used in the literature. However, in order to avoid the time connotation implied by these adjectives, the SCMR prefers the terms 'shallow' and 'deep'.

Deep diagenesis is the term recommended by the SCMR, as the equivalent to the three terms

Metamorphic Rocks: A Classification and Glossary of Terms. Recommendations of the International Union of Geological Sciences, eds. Douglas Fettes and Jacqueline Desmons. Published by Cambridge University Press. © Cambridge University Press 2007.

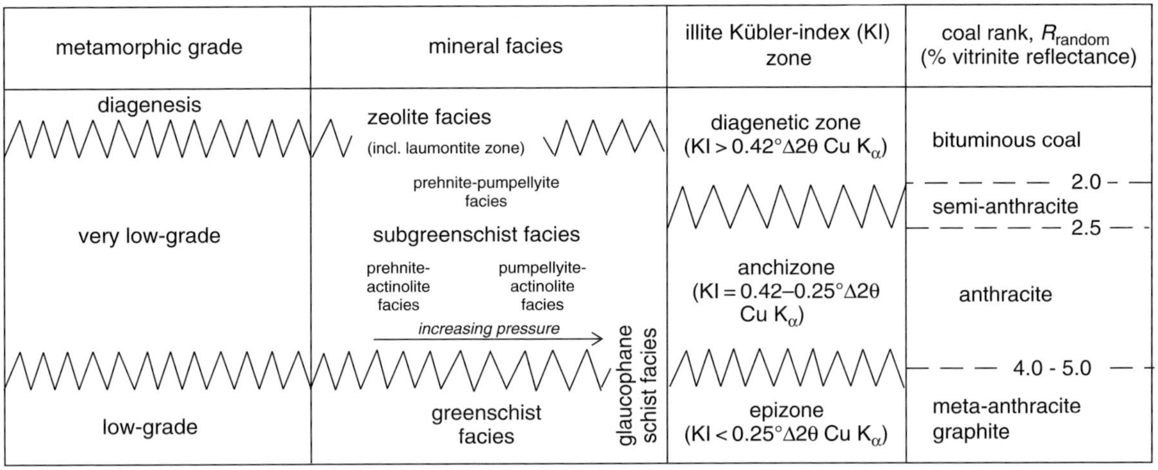

Fig. 2.5.1 Comparison of mineral facies, illite Kübler index (KI) 'crystallinity' zones and coal rank in the diagenetic, very low- and low-grade metamorphic realms. Zigzag lines represent uncertainties of correlation. Scheme simplified after Frey (1987b), Kisch (1987) and Merriman and Frey (1999).

epigenesis, katagenesis and catagenesis of Russian authors, and middle + late or deep burial stage diagenesis of Müller (1967) and Dunoyer de Segonzac (1970).

VERY LOW-GRADE METAMORPHISM: THE TRANSITION ZONE

The transition zone between diagenesis (sensu lato) and metamorphism, effectively the field of very low-grade metamorphism, is characterized by the gradual change of various features, which affect the rocks through this zone up to their partial or complete alteration into metamorphic rocks.

However, in the context of rock nomenclature, the main problem is that the most critical changes, marking this transition, are not visible with the naked eye because they commonly occur only on the microscopic or submicroscopic scale. Thus, the related nomenclature requires an important exception to the SCMR principle (Schmid *et al.*, this vol.), which states that, wherever possible, definitions are based on features that can be seen with the naked eye.

Indeed, on a mesoscopic scale these transitional rocks commonly display characteristics identical with, or very similar to, those of their non-metamorphic equivalents. Some of these very low-grade rocks can be recognized under the microscope but most can only be identified using other instrumental techniques. These techniques include measurements of the illite Kübler index (formerly called 'crystallinity') and coalified organic matter order–disorder by means of X-ray powder diffraction, measurements of vitrinite reflectance by means of optical microscopy, fluid inclusion thermobarometry, etc.

Different criteria have been used for the characterization and subdivision of the transitional field between diagenesis and low-grade metamorphism in various rock types. Consequently, various systems and nomenclatures have been established (e.g. specific mineral associations, illite Kübler index zones, microstructural zones, coal rank scales). See Fig. 2.5.1 for a schematic comparison of these systems. Details and comprehensive interpretations can be found in the textbooks edited by Frey (1987a) and Frey and Robinson (1999), including the studies of Frey (1987b), Liou *et al.* (1987), Mullis (1987), Teichmüller (1987), Alt (1999), Merriman and Frey (1999), Merriman and Peacor (1999), and Robinson and Bevins (1999).

The temperature (and pressure) boundaries related to the sequence of stages or zones of each nomenclature scheme are not defined exactly, and in most cases they do not coincide with those of the other schemes. The correlation between the nomenclature schemes and absolute temperature is full of uncertainties, mostly because of the greatly differing nature of the disequilibrium processes considered, the transitional open- to semi-closed systems, and in consequence, the great variations in the chemical effects of the fluids present.

Because the transition between diagenesis and metamorphism is gradual, any boundary between these two fields can only be arbitrarily defined. In practice, it is not possible to establish an isothermal boundary that can be applied to all, or even to the majority, of the rock types. Furthermore, rocks having different bulk composition, different grain size, different amount of strain, different porosity and permeability, etc., may react at relatively higher or lower temperatures, so that at a given depth or temperature a given rock may react whilst others may remain unchanged. Thus, the temperature of the beginning of metamorphism may vary considerably between rock types.

THE PROBLEM OF MIXED SEQUENCES

The problem of different rock types, as discussed above, raises a related and important question. If the very low-grade character of a rock in a lithologically mixed sequence can be determined (e.g. as the prehnite-pumpellyite facies (see below) in basic volcanic rocks), can that classification be extrapolated to other rock types in the sequence even though they do not display any measurable alteration in hand specimen (and, commonly, not even at the microscopic scale)? Alternatively, should different grade terms be used for the different rock types?

To put the question another way. Within the same rock pile, depending on the chemical, mineralogical and physical properties of the different rocks, some (e.g. carbonate rocks) may show hardly any signs of alteration, whereas others that underwent the same T–P_l–P_f conditions[7] are clearly recrystallized, and therefore must be classified as metamorphic rocks, and receive metamorphic rock names. In these cases, should the metamorphic name/grade be extrapolated to the unaltered rocks, or should they be given the most appropriate, non-metamorphic name, disregarding that of their surroundings?

In response to this very difficult question the SCMR recommends the following.

1. Those rocks that display at least one observable or measurable sign of very low-grade metamorphism (i.e. diagnostic minerals or mineral assemblages, appropriate values of illite Kübler index, appropriate values of vitrinite reflectance, etc.) should be given metamorphic rock names.
2. Those rocks that do not show any sign of very low-grade metamorphism should be given the appropriate non-metamorphic rock name, regardless of their close spatial and genetic relationship to the 'metamorphic' rocks.

This implies that metamorphic rock names may alternate with sedimentary or magmatic ones in a geologic profile or map. This might give rise to cartographic boundaries that would require specific explanations.

2.5.4 The various nomenclature systems of very low- and low-grade metamorphism

The possibilities for discrimination between diagenesis, very low- and low-grade metamorphism are schematically shown in Fig. 2.5.1.

ROCKS WITH DIAGNOSTIC MINERALS AND MINERAL ASSEMBLAGES (MAINLY BASIC ROCKS)

(a) Zeolite and subgreenschist facies

In rocks of basic to intermediate compositions, the occurrence of zeolites is attributed to alterations at low temperatures in the presence of CO_2-poor or CO_2-absent aqueous fluids. The SCMR defines the zeolite facies as follows:

> **Zeolite facies**: a facies (in the sense of Eskola, 1920) embracing all mineral assemblages that include various zeolites plus quartz, irrespective of the mode of origin, whether metamorphic (including hydrothermal) or diagenetic (Coombs *et al.*, 1959; Coombs, 1971; Boles & Coombs, 1977).

[7] P_l – lithostatic pressure; P_f – fluid pressure

In silicate rocks (see Bucher & Frey, 1994), the first appearance of one or some of the following minerals, Fe-Mg-carpholite, glaucophane, lawsonite, laumontite, paragonite, prehnite, pumpellyite or stilpnomelane, is commonly regarded as the beginning of metamorphism (although the status of laumontite is questionable).[8]

The SCMR recommends that only a few metamorphic facies should be used as a general rule (Smulikowski *et al.*, Section 2.2), but leaves open the possibility of using other facies or subfacies if required to describe a specific region, provided that these additional facies or subfacies are clearly defined. Although the very low- and low-grade *P–T* field covers quite a small part of the whole metamorphic field, various authors have suggested that it should be subdivided into several facies or subfacies to cover the situation with particular lithologies. They also suggest different methods for defining the additional facies or subfacies.

The following are a few of these suggestions and the recommended SCMR definitions. As grade is taken as equivalent to temperature (Smulikowski *et al.*, Section 2.2) it follows that metamorphic rocks formed at temperatures lower than the low-temperature limit of the greenschist facies belong to the very low-grade field. The high *P/T* and high-*P* part of this field includes a small part of the glaucophane-schist (blueschist) facies (Smulikowski *et al.*, Section 2.2).

The part of the field of very low-grade metamorphism characterized by pressures lower than those of the glaucophane-schist facies has been called the **subgreenschist facies** (e.g. Bucher & Frey, 1994; Merriman & Frey, 1999). The subgreenschist facies embraces various metabasite facies/subfacies (Fig. 2.5.1) characterized by the diagnostic mineral assemblages of prehnite + pumpellyite, prehnite + actinolite, pumpellyite + actinolite, and also, according to certain authors, laumontite.

[8] Coombs *et al.* (1959), and Boles and Coombs (1977) argue that the zeolite facies should be regarded as a *mineral facies*, irrespective of the origin of the mineral assemblage. They regarded it as unrealistic to arbitrarily interpret laumontite as metamorphic and heulandite or analcime (plus quartz) as diagenetic.

The SCMR defines these facies/subfacies as follows.

The **prehnite-pumpellyite facies** is characterized in metasandstones and metavolcanic rocks of appropriate composition by the presence of prehnite and/or pumpellyite in the absence of zeolites, lawsonite or jadeite. Quartz-albite-chlorite-prehnite and/or pumpellyite may coexist stably (Coombs, 1960).

The **pumpellyite-actinolite facies** is characterized by the mineral association of pumpellyite-actinolite-quartz (± chlorite, albite and epidote) and by the lack of prehnite (Hashimoto, 1966).

The **prehnite-actinolite facies** is characterized by the mineral association of prehnite-actinolite-epidote (± chlorite, albite, quartz and titanite) and by the absence of pumpellyite in rocks of appropriate bulk composition (mostly metabasic rocks and their clastic derivates) (Liou *et al.*, 1985).

(b) The transition to the greenschist facies

In the case of rocks of suitable composition (mostly of basic to intermediate composition), the first appearance of the diagnostic mineral assemblage actinolite + epidote + chlorite + albite + quartz in the absence of pumpellyite and/or prehnite indicates the onset of low-grade metamorphism, that is, the transition from very low- to low-grade metamorphism, or from subgreenschist to greenschist facies.

The first appearance of the lawsonite + chlorite + albite association or of sodic amphibole lies within the very low-grade field, indicating high *P/T*.

Although very fine-grained chloritoid may appear in very low-grade (anchizonal) metapelites (commonly hard to detect by optical microscopy), chloritoid typically occurs in low-grade (greenschist facies) rocks.

ROCKS DEVOID OF DIAGNOSTIC MINERALS AND MINERAL ASSEMBLAGES

In many common rocks, such as 'normal' marine pelites and carbonate rocks, no diagnostic minerals and mineral assemblages form in the very low-grade field. In these rocks, the transitions from non-metamorphic to very low-grade and from very low-grade to low-grade metamorphic domains take place through the diagenetic zone, the anchizone and the epizone (Fig. 2.5.1), each zone being characterized by specific values of the illite Kübler index (KI), which is measured on the <2 μm fraction of clay-rich clastic rocks following the recommendations on sample preparation, X-ray diffraction settings and inter-laboratory standardization summarized by Kisch (1991). For comparing the various sample preparation techniques, the procedure and standards suggested by Warr and Rice (1994) are useful. Concerning the proper nomenclature of indices (formerly called 'crystallinity') expressing the reaction progress of illite-muscovite and chlorite in diagenetic and low-temperature metamorphic conditions, the authors refer to the recommendations of Guggenheim *et al.* (2002).

Diagenetic zone [or more precisely, diagenetic illite Kübler index zone]: zone characterized by illite Kübler index (KI) values greater than $0.42°\Delta 2\theta CuK_\alpha$ (after Kübler, 1967, 1968, 1984).

Anchizone: transitional zone between the diagenetic zone and the epizone characterized by illite Kübler index (KI) mean values between 0.42 and $0.25°\Delta 2\theta CuK_\alpha$ (after Kübler, 1967, 1968, 1984).

Metamorphism in this zone is consistently called **anchimetamorphism**, which roughly corresponds to very low-grade metamorphism. Note: the term 'anchimetamorphism' was originally introduced by Harrassowitz (1927) to indicate changes in mineral content of rocks under pressure and temperature conditions prevailing between the Earth's surface and the zone of metamorphism.

Epizone: zone of low-grade metamorphic rocks characterized by illite Kübler index (KI) mean values less than $0.25\,°\Delta 2\theta\ CuK_\alpha$ (after Kübler, 1967, 1968, 1984).

Note: the term epizone was originally proposed by Grubenmann (1904) to indicate shallow depth of metamorphism. At present, however, this term is mainly used in the context of illite Kübler index investigations.

In addition to the illite Kübler index, some other characteristics such as vitrinite reflectance (Fig. 2.5.1), chlorite Árkai index (see Gugenheim *et al.*, 2002) or Conodont Colour Alteration Index (CAI) can also be used for determining diagenetic and metamorphic zones. Note, however, that only approximate correlations can be established between these parameters. Examples and explanations of common deviations from the generalized scheme are given by Kisch (1987) and Merriman and Frey (1999).

2.5.5 Protolith names and definitions of specific rock names

Very few specific rock terms exist in the realm of very low-grade metamorphism. This can be explained by the fact that, with the exception of the structure of slates, mesoscopic features characteristic of this metamorphic grade do not exist.

As the original (sedimentary or magmatic) features of the protoliths can usually be easily recognized in this realm, the general use of the sedimentary or magmatic name of the protolith prefixed with '*meta*' or '*meta-*' is highly recommended (e.g. meta-andesite, metasandstone), excluding the very few cases when *specific rock terms* are available (e.g. slate). The SCMR recommends the use of the prefix *meta or meta-* in combination with a protolith-based name only when the protolith is easily identifiable or obvious. It must never be used for former metamorphic rocks (for example, 'meta-phyllite' is not an acceptable term) (Schmid *et al.*, Section 2.1).

The *specific rock terms* are defined below.

Slate: an ultrafine- or very fine-grained rock displaying slaty cleavage.

Slate is usually of very low metamorphic grade, although it may also occur in low-grade conditions. The definition of *slaty cleavage* is given by Brodie *et al.* (Section 2.3) as: 'a type of continuous cleavage in which the individual grains are too small to be seen by the unaided eye.' The definition of *continuous cleavage* is given by Brodie *et al.* (Section 2.3) as: 'A type of cleavage characterized by the preferred orientation of all inequant mineral constituents of a rock, and in which the cleavage planes are developed at the grain-size scale.'

Phyllite: a fine- to medium-grained rock characterized by a lustrous sheen and a well-developed schistosity resulting from the parallel arrangement of phyllosilicates. Phyllite is usually of low metamorphic grade.

Palaeovolcanite rock terms: the SCMR recommends following the definitions for volcanic rock terms given by Le Maitre (1989, 2002) and that the recommendations given in the present paper are used for rocks which have undergone any kind of very low-grade metamorphic alteration.

Spilite: an altered basic to intermediate, volcanic or subvolcanic rock in which the feldspar is partially or completely composed of albite and is typically accompanied by chlorite, calcite, quartz, epidote, prehnite, or other low-temperature hydrous crystallization products.

Preservation of eruptive (volcanic and subvolcanic) features is an important characteristic of spilites. The term spilite may be classified as metabasalt or meta-andesite, as appropriate, regardless of its origin.

The 'spilite problem', debated in the geological literature for many decades, is of special interest for researchers working in the field of very low- and low-grade metamorphism. This is the reason why the term 'spilite' has been considered by the SCMR, despite the fact that the Subcommission on the Systematics of Igneous Rocks include this term in their glossary (Le Maitre, 1989, 2002).[9] The SCMR recommends that the definition given by Le Maitre should be expanded to include rocks of intermediate composition, because many spilites have an andesitic rather than a basic composition.

Greenschist: schist whose greenish colour is due to the presence of minerals such as actinolite, chlorite and epidote. More precise terms should be used whenever possible (e.g. epidote-bearing actinolite-chlorite schist).

[9] Note: the term 'spilite' is not a root or recommended name in the IUGS classification of igneous rocks, although it is included in the glossary (Le Maitre, 2002).

Greenstone: a granofels whose greenish colour is due to the presence of minerals such as actinolite, chlorite and epidote. More precise terms should be used whenever possible (e.g. chlorite-epidote granofels).

2.5.6 *Terms not recommended for general use*

In addition to the terms listed above, the Study Group also discussed some other terms, even though they are not recommended for international use. They are listed below, with the corresponding explanations.

Brownstone facies: a low-temperature mineral facies encompassing ocean-floor weathering and low-temperature hydrothermal alteration of the ocean floor.

The most widespread secondary minerals present under oxidizing conditions include a K-rich dioctahedral iron illite resembling celadonite. This replaces olivine, occupies vesicles, replaces interstitial glass, and eventually replaces augite. Under reducing conditions its place is taken by saponite, and pyrite is also characteristic. Plagioclase may be replaced by clay minerals or potassium feldspar. Glassy rinds of basaltic pillow lava alter to palagonite in association with phillipsite and other low-temperature zeolites and calcite.

Catagenesis = katagenesis: terms used, especially by Russian authors (see above), to indicate changes occurring in (an already lithified) sedimentary rock buried under a distinct covering layer, characterized by P–T conditions that are significantly different from those of both deposition and metamorphism.

This term is equivalent to 'epigenesis' or to 'middle and deep diagenesis'. It is subdivided into early catagenesis (= middle diagenesis = shallow epigenesis) and late catagenesis (= deep diagenesis = deep epigenesis).

Cryptic metamorphism: a term for metamorphism that can be detected only by special study, for example, vitrinite reflectance, illite Kübler index, etc., and not by ordinary hand-specimen or microscopic study.

Epigenesis: a term used, especially by Russian authors, to indicate changes, transformations or processes, occurring at low temperatures and pressures, that affect sedimentary rocks after their compaction, excluding superficial alteration (weathering) and metamorphism (= catagenesis = middle and deep diagenesis). Subdivided into early and deep or late epigenesis.

Incipient regional metamorphism: a general term for the stages of mineral modifications as characterized by the appearance of the attributes of the anchizone (anchimetamorphism).

It is approximately equivalent to the early metagenesis or with the greater part of Winkler's (1974) 'very low-grade metamorphism', including the higher-T part of the 'pumpellyite-prehnite-quartz facies', and probably, all of the 'lawsonite-albite schist' and 'glaucophane-lawsonite schist facies' in the sense of Winkler (1974).

Metagenesis: a term used, especially by Russian authors (see above), to indicate a more advanced stage of post-diagenetic alteration than epigenesis or catagenesis.

It is subdivided into 'early metagenesis' (which roughly corresponds to the anchizone) and 'late or deep metagenesis' (which is more or less equivalent to the epizone or the chlorite zone of the greenschist facies).

2.6 Migmatites and related rocks

Wolfhard Wimmenauer and Inge Bryhni

2.6.1 *Introduction*

A Study Group under the leadership of W. Wimmenauer was set up to look at the nomenclature of migmatites and related rocks. In addition to that small group, a number of worldwide specialists were consulted on their views on the more important terms being considered. Their answers were greatly appreciated and provided valuable contributions to the deliberations. This paper presents the definitions together with some notes explaining the reasoning.

In the discussions it turned out that the definition of migmatites and their subgroups is not an easy task. Rosenbusch's statement: 'The essence of rocks is transition' is particularly valid for migmatites. They form, in their total spectrum, a continuous transition from metamorphic to plutonic rocks. The establishment of limits within such a continuum is very difficult and the application of quantitative criteria virtually impossible. Thus, many of the 'definitions' presented below are characterizations of certain prominent rock types rather than definitions sensu stricto. Their application to a natural rock will often demand some scientific experience.

It should also be noted that the scale of migmatite structures is such that they mainly require definitions which refer to rock masses greater than the preferred hand-specimen size.

The work was greatly aided by the existence of two comprehensive glossaries of terms bearing on migmatites, namely those of Dietrich and Mehnert (1960), and Mehnert (1968).

The following pages present the definitions proposed by the SCMR; where appropriate, some comments on the reasoning leading to them are added.

2.6.2 *Definitions of terms*

Migmatite: a composite silicate metamorphic rock, pervasively heterogeneous on a meso- to megascopic scale. It typically consists of darker and lighter parts. The darker parts usually exhibit features of metamorphic rocks whereas the lighter parts are of igneous appearance (see also leucosome, melanosome, mesosome, neosome, palaeosome). Wherever minerals other than silicates and quartz are substantially involved, it should be explicitly mentioned.

The essential elements of the above definition received wide approval from the Working Group. The last sentence makes allowance for the comments of some contributors who pointed out that migmatitic structures, as described in the definition, may also occur in non-silicate rocks. For the sake of clearness and simplicity, other versions of the definition, which might also cover very rare and uncommon varieties of migmatites, were eventually abandoned.

Anatexis: melting of a rock.

The term is used irrespective of the proportion of melt formed, which may be indicated by adjectives such as *initial, advanced, partial, differential, selective, complete*, etc.

Migmatization: process of formation of a migmatite.

Leucosome: the lightest-coloured parts of a migmatite.

Mesosome: the part of a migmatite that is intermediate in colour between leucosome and melanosome. If present, the mesosome is mostly a more or less unmodified remnant of the parent rock (protolith) of the migmatite.

In spite of the near-identity of most mesosomes with the palaeosome, a purely descriptive term for the intermediate parts of a migmatite appeared desirable.

Melanosome: the darkest parts of a migmatite, usually with prevailing dark minerals. It occurs between two leucosomes, or, if remnants of the more or less unmodified parent rock (mesosome) are still present, it is arranged in rims around these remnants.

Palaeosome: part of a migmatite representing the parent rock (cf. mesosome).

Metamorphic Rocks: A Classification and Glossary of Terms. Recommendations of the International Union of Geological Sciences, eds. Douglas Fettes and Jacqueline Desmons. Published by Cambridge University Press. © Cambridge University Press 2007.

Neosome: The newly formed parts of a migmatite (metatects and restites).
Restite: remnant of a metamorphic rock from which substantial amounts of the more mobile components have been extracted without being replaced.
Resister: rock offering greater resistance to the processes of granitization than another by virtue of its composition or its 'impenetrable' fabric.

Whereas 'restites' are rock portions that have undergone essential changes of their earlier composition, 'resisters' are rocks that have survived the formation of the surrounding migmatite (or granite) without significant changes to their mineralogical and chemical composition. Although the term 'resister' is not widely used, Mehnert's definition is proposed here as a restricted term. Its meaning is not easily covered by another, more frequently used term.

Although a few members of the Working Group would like to confine the definition to crustal processes, the SCMR recommendation is that it should also be applicable to mantle processes.

Anatexite: two versions of the definition were discussed in the Subcommission; the first one (a, below) was preferred because the second version could be applied to any magmatic rock believed to be of anatectic origin.

(a) Rock still showing the evidence of *in situ* formation by anatexis.
(b) Any rock showing structural and/or compositional evidence of formation by anatexis.

Metatexis: initial stage of anatexis where the parent rock (palaeosome) has been partly split into a more mobile part (metatect) and a non-mobilized (depleted) restite (cf. palaeosome, metatect, restite).
Metatexite: a variety of migmatite with discrete leucosomes, mesosomes and melanosomes (cf. leucosome, mesosome, melanosome).

Although the term refers directly to the genetic term 'metatexis', a descriptive definition of the rock type is also required.

Metatect: discrete, mostly light-coloured body in a migmatite formed by metatexis.
Arterite: a type of migmatite where the darker parts are injected by veins of lighter material (leucosomes) introduced from outside.

Wherever the introduced material is not lighter than the surrounding rock, it should be explicitly mentioned.

Venite: a type of migmatite in which the material of the lighter veins (leucosomes) is extracted from the parent rock.

As with 'arterite', the definition is explicitly a genetic one. As a non-genetic name for veined rocks, Scheumann's (1936a) 'phlebite' is proposed (see below).

Phlebite: a veined rock; the veins may have been injected from outside or exuded *in situ*.
Diatexis: advanced stage of anatexis where the dark-coloured minerals are also involved in melting; the melt formed has not been removed from its place of origin (cf. metatexis).
Diatexite: a type of migmatite where the darker and the lighter parts form schlieren and nebulitic structures which merge into one another (cf. diatexis).

Although the term refers directly to the genetic term 'diatexis', a descriptive definition of the rock type is also required.

Nebulite: migmatite with diffuse relics of pre-existing rocks or rock structures.
Agmatite: migmatite with breccia-like structure.
Palimpsest structure: structure in a migmatite or granitized rock that can be recognized as pre-migmatitic (or pre-granitic).

Definition of Mehnert (1968) unchanged. The term is also used in a more general meaning for relict features.

Dictyonite: a type of migmatite with a reticulated structure formed by a network of small veins.
Schollen: in a migmatite, blocks or rafts of palaeosome within the neosome; the structure is similar to agmatite but the

neosome is more abundant so that the disrupted blocks float like rafts.

Stromatite: a type of migmatite with regular layers, the layers having two or more different compositions or appearances, for example the alternation of mesosome and leucosome.

Palingenesis: formation of a new magma by complete or nearly complete melting of pre-existing rocks.

Granitization: a comprehensive term for processes by which pre-existing rocks are converted to granitoids (melting, pervasive influx of chemical components such as silica, potash, soda or other means of pervasive transformation).

On the questionnaire sent to the worldwide Working Group, 79% of the answers agreed that the term should be defined. Some contributors proposed to abolish the term, but it is widely used, comprehensive, and cannot be replaced by a better one. Some suggested that the term should be restricted to metasomatic processes. However, we consider the term to have a general meaning, and special cases can easily be specified by an appropriate adjective (e.g. metasomatic granitization, anatectic granitization).

Degranitization: a process by which a rock is depleted in chemical components that are significant in making up a granitoid, essentially silica and potash ± soda.

Several members of the Working Group proposed to abolish the term. However, it has been used since 1955 and designates a process demonstrated in several well-studied areas, for example the granulites of the Ivrea Zone and Calabria.

Feldspathization: formation of feldspar, due to metasomatism.

Metablastesis: Preferred crystallization and growth in size of a mineral or a group of minerals by metamorphic (including metasomatic) processes.

'Preferred' is used to emphasize the fact that certain minerals grow to larger sizes than others.

Metamorphic differentiation: mechanical redistribution of minerals by species and/or segregation of chemical components during metamorphism to form an inhomogeneous structure of two or more species within a rock body.

The wording: 'to form an inhomogeneous structure' was chosen because redistribution may also result in forming a more homogeneous body. Particular attention was given to the distinction between 'redistribution' of solid minerals and 'segregation' of chemical components, transported in solution.

Ultrametamorphism: metamorphism under the extreme upper range of temperatures and pressures as a result of which rocks suffer complete or almost complete *in situ* melting.

Ptygmatic folding: originally defined by Sederholm (1907) to describe contorted and folded granitic veins which characteristically occur in migmatites. The term is now used more widely to describe a form of folding where single isolated layers of relatively high competence material are enclosed in a matrix of lower competence and strongly shortened (see Ramsay & Huber, 1983).

2.6.3 Concluding remarks

The members of the Study Group feel that the terms presented above are useful and even necessary, including rather new ones like 'mesosome'. They believe that the terms defined above are the minimum required to describe and classify common natural phenomena and well-known processes associated with migmatites. Additional information on the category and source of the terms is contained in the glossary. Various other terms related to migmatites which were considered unnecessary by the Study Group are also given in the main glossary and the glossary lists.

2.7 Metacarbonate and related rocks

OLEG ROSEN, JACQUELINE DESMONS
and DOUGLAS FETTES

2.7.1 *Introduction*

A Study Group under the leadership of O. Rosen was set up to prepare the nomenclature of metacarbonate and related rocks. Questionnaires were sent to a worldwide panel of specialists on metasediments. Eleven agreed to cooperate with the Study Group. This paper presents the results of this work.

2.7.2 *Basis of the classification scheme*

SCOPE OF THE CLASSIFICATION

The following discussion is based on the classification of metacarbonate rocks (here taken to include metacarbonate-bearing rocks): the derived nomenclature is applicable to all carbonate and carbonate-bearing metamorphic and metasomatic rocks including all calc-silicate rocks (which may be carbonate-free).

DIFFICULTIES IN ERECTING
A CLASSIFICATION SCHEME

In considering a nomenclature scheme for metacarbonate rocks the SCMR noted the lack of any systematic data on which to base their definitions. This situation existed even though the extent of metacarbonate rocks in regional fold belts is comparable to many other well-documented rock types (e.g. basic rocks). A brief summary of existing terminologies is given by Rosen *et al.* (2005). This highlights the great diversity of terms used by authors and the variable use of the same terms between authors.

In addition, the wide variety of mineral assemblages, even in a simple prograde metamorphic sequence of a single impure-carbonate rock type, is a typical feature of metacarbonate rock sequences and creates a major difficulty in erecting a systematic classification based on modal content.

ERECTION AND ANALYSIS OF A DATABASE

In order to address these difficulties a sizeable database of chemical and mineral compositions was collected and used to define systematic variations in the composition of metacarbonate rocks. This database is representative of almost all known compositional varieties, from silicate-free to carbonate-free types. The data were derived from regional metamorphic belts, and comprise rocks from a wide geographical range and representatives of greenschist, amphibolite and granulite facies terrains.

Although the data were selected to be broadly representative of metacarbonate rocks, it should be noted that the study is only a preliminary attempt to produce such a systematic approach.

An analysis of this database was then used to underpin the nomenclature divisions presented below.

The proposed classification is based on an analysis of modal carbonate mineral contents supported by chemical compositions. The database, its analysis and discussion are published separately (Rosen *et al.*, 2005).

VARIETY OF METACARBONATE ROCKS

Metacarbonate rocks form a large and complicated group of rocks ranging from pure carbonate to almost carbonate-free calc-silicate varieties. The metacarbonate rocks discussed in the database referred to above were derived from sedimentary basin sequences overprinted by near-isochemical regional metamorphism. Limestone and dolostone are precursors of marbles. Carbonate-bearing mudstones, sandstones, tuffaceous and evaporitic sediments and marlstones are the precursors of carbonate-silicate and calc-silicate rocks. The latter rock types are more widespread than marbles, reflecting their relative abundance in the sedimentary sequence. The metamorphism of the impure carbonate rocks is characterized by decarbonization reactions, whose net result may be the total consumption of the original carbonate minerals. In this way an original impure carbonate rock may be converted into a carbonate-silicate rock while a carbonate-bearing mudstone may be converted into a calc-silicate rock. By contrast, decarbonization reactions in pure carbonate rocks (limestones and dolostones) are rare because of the high thermal stability of carbonate minerals under most metamorphic conditions. Therefore

Metamorphic Rocks: A Classification and Glossary of Terms. Recommendations of the International Union of Geological Sciences, eds. Douglas Fettes and Jacqueline Desmons. Published by Cambridge University Press. © Cambridge University Press 2007.

marbles are found in virtually all metamorphic environments.

ISOCHEMICAL METAMORPHISM, CONTACT METAMORPHISM AND CONTACT METASOMATISM

A distinction may be drawn between metamorphic carbonate rocks produced under isochemical metamorphism and contact metasomatism. The database and associated analytical results mentioned above are concerned only with rocks that are the products of regional (isochemical) metamorphism. The classification scheme is, therefore, supported by a study of the systematics of regionally metamorphosed carbonate rocks. However, the recommended names given below apply to all metamorphic carbonate rocks regardless of their genesis.

In addition, certain specific terms given in the associated SCMR papers on 'Metasomatism' (Zharikov *et al.*, Section 2.9) and 'Contact metamorphism' (Callegari & Pertsev, Section 2.10) also relate to metacarbonate rocks (e.g. **skarn**, **contact marble**, **predazzite**). The user is referred to these papers or the SCMR glossary for the relevant definitions.

2.7.3 The classification scheme

Metacarbonate rocks may be named using the systematic compound names with structural root terms and mineral prefixes (e.g. calcite-diopside granofels) as set out in the SCMR recommended scheme (Schmid *et al.*, this vol.). This section, however, is mainly concerned with the specific names (see Schmid *et al.*) related to metacarbonate rocks. These are defined below.

SUBDIVISIONS OF METAMORPHOSED Ca- AND Mg- CARBONATE ROCKS

The SCMR classifications are based on features visible or measurable at the hand-specimen scale (Schmid *et al.*, Section 2.1). In metacarbonate rocks the modal content of carbonate minerals is the most evident and reliable criterion and may be selected as the basis of classification. The analysis of the database discussed above (Rosen *et al.*, 2005) showed significant boundaries at 95%, 50% and 5% of modal Ca- and Mg-carbonate minerals and these boundaries are used to define the main nomenclature fields. Thus, the 95% level separates pure marble from impure marble; the 50% level separates marble from carbonate-silicate rock; and the 5% level separates carbonate-silicate rock from calc-silicate rock and carbonate-bearing silicate rock. These divisions are shown on Fig. 2.7.1.

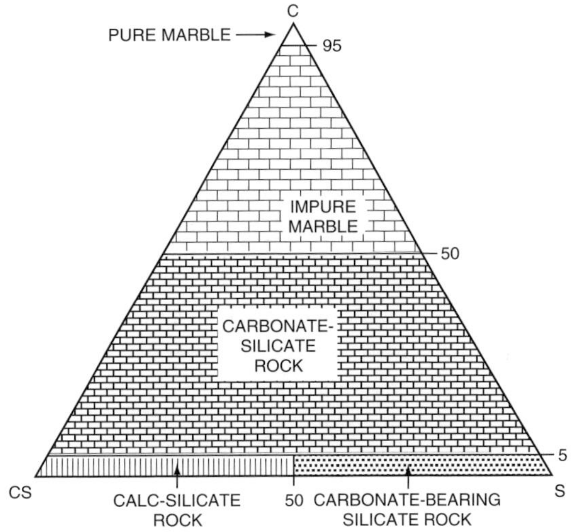

Fig. 2.7.1 Descriptive diagram for metacarbonate and related rocks. C, carbonate minerals: calcite, dolomite and aragonite. CS, calc-silicate minerals; see text for list of minerals. S, all other silicate minerals. 'Rock' is used as a non-specific root term; schist, gneiss or granofels may be used instead, as appropriate.

In these names 'rock' is used as a general term; the more specific structural terms *schist, gneiss* and *granofels* may be used as appropriate.

The SCMR considered various possible fields for calc-silicate rocks but concluded that, in general, these rocks only have carbonate minerals present as minor phases. Therefore, the calc-silicate field is restricted to rocks with less than 5% modal carbonate minerals.

The name *carbonate-bearing silicate rock* is determined by the SCMR rules on mineral prefixes.

A modal value of 50% or greater of calc-silicate minerals separates calc-silicate rocks from carbonate-bearing silicate rocks (Fig. 2.7.1).

The SCMR noted that there was no precise definition of **calc-silicate minerals** that could be used to define this latter boundary. Consideration

Table 2.7.1 *List of Ca-silicates arranged according to their CaO wt% contents (on the base 100% including volatile components such as H_2O, CO_2, C, F, etc.).*

This compilation by E. Callegari is based on data from the *Handbook of Mineralogy* (Bideaux *et al.*, 1995). The minerals given in plain letters have been found as main components of metacarbonate rocks, whereas those in italics are secondary or rare minerals (for detailed references see Bideaux *et al.*, 1995).

CaO > 60%	Larnite/Bredigite, Spurrite/Paraspurrite, Cuspidine, Hatrurite, Jassmundite, Nagelschmidtite, Reinhardbraunsite, Rustumite, *Defernite*
CaO 50 to 60%	Tilleyite, Rankinite, Merwinite, Hillebrandite, Bultfontainite, Fukalite, Hydroxylellestadite, Chlorellestadite, *Killalaite, Foshagite, Trabzonite*
CaO 40 to 50%	Gehlenite/Åkermanite, Wollastonite, Katoite, Baghdadite, Götzenite, Gugiaite, *Xonotlite, Scawtite, Harkerite, Afwillite, Aminoffite, Jennite, Oyelite, Riversideite, Rosenhanite, Zeophyllite*
CaO 30 to 40%	Grossular, Andradite, Uvarovite, Hydrogrossular, Schorlomite, Vesuvianite, Datolite, Pectolite, Fe-Bustamite, Hardystonite, Jeffreyite, Kirschsteinite, Sarcolite, Monticellite, *Cebollite, Gyrolite, Bicchulite, Tobermorite/Clinotobermorite, Plombierite, Kimzeyite, Juanite, Kamaishilite*
CaO 20 to 30%	Diopside, Hedenbergite, Zoisite/Clinozoisite, Epidote, Titanite, Meionite, Pumpellyite, Prehnite, Esseneite, Charoite, Latiumite, Bavenite, Bustamite, Danburite, Glaucochroite, Johannsenite, Axinite, Mukhinite, Malayaite, *Carletonite, Chantalite, Natroapophyllite, Nekoite, Strätlingite, Ruizite*
CaO 15 to 20%	Bytownite, Anorthite, Lawsonite, Mizzonite, Piemontite, Augite, Babingtonite, Omphacite, Partheite, Serendibite, *Junitoite, Liottite*
CaO 10 to 15%	Labradorite, Tremolite, Actinolite, Fe-Actinolite, Ilvaite, Margarite, Clintonite, Britholite, Calderite, Dorrite, Ekanite, Rhönite, Welshite, *Thompsonite, Gismondine, Laumontite, Afghanite, Allanite, Franzinite, Trimerite, Wenkite, Chabazite*
CaO < 10%	Sodic Plagioclases, Sodic Scapolites, Lazurite, *Heulandite*

was given to selecting minerals with ≥20% CaO content (see Table 2.7.1) but this boundary excluded, inter alia, calcic plagioclase which was considered to be important; similarly, other values of CaO% presented other anomalies. The SCMR therefore proposes the following list as the **main** calc-silicate minerals: calcic garnet (ugrandite), calcic plagioclase, calcic scapolite, diopside-hedenbergite, epidote group minerals, hydrogrossular, johannsenite, prehnite, pumpellyite, titanite, vesuvianite, wollastonite.

Note: in rare cases where other calc-silicate minerals (for example cuspidine, danburite, larnite, melilite, merwinite, monticellite, rankinite, scawite, spurrite, xonotlite) are present in significant amounts and are used in determining the rock name, this should always be indicated. Similarly, if it is desired to include other minerals as calc-silicate minerals (e.g. calcic amphibole, laumontite, lawsonite) in calculating the rock name, then this should always be stated.

It is characteristic of metacarbonate rocks, as with other metamorphic rocks, that with increasing grade an individual rock may move across fields and field boundaries (see above).

SIDERITE AND MAGNESITE ROCKS

As discussed by Rosen *et al.* (2005), siderite and magnesite rocks can occur in substantial deposits. However, it is recommended that such rocks, if known to be metamorphic, should not be given a specific name but should be defined using compound names with structural root terms and appropriate mineral prefixes (e.g. mica-siderite schist/gneiss).

BOUNDARIES WITH NON-CARBONATE ROCKS

In terms of quantitative mineral composition, marbles show a transition into carbonate-silicate rocks, which, in turn, pass into calc-silicate rocks and carbonate-bearing silicate rocks. The latter two, when free of carbonate, pass into the area of other felsic and mafic metamorphic rocks. Rosen *et al.* (2005) propose the following boundaries, based on empirical evidence: a maximum modal content of 35% of clinopyroxene and amphibole separates metacarbonate rocks from mafic silicate rocks and a maximum modal content of 50% of feldspar separates metacarbonates from feldspar-rich rocks. The SCMR definition of amphibolite also states that the boundary between a metacarbonate rock and an amphibolite is fixed at 75% of modal amphibole + plagioclase.

STRUCTURAL VARIATION

As with other varieties of metamorphic rocks the degree of schistosity in metacarbonate rocks varies with the degree of strain, the mineral content and the metamorphic grade. For example, high degrees of strain or a high content of phyllosilicates will result in a well-developed schistosity, whereas lower strain values or the presence of more equant mineral grains will result in a poorly developed or absent schistosity. In each case the appropriate structural root term (schist, gneiss, granofels) may be used (e.g. carbonate-silicate gneiss).

The structure can also vary with the metamorphic grade. For example, metacarbonate rocks in the greenschist facies are more likely to have a relatively high modal content of phyllosilicates than at higher grades and may therefore display a well-developed schistosity (e.g. chlorite marble or carbonate-chlorite-mica schist). At higher grades, on the other hand, the phyllosilicate (muscovite, chlorite and/or biotite) content or degree of preferred orientation is likely to be lower and the rocks characteristically develop a granofelsic structure (e.g. calc-silicate granofels). Exceptions to this generality may of course occur; abundant phyllosilicate minerals may exist in a rock but have a random orientation and also some phyllosilicate minerals may persist to relatively high grades.

2.7.4 *Definitions*

RECOMMENDED NAMES

Marble: Metamorphic rock containing more than 50% vol. of carbonate minerals (calcite and/or aragonite and/or dolomite). Pure marble contains more than 95% vol. of carbonate minerals; a marble containing less than 95% of carbonate minerals is classified as impure marble.

The specific carbonate mineral(s) may be given as an informal adjective (e.g. calcitic marble). Impure marbles may carry prefixes denoting the non-carbonate mineral components and structural state (e.g. diopside-grossular marble; tremolite-bearing quartz-diopside marble; gneissose phengite-omphacite marble).

Carbonate-silicate rock:[10] metamorphic rock mainly composed of silicate minerals (including calc-silicate minerals) and containing between 5 and 50% vol. of carbonate minerals (calcite and/or aragonite and/or dolomite).

Calc-silicate rock: metamorphic rock mainly composed of calc-silicate minerals and containing less than 5% vol. of carbonate minerals (calcite and/or aragonite and/or dolomite).

COMMON NAMES WITH RESTRICTED OR UNNECESSARY STATUS

Calciphyre (Brongniart, 1813): a metacarbonate rock containing a conspicuous amount of calcium-silicate and/or magnesium-silicate minerals.

The term '*carbonate-silicate rock*' should be used in preference if the non-carbonate mineral content is higher than 50% vol. and the term '*impure marble*' should be used in preference if the non-carbonate mineral content is lower than 50% vol.

[10] Carbonate-silicate and calc-silicate rocks largely arise from the isochemical metamorphism of carbonate-sandstone and/or carbonate-mudstone sediments. These rocks, as defined above, may also be derived from these and other lithologies as a result of commonly complex processes associated with contact metasomatism. It is desirable when dealing with such rocks to make their context clear.

More specific names may be given using mineral-structural root names, for example carbonate-zoisite-clinopyroxene-plagioclase granofels.

Calc-mica schist (Cotta, 1855): a metacarbonate rock with a schistose structure and composed of calcite, oriented mica and quartz.

The compound term 'quartz-mica-carbonate schist' should be used in preference.

Calc-schist (Brongniart, 1813): a metamorphosed argillaceous limestone containing calcite as a substantial component and with a schistose structure produced by parallelism of platy minerals.

The term '*carbonate-silicate schist*' should be used in preference if the non-carbonate mineral content is more than 50% vol. and the term '*schistose impure marble*' should be used in preference if the non-carbonate minerals constitute less than 50% vol.

Cipolin (Brongniart, 1813; Cordier, 1868): a metacarbonate rock rich in chlorite and other phyllosilicates and displaying a saccharoidal structure.

The term was also used in France for any crystalline limestone but this use is now obsolete. Terms such as *carbonate-chlorite schist* should be used in preference if chlorite and other phyllosilicates constitute more than 50% vol. of the rock and the term 'impure marble' should be used if chlorite and other phyllosilicates constitute less than 50% vol.

Crystalline limestone (Daubrée, 1867): metamorphosed limestone; a marble formed by recrystallization of limestone as a result of metamorphism.

The term '*pure marble*' should be used in preference if the non-carbonate content is lower than 5% vol. and the term '*impure marble*' should be used in preference if the non-carbonate content is higher than 5% vol.

OTHER NAMES

Ophicarbonate: a rock consisting of serpentinite and carbonate; the serpentinite is commonly fragmented or brecciated, and veined and impregnated by the carbonate material (calcite, dolomite or magnesite). It forms by the serpentinization of ultramafic rocks and their reaction with CO_2 solutions. Hence ophimagnesite (where the carbonate is predominantly magnesite), etc.

Ophicalcite: strictly a form of *ophicarbonate* in which calcite is the predominant carbonate. However, the term has traditionally been used in the meaning of *ophicarbonate*, that is, it is taken to include rocks containing a variety of different carbonates.

Other names that relate to metacarbonate rocks produced by contact metamorphism or metasomatism (for example, ***skarn***) are given in the relevant SCMR papers and in the glossary.

2.8 Amphibolite and Granulite

José Coutinho, Hans Kräutner, Francesco Sassi, Rolf Schmid and Sisir Sen

2.8.1 Introduction

This paper summarizes the results of the discussions by the SCMR, and the response to circulars distributed by the SCMR, which demonstrated the highly variable usage of the terms granulite and amphibolite. It presents those definitions which currently seem to be the most appropriate ones. Some notes are included in order to explain the reasoning behind the suggested definitions. In considering the definitions it is important to remember that some of them are a compromise between very different, even conflicting, opinions and traditions.

The names *amphibolite* and *granulite* have been used in geological literature for nearly 200 years: amphibolite since Brongniart (1813), granulite since Weiss (1803). Although Brongniart described amphibolite as a rock composed of amphibole and plagioclase, in those early days the meaning of the term was variable. Only later (e.g. Rosenbusch, 1898) was it fixed as a *metamorphic* rock consisting of hornblende and plagioclase, and of medium to high metamorphic grade. In contrast, the use of the name granulite was highly variable, a position which was further complicated by the introduction of the facies principle (Eskola, 1920, 1952) when the name granulite was proposed for all rocks of the granulite facies.

The aim of the SCMR was to define amphibolite and granulite in accordance with its general principles (Schmid *et al.*, Section 2.1), that is principally according to their mineral composition and macroscopic characteristics, without, as far as possible, any genetic connotation.

2.8.2 Amphibolite

EVOLUTION OF THE MEANING OF THE TERM

It is generally accepted that amphibolite is a metamorphic rock consisting mostly of plagioclase and hornblende. However, different opinions have been presented concerning the modal proportions of hornblende and plagioclase, and the presence or absence of other minerals in the rock. Several classifications have been proposed, commonly based on triangular classification diagrams with plagioclase and amphibole (hornblende) at two apices. Quartz has frequently been chosen for the third apex. These three components (expressed in volume %) were recalculated to 100%, and other minerals were either not considered in the calculation, or combined with quartz, plagioclase or hornblende. When the transition of amphibolite to quartzofeldspathic rock (para- or ortho-) was described, the main candidate for the third apex was quartz, sometimes with biotite and K-feldspar. In the classification of Matthes and Kramer (1955), quartz is the critical mineral: these authors define amphibolite as essentially composed of amphibole and plagioclase with only traces of quartz, rocks with less than 5% quartz are called quartz-bearing amphibolites, rocks with 5 to 20% quartz are quartz amphibolites, and rocks with more than 20% quartz are hornblende gneisses. According to Oen (1962) amphibolite should contain less than 70% hornblende, rocks with more hornblende are classified as hornblendic rocks and hornblendites; the lower limit of hornblende content is not given. According to Cannon (1963) the amount of hornblende as the dominant mafic mineral should normally exceed 50% of the total mineral content; quartz and plagioclase should be present in equal proportions or plagioclase should predominate. In the case when the amount of quartz is greater than plagioclase, the rock is classified as quartz amphibolite.

A double triangle was used by Fritsch *et al.* (1967) for classifying rocks of the amphibolite (metabasite) group; pyroxene, plagioclase, amphibole and zoisite (epidote group) were placed at the apices. According to the definition of Fritsch *et al.* amphibolite should contain more than 40% amphibole; rocks with less amphibole (20–40%) were called leucoamphibolites.

Metamorphic Rocks: A Classification and Glossary of Terms. Recommendations of the International Union of Geological Sciences, eds. Douglas Fettes and Jacqueline Desmons. Published by Cambridge University Press. © Cambridge University Press 2007.

Other varieties of the classification triangle were proposed and used with different groupings of the minerals at the apices, for example:

Fsp – Am – Qtz (Lorenz, 1980);
Pl (Kfs) – Hbl – Qtz (Pešková, 1973);
Pl (Kfs) – Hbl – Grt + Cpx (+ Ep) (Tonika, 1969);
Pl (Kfs) – Hbl (+ Cpx, Grt, etc.) – Qtz (Fišera, 1968).

There are no great differences in the principles used by these classifications, and in all of them most samples commonly called amphibolite plot in the respective amphibolite fields.

Wholly different is the proposal of Berthelsen (1960), who includes in the classification the transition of amphibolite to basic granulite (trappgranulite, pyriclasite). The presence of orthopyroxene in the rock is critical, and the classification is based on the relation of hornblende to clinopyroxene + orthopyroxene (the corresponding triangle was constructed by Lorenz, 1981).

To determine the range in the modal composition of amphibolites, Fišera (1973) collected about 260 modal analyses from the literature (Fig. 2.8.1a). More than 80% of them contain less than 10% of quartz, and more than 50% have no quartz at all, or contain quartz only as an accessory mineral. It may be concluded that most of the amphibolites described in the literature correspond to the common definition of amphibole + plagioclase rocks. The same conclusion may be drawn from the descriptions and modal compositions of about 130 amphibolites from the Moldanubicum of the Bohemian Massif, as illustrated in Fig. 2.8.1a and b. From these collected data, which serve only as an example of the extensive literature, it follows that in the triangular plots the amphibolites cluster mostly in the field between 50 and 90% amphibole. The field between 30% and 50% amphibole is also densely populated, and contains no more than 10% quartz. K-feldspar is generally absent or only present in small quantities, and has no influence on amphibolite nomenclature.

THE SCMR DEFINITION

The SCMR arrived at a definition of amphibolite after extensive discussion, and consideration of the questionnaire results. The wording is based on the following statements.

1. Since the beginning of petrography, the term amphibolite has been used for a rock composed of amphibole and plagioclase. This name is therefore recommended for further use, even if it does not fit the general SCMR rules, for use of the suffix 'ite' (see Schmid *et al.*, Section 2.1).
2. The modal compositions of amphibolites show that most of them contain more than 50% amphibole, but those with 30 to 50% are not unusual. The content of amphibole and plagioclase together is mostly higher than 90%, and may be as low as 75%. The value of 75% is therefore taken as the lowest boundary of Pl + Am.
3. The colour of amphibole is green, brown or black in hand specimen and green or brown in thin section. The common varieties are tschermakitic and magnesio- and ferro-hornblende as defined in the IMA classification (Leake, 1997). Other types of amphibole, although less common, for example cummingtonite or anthophyllite, may also be constituents of amphibolites.
4. Plagioclase is the prevalent light-coloured constituent; the quantity of quartz or epidote or scapolite should be lower than that of plagioclase.
5. Clinopyroxene, where present, should be less abundant than amphibole (hornblende). When pyroxene prevails, the rock should be named hornblende-pyroxene rock or calc-silicate rock, depending on its composition and on the composition of the clinopyroxene. The name *pyribolite*, proposed for Pl-Hbl-Px (Cpx + Opx) metamorphic rocks (Berthelsen, 1960) is not recommended for use by the SCMR (see below).
6. The presence of other major mineral constituents (>5%) is expressed by the corresponding prefix according to general SCMR rules (e.g. garnet amphibolite, pyroxene amphibolite, quartz amphibolite).
7. The amphibolite is characterized by the presence of hydroxyl-bearing minerals (amphibole, biotite), which prevail over the

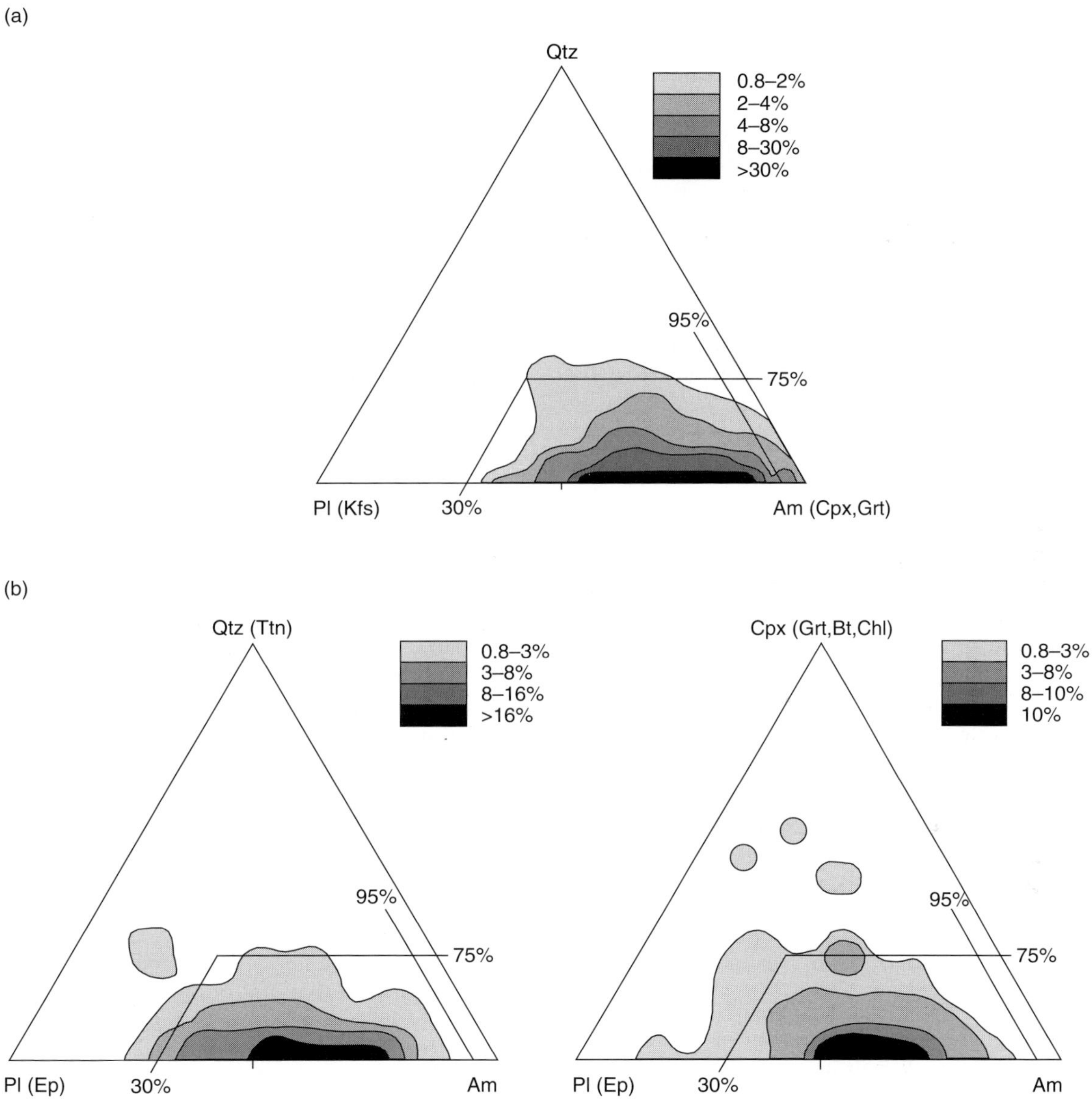

Fig. 2.8.1 **a**, Qtz – Pl (Kfs) – Am (Cpx, Grt) diagram with contoured plots of 260 rocks described as amphibolites from various localities. Straight lines at 30%, 75% and 95% indicate limits of the amphibolite field recommended by the SCMR. Modal data from Fišera (1968). **b**, Qtz (Ttn) – Pl (Ep) – Am and Cpx (Grt, Bt, Chl) – Pl (Ep) – Am diagrams with contoured plots of 127 rocks described as amphibolites from the Moldanubicum of the Bohemian Massif. Straight lines at 30%, 75% and 95% indicate limits of the amphibolite field recommended by the SCMR. Modal data from Fišera (1973), Pešková (1973), Klápová (1977) and Šichtářová (1977).

hydroxyl-free ones (garnet, diopside). The boundary with the higher grade, granulite-facies metamorphic rocks is determined by the appearance of orthopyroxene.

8. The names ortho-amphibolite, para-amphibolite and meta-amphibolite may be found occasionally in the literature. The prefixes ortho-, para- and meta- have been used for the description of very different properties of rocks or processes, and in very different meanings (e.g. ortho- has been used for the following rocks: metamorphic rocks of igneous origin, rocks containing K-feldspar, SiO_2-saturated rocks, sediments with defined contents of matrix, contents of carbon in coals, type of cumulates, etc.). The use of the prefixes ortho- and para- is accepted by SCMR in the classical concept of Rosenbusch for metamorphic rocks derived from igneous and sedimentary parents respectively, but difficulties may exist in determining the nature of the protolith by directly observable features. The prefix meta- must never be used for repeatedly metamorphosed rocks according to SCMR guidelines. If in amphibolites, amphibole is partly replaced by chlorite, and plagioclase by albite ± epidote ± calcite, owing to alteration by retrograde metamorphism (retrogressive overprint), the term retrograde (retrogressive) amphibolite may be used in accordance with the SCMR rules.

The recommended definition of amphibolite reads:

> **Amphibolite** is a gneissose or granofelsic metamorphic rock mainly consisting of green, brown or black amphibole and plagioclase (including albite), which combined form ≥75% of the rock and both of which are present as major constituents; the amphibole constitutes ≥50% of the total mafic constituents and is present in an amount of ≥30%. Other common minerals include quartz, clinopyroxene, garnet, epidote-group minerals, biotite, titanite and scapolite.

The boundaries of the amphibolite field as defined here are presented on Fig. 2.8.2.

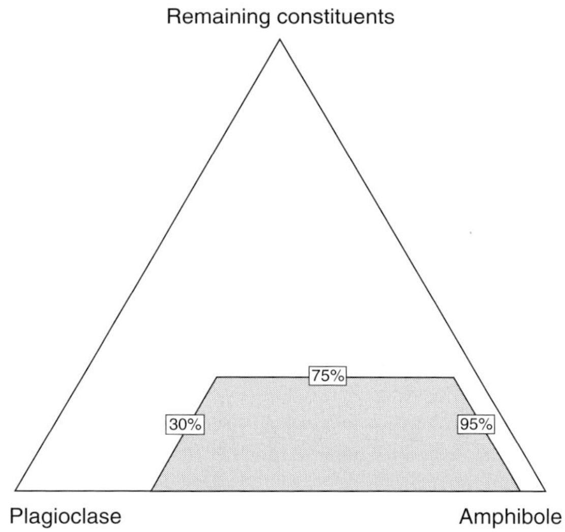

Fig. 2.8.2 Amphibolite field according to the SCMR definition.

RELATION TO OTHER ROCK GROUPS

Transition amphibolite – quartzofeldspathic rock
Transitions of amphibolite to biotite-quartz-feldspar rocks are relatively common. The boundary between the amphibolite and the quartzofeldspathic rock is set at a modal content of 25% of minerals other than amphibole and plagioclase.

Transition amphibolite – metacarbonate rock
The transition from amphibolite to *metacarbonate* rock (calc-silicate rock and carbonate-bearing silicate rock, sensu Rosen *et al.*, Section 2.7) is generally marked by increasing amounts of calcic pyroxene and plagioclase in the rock. The critical boundary is again the 25% content of minerals other than amphibole and plagioclase.

In some metamorphic rocks there are alternating millimetre to decimetre thick layers of amphibolite and biotite-quartz-feldspar rock or layers of amphibolite and *calc-silicate* rock (or even marble). For such rock associations the name banded amphibolite has been used by some authors. Because rock names, even with adjectives, cannot be used for mixed or juxtaposed rocks of different petrographic types, these complex rock associations may be called an ***interlayered amphibolite and quartzofeldspathic rock***, or ***interlayered amphibolite and calc-silicate rock*** (or ***interlayered amphibolite and marble***).

Rocks composed entirely of amphibole, other minerals being present only as accessories, set another nomenclature problem. Under the SCMR guidelines (Schmid *et al.*, Section 2.1), the term hornblendite is reserved for rocks with 90% hornblende. The terms ***hornblende granofels*** or ***hornblende gneiss*** may be used as an alternative. However, as may be seen from the triangular diagrams, rocks containing more than 90% amphibole are rare.

2.8.3 Granulite

EVOLUTION OF THE MEANING OF THE TERM

In contrast to the name amphibolite, the name granulite is burdened by many ambiguities and has been used with different meanings in different countries. In France it was applied to fine-grained granitic rocks (Michel-Lévy, 1874; Cogné & von Eller, 1961), but this use did not find common acceptance. In Scotland and England the term granulite was applied to high-grade metamorphic products of psammitic rocks. Most widespread is the use of the term granulite for light-coloured, quartzofeldspathic, high-grade metamorphic rocks. This meaning for the name was introduced by Weiss (1803) and later defined by many authors, mainly in consideration of the Central European fine-grained granulites of the Granulitgebirge in Saxony and of the Bohemian Massif (among others, Lehmann, 1884; Scheumann, 1961; Scharbert, 1963). The introduction of the facies principle (Eskola 1920, 1952) further complicated the terminology, the name granulite being proposed and used for all rocks of the granulite facies, even for those with intermediate and basic compositions. To avoid these ambiguities, Winkler and Sen (1973) proposed the name *granolite* instead of granulite for rocks with mineral associations diagnostic of the granulite facies (regional hypersthene zone). But this suggestion was not widely accepted.

By the end of the 1960s, an international group, referred to as the Granulite Commission,[11] tried to define the term granulite and published two reports (Behr *et al.*, 1971; Mehnert, 1972). The results of the discussion were excellently summarized by Mehnert (1972, pp. 148–9) and the proposed definition of granulite was:

Granulite is a fine- to medium-grained metamorphic rock composed essentially of feldspar with or without quartz. Ferromagnesian minerals are predominantly anhydrous. The texture is mainly granoblastic (granuloblastic), the structure is gneissose to massive. Some granulites contain lenticular grains or lenticular aggregates of quartz ('disc-like quartz'). Granulite is the type rock of the granulite metamorphic facies. The composition of the minerals corresponds to granulite *P–T* conditions. The following rock types should not be included in the definition of 'granulite': medium- to coarse-grained rocks (>3 mm) of corresponding composition and origin should be termed **granofelses**. Granulites that are rich in ferromagnesian minerals (>30%) should be termed pyriclasites, pyribolites, or pyrigarnites, depending on their composition.

THE SCMR DEFINITION

The definition given by the Granulite Commission was an important basis for discussions in the SCMR and for the questionnaires. The comments on the proposed definition were strongly contrasting; a clear distinction from ***leptynite*** was also required (proposed mainly by French geologists). The results of the discussion are summarized as follows:

1. The main constituents of granulites are feldspars (perthitic alkali feldspar, plagioclase) and quartz. Typical mafic minerals are garnet (pyralspite), orthopyroxene (commonly with a high alumina content) and clinopyroxene

[11] H. J. Behr, GDR; E. den Tex, the Netherlands; D. de Waard, USA; K. H. Mehnert, Berlin; H. G. Scharbert, Austria; V. S. Sobolev, USSR; A. Watznauer, GDR; V. Zoubek, Czechoslovakia; H. J. Zwart, the Netherlands.

(diopsidic); other typical constituents are kyanite, sillimanite, rutile and ilmenite. Many granulites contain biotite (Mg- and Ti-rich) and hornblende (with high Ti-contents). In some granulites cordierite may also be present.

2. In some types of granulites the fine grain size and the presence of platy quartz crystals ('Diskenquarze') are very typical (e.g. Scheumann, 1961; Behr *et al.*, 1971). However, the SCMR proposes to omit the structural and grain-size criteria from the granulite definition. The structural criteria were based on the classical Central European terrains, whereas in the basement complexes of the world, the granulite facies rocks are prevalently coarser-grained and have a massive microstructure/random fabric.
3. The main problem of the granulite definition is the nomenclature of mafic high-grade metamorphic rocks ('basic granulites') and their relation to granulites. These rocks, of mostly basaltic composition, were named in older papers *pyroxene granulites* (e.g. Lehmann, 1884) and this name remained the only one generally used, even though it was considered to be not wholly appropriate. The necessity of a special name for 'basic granulites' was mentioned by many authors (Lehmann, 1884; Scheumann, 1954), and if a special name could be adopted, 'one of the main difficulties of the granulite nomenclature could be settled' (Mehnert, 1972). But the proposed and practical new term pyriclasite (Berthelsen, 1960) did not find general acceptance during the SCMR discussions. The informal name ***mafic granulite*** (Harley, 1989) seems, therefore, to be the most appropriate at present.
4. Contrasting opinions were expressed by the SCMR Study Group about the facies characterization of the term granulite. Because it is difficult in the field to determine the metamorphic facies of rocks, especially where the critical mineral (orthopyroxene) is missing, the SCMR does not preclude granulites belonging to the amphibolite facies, because the characteristic mineral composition may be strongly influenced by the rock bulk chemistry.

Considering the above comments the recommended definition of granulite reads:

> **Granulite** is a high-grade metamorphic rock in which Fe-Mg-silicates are dominantly hydroxyl-free; the presence of feldspar and the absence of primary muscovite are critical, and cordierite may also be present. The mineral composition is to be indicated by prefixing the major constituents.
> The rocks with >30% mafic minerals (dominantly pyroxene) may be called **mafic granulites**, those with <30% mafic minerals (dominantly pyroxene) may be called **felsic granulites**. The term should not be applied to ultramafic rocks, calc-silicate rocks, marbles, ironstones or quartzites.

Detailed names and subdivisions may be given using mineral-root names, for example, garnet-clinopyroxene-plagioclase granulite, biotite-garnet-plagioclase granofels.

RELATED ROCK NAMES

Several rock names are associated with the nomenclature of granulite rocks. They are briefly mentioned below with the SCMR recommendations on their use.

Charnockite, **enderbite**, etc. The nomenclature of these orthopyroxene-bearing granitoid rocks was given by Le Maitre (1989, 2002).

Hälleflinta (Cronstedt, 1758). Obsolete term, used mainly in Sweden and Finland, for a fine-grained compact quartzofeldspathic rock of horny aspect, which may be banded and/or blastoporphyritic. It is derived from acid igneous rocks or acid tuffs and is partly synonymous with *leptynite*. Not recommended for further use.

Leptynite (Haüy, 1782, as recorded in Cordier, 1868). Name created by Haüy and probably first published by Brongniart, initially applied to a fine-grained granulite-facies rock, predominantly consisting of alkali feldspar, containing minor quartz, white mica, garnet and tourmaline, and with a planar gneissose structure. Later used for any white-coloured, quartzofeldspathic rock, typically forming bands alternating with metabasic rock, irrespective of the intensity of metamorphism (Cogné & von Eller, 1961). It is not recommended for further use.

Leptite (Sederholm, 1897). Old term used by Swedish geologists for fine-grained gneissose to granulose metamorphic rocks of sedimentary origin mainly composed of feldspar and quartz with subordinate mafic minerals. The grade of metamorphism is higher than that of hälleflinta. Partly synonymous with *leptynite*; not recommended for further use.

Náměšter Stein (von Justi, 1757). An obsolete name for granulite from the locality Náměšt in the Bohemian Massif, from where this rock was initially described.

Pyribolite (Berthelsen, 1960). According to the original definition, a high-grade metamorphic rock composed of plagioclase, hornblende, clinopyroxene, orthopyroxene + garnet. The presence of orthopyroxene is essential, according to the original definition, hornblende and pyroxene being present in approximately equal amounts. The SCMR does not recommend the use of this term, which may be substituted by *pyroxene amphibolite* or *hornblende mafic granulite* according to the quantities of the respective minerals present.

Pyriclasite (Berthelsen, 1960). High-grade metamorphic rock consisting mainly of feldspar (plagioclase) and pyroxene (Cpx and/or Opx) with or without garnet. The presence of Opx is essential according to the original definition. As redefined by the Granulite Commission, the contents of mafic constituents should be higher than 30% (in vol.). As this name was not generally accepted by the SCMR, the term *mafic granulite* is most appropriate at present.

Pyrigarnite. Initially defined by Vogel (1967) as a high-grade metamorphic rock composed of pyroxenes and garnet, in which the presence of plagioclase may be expressed by a prefix (plagio-pyrigarnite). This definition was modified by Mehnert (1972), and plagioclase was added to the characteristic constituents of pyrigarnite. According to this revised definition pyrigarnite is composed of plagioclase, garnet and pyroxenes (Cpx and/or Opx), the contents of mafic constituents being higher than 30% vol. The SCMR does not recommend the use of this term, which may be replaced by *garnet-rich mafic granulite* or, according to the general rules, by *Px-Grt-Pl-granofels* (Schmid *et al.*, Section 2.1).

Pyroxene granulite. Name for mafic granulites, mostly of basaltic composition. The term *mafic granulite* is recommended instead.

Stronalite (Artini & Melzi, 1900). Regional term for high-grade metamorphic rock mainly composed of garnet, feldspar and quartz. Biotite and cordierite may be present, as well as kyanite and sillimanite. The name may be replaced by *granulite*, or according to the SCMR general rules, by *Grt-Qz-Pl gneiss/granofels* (see also Schmid, 1968; Schmid *et al.*, Section 2.1).

Trappgranulite (Stelzner, 1871). Obsolete term for *mafic granulites* ('pyroxene granulites') mostly of basaltic composition and consisting of plagioclase, quartz, pyroxene (originally described as a micaceous mineral), pyrrhotine and garnet. Not recommended for further use.

Weissstein (Engelbrecht, 1802). Obsolete term for a granulite from the earliest descriptions of these rocks in Saxony. Not recommended for further use.

2.9 Metasomatism and metasomatic rocks

VILEN ZHARIKOV, NICOLAÏ PERTSEV, VLADIMIR RUSINOV, EZIO CALLEGARI and DOUGLAS FETTES

2.9.1 Introduction

A Study Group was set up under the leadership of N. Pertsev to consider the nomenclature of metasomatism and metasomatic processes. This paper reports the conclusions of that group.[12]

2.9.2 Definition of metasomatism

The term metasomatism was introduced by Naumann (1826). *Metasomatism, metasomatic process* and *metasomatose* are synonyms although some authors use *metasomatose* as a name for specific varieties of metasomatism (e.g. Na-metasomatose, Mg-metasomatose). Metasomatism is defined as follows:

Metasomatism: a metamorphic process by which the chemical composition of a rock or rock portion is altered in a pervasive manner and which involves the introduction and/or removal of chemical components as a result of the interaction of the rock with aqueous fluids (solutions). During metasomatism the rock remains in a solid state.

Metasomatic rocks, in general, have a granofelsic or granoblastic structure. They may be coarse- or fine-grained and may sometimes exhibit banding which may be rhythmic. They may demonstrably overprint earlier structures.

Metasomatism is separated from other endogenic processes by the following features (Zharikov *et al.*, 1998):

1. *From the ion-by-ion replacement in minerals* (e.g. zeolites) by mechanisms in which the dissolution of minerals occurs synchronously with the precipitation of new minerals thus maintaining a constant volume and conforming with Lindgren's (1925) rule of constant volume during metasomatism. A good example of metasomatism is the pseudomorphic replacement of a mineral crystal by another mineral (or by a mixture of other minerals) with preservation of the former shape and volume.
2. *From the group of processes including the in-filling of cavities or cracks, magma crystallization, and magma–rock interactions*, by the preservation of rocks in the solid state during replacement (the volume of solution filling pores is negligible in comparison with the total rock volume). The crystallization of melts follows the eutectic law. However, the main trend of metasomatic replacement, that is with a decreasing number of mineral phases from the outermost zone to the solution conduit, is incompatible with the eutectic law. The chemical and mineral compositions of magmatic rocks are uniform throughout the greater part of a magmatic body in contrast to the zonation pattern of metasomatic rocks. Magmatic rocks, particularly residual melts, are typically polymineralic, in contrast to metasomatic rocks, thus terms like *metasomatic granites* are incorrect according to metasomatic theory.
3. *From isochemical metamorphism* by substantial changes in the chemical composition by either the addition or subtraction of major components other than H_2O and CO_2. Changes in the water and/or carbon dioxide concentrations are allowed in isochemical metamorphism, so hydration/dehydration or carbonation/decarbonation reactions are not specific to metasomatism and terms such as *carbonate metasomatism* or *hydrometasomatism* are undesirable. Only H_2O and CO_2 are perfectly mobile (in a thermodynamic sense) during metamorphic processes, whereas during

[12] This paper is based on the report of the Study Group on Metasomatic rocks; it presents a systematic scheme and defines related terms. It is noted, however, that several of the terms have not been widely used internationally and are therefore, under the SCMR rules, classified as restricted. Therefore the results of the paper are presented as provisional recommendations.

It is with sadness that we record the death of Vilen Zharikov on 29th July 2006, before publication of this book.

Metamorphic Rocks: A Classification and Glossary of Terms. Recommendations of the International Union of Geological Sciences, eds. Douglas Fettes and Jacqueline Desmons. Published by Cambridge University Press. © Cambridge University Press 2007.

metasomatic reactions other rock- or ore-forming components may become perfectly mobile. The number of coexisting minerals in metasomatic zones is usually less than in a replaced rock unless the former rock was monomineralic.

4. *From magmatism and metamorphism* by the formation of a regular set of zones. These zones form a characteristic pattern (*metasomatic column*) across the metasomatic body. The zonal pattern represents chemical equilibration between two rocks or between a rock and a filtrating fluid (solution). In the case of diffusional metasomatism mineral changes across the zones are transitional whereas in the case of infiltrational metasomatism changes across zones occur in steps. The number of metasomatic zones in the column depends on the physico-chemical conditions of the interacting media. In the simplest cases it can be represented by a single zone. All zones in a metasomatic column are generated and grown contemporaneously, increasing their thickness along the direction of mass transport.

A **metasomatic column** (or metasomatic zone pattern) is the complete sequence of metasomatic zones characterizing an individual metasomatic facies.

2.9.3 Types of metasomatic processes

There are two main types of metasomatism, namely, *diffusional* and *infiltrational* as determined by the prevailing nature of the mass transport (Korzhinskii, 1957). They are defined as follows:

Diffusional metasomatism is a type of metasomatism that takes place by the diffusion of a solute through a stagnant solution (fluid). The driving force of diffusion is the chemical potential (or chemical activity) gradients in the rock-pore solution.

Infiltrational metasomatism is a type of metasomatism that takes place by the transfer of material in solution, infiltrating through the host rocks. The driving force is the pressure and concentration gradients between the infiltrating and rock-pore solutions.

Diffusional metasomatic rocks form rather thinly zoned bodies (rims) along cracks, veins and contact surfaces, and the composition of minerals may vary gradually across each metasomatic zone. Infiltrational metasomatic rocks generally occupy much greater volumes and the composition of minerals is constant across each of the metasomatic zones.

The term *metasomatism* is conventionally limited by hydrothermal conditions (both sub- and supercritical) related to endogenic processes. Rock alteration under hypergenic (exogenic) conditions is strongly dependent on chemical kinetics, surface forces and microbiological activity, which are less noticeable in the hydrothermal environments. Therefore hypergenic processes have to be related to *hypergenic metasomatism*.

Alteration at high pressures and temperatures in the mantle is connected with concentrated liquids whose properties are intermediate between fluids and magmas. The processes are quite specific, although the mechanisms are not clear. These processes are referred to as *mantle metasomatism*.

Korzhinskii (1953) stressed the relation of metasomatism to magmatism and distinguished two metasomatic stages, namely *metasomatism of the magmatic stage* and *metasomatism of the postmagmatic stage*. The first stage is connected with fluids emanating from a liquid magma body and metasomatizing the solid host rocks. Metasomatic processes of the postmagmatic stage are retrogressive and are connected with hydrothermal solutions emanating from the cooling magma and/or other heated exogenic sources, caused by, for instance, the mixing of juvenile water with meteoric water.

The following common types of metasomatism can be recognized according to their geological position: *autometasomatism*, *boundary metasomatism*, *contact metasomatism*, *near-vein metasomatism* and *regional metasomatism*.

Autometasomatism is a type of metasomatism that occurs at the top of magmatic bodies during the early postmagmatic stage. Typical autometasomatic processes, for example, are albitization in granitic plutons and serpentinization of ultramafic rocks.

Boundary metasomatism is a type of metasomatism that occurs at the contact between two rock types.

Contact metasomatism is a type of metasomatism that occurs at or near to the contact between a magmatic body and another rock. It may occur at various stages in the magmatic evolution.

Endocontact zones develop by replacement of the magmatic rocks and *exocontact* zones are formed by replacement of the host rocks.

Bimetasomatism is a variety of boundary metasomatism, which causes replacement of both the rocks in contact due to two-way diffusion of different components across the contact.

Near-vein metasomatism is a type of diffusional metasomatism, which forms symmetrical metasomatic zonation on either side of an infiltrational metasomatic vein (or a vein in-filling).

Regional metasomatism occupies great areas developing in various geological situations. It commonly forms alkaline metasomatic rocks during the magmatic and early postmagmatic stages. Regional metasomatic rocks at moderate and even shallow depths form the outer zones of metasomatic rock associations accompanying ore deposits (*near-ore*) such as greisens, quartz-sericite rocks, propylites and some others. The inner zones marked by intensive metasomatism are commonly rimmed by zones of weak alteration extending over great areas (Zhdanov *et al.*, 1978; Plyushchev, 1981; Gryaznov, 1992). The weak alteration zones can be enclosed in turn by regional metamorphic rocks especially since the processes of metamorphism, metasomatism, and magmatism are intimately related.

2.9.4 *Systematics of metasomatic rocks*

GENERAL APPROACH

The classification of metasomatic rocks is a much more difficult problem than that of magmatic rocks. Knowledge of the mineral and chemical composition as well as the structural features is insufficient for determining the type of metasomatic process, because the composition of a metasomatic rock depends not only on temperature, pressure and replaced rock composition, but also on the type and stage of the metasomatic process and the composition of the fluid or solution.

So, for a genetic interpretation a hierarchical system may be constructed.

1. The basic unit is a rock with its mineral and chemical composition and structure (a metasomatic zone is taken as a specific rock with a single paragenesis).
2. An assemblage of rocks formed under specific physico-chemical conditions determines a metasomatic facies. Each facies is characterized by a specific type of metasomatic zonation or zonal assemblage.
3. A petrogenetic process (including magmatic, tectonic and hydrothermal activity) commonly forms a number of metasomatic facies characteristic of that process. This facies-association forms a metasomatic family.

Some difficulties arise here because metasomatic rocks of similar composition can be produced by different hydrothermal metasomatic processes with different metallogenic associations, and so they have to be separated.

Thus it becomes important to determine the distinguishing features of each petrogenetic process that results in the formation of a specific association of metasomatic rocks or facies. The nature of the 'metasomatizing' solutions is one of the most important of these features.

THE EVOLUTION OF METASOMATIC SYSTEMATICS

Metasomatic rock classifications were initially based on either the newly formed mineral assemblage (Lindgren, 1925; Schwartz, 1939) or on the input–loss chemical mass balance (Goldschmidt, 1921; Eskola, 1939; Turner, 1948; Barth, 1952). The facies approach was developed subsequently by Burnham (1962). However, neither of these approaches took account of the geological environment of the metasomatic rocks nor the nature of the metasomatic processes. The concept of associated groups of metasomatic facies or *families* ('metasomatic formations' in Russian publications) allowed the development of a more geologically based classification in which each metasomatic family (e.g. skarns, fenites, greisens, beresites, propylites, secondary (or hydrothermal) quartzites, argillisites) includes an association of rocks tied to a specific petrogenetic process. These ideas were first proposed in the 1950s (Korzhinskii, 1953; Nakovnik, 1954; Zharikov, 1959). However, the facies approach remained the common method of classification for some time thereafter, although proposals to base classifications on the family ('formation') concept continued in Russia during the period from the 1960s to the 1990s (Rundquist & Pavlova, 1974; Belyaev & Rudnik, 1978; Zharikov & Omel'yanenko, 1978; Zhdanov *et al.*, 1978; Plyushchev, 1981; Shlygin & Gukova, 1981; Gryaznov, 1992; Zharikov *et al.*, 1992). A recent summary of some of this work is given in Zharikov *et al.* (1998).

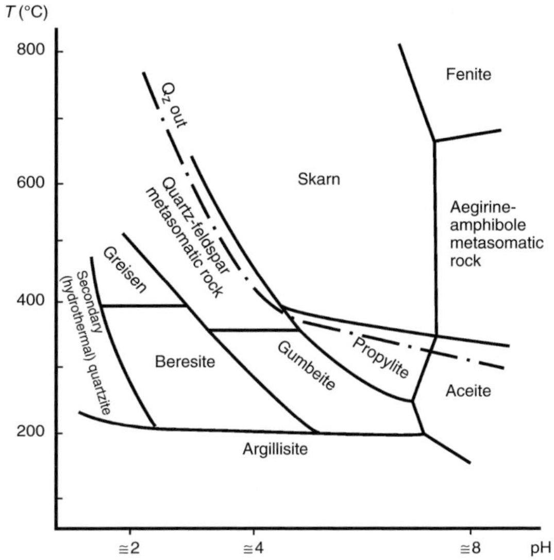

Fig. 2.9.1 Diagram illustrating the general *T* versus pH fields of the metasomatic families. The dot-dashed line separates acidic and neutral-alkaline families (with and without quartz).

THE SYSTEMATIC SCHEME

Metasomatic zone is a metasomatic rock defined by a specific mineral paragenesis.

Metasomatic facies is a regular set of metasomatic zones (a metasomatic column) developed under similar physico-chemical conditions (the compositions of the original rock and the metasomatized rock, the temperature and pressure conditions and the composition of metasomatizing solutions or fluids).

Metasomatic family is the totality of related metasomatic facies typical of a given petrogenetic process. The facies differ from each other by the appearance or disappearance of mineral parageneses or metasomatic zones reflecting a difference in one (or more) of the physico-chemical parameters.

The underlying concept of the 'metasomatic families' is the acid–basic interaction of hydrothermal fluids and their general evolution as they cool through time from an ***early alkaline*** to an ***acidic*** and then to a ***late alkaline*** substage (Korzhinskii, 1953; Zharikov, 1976). The related metasomatic process involves a mass balance change as a result of input/loss of acid (SiO_2, F, Cl, SO_3, CO_2, etc.) and alkaline/basic (K_2O, Na_2O, CaO, MgO, etc.) components. The acidic metasomatic rocks (or the products of acid leaching) show enrichment with acid components and depletion of alkalis and alkali earths relative to the initial rocks. Alkaline and basic metasomatism produces the opposite mass exchange. Each of the metasomatic families that have been distinguished relate to a specific *T*–pH field: the most common metasomatic families are shown on Fig. 2.9.1. All the fields plotted on Fig. 2.9.1 may be divided into two general groups, namely: those with and those without quartz (acidic and neutral-alkaline). The acidic group may be further divided into three subfields forming bands of approximately equal acidity according to the stability of (1) clay minerals, (2) muscovite-hydromicas and (3) feldspars. The alkaline group is divided in the same way into subfields with stable parageneses containing (1) feldspars and (2) feldspathoids (nepheline).

Within most of the metasomatic families some facies may be distinguished on the basis of variations in the mineral assemblages related to the component activities in the hydrothermal solutions, the most important of which are variations of the F, Cl, S, CO_2, K_2O, Na_2O, MgO and CaO chemical activities.

THE METASOMATIC FAMILIES

The following is a brief description of some of the metasomatic families shown in Fig. 2.9.1.

FENITE FAMILY

Fenites are high-temperature metasomatic rocks composed mainly of K-Na-feldspars (perthite or antiperthite), albite, nepheline, alkali pyroxenes (aegirine, aegirine-diopside, aegirine-augite), and alkali amphiboles (arfvedsonite, riebeckite, hastingsite, richterite); subordinate minerals include biotite-phlogopite micas, magnetite and ilmenite, and the most common accessories are titanite and apatite (Brögger, 1921; Evdokimov, 1982; Bardina & Popov, 1994). In some cases fenites contain calcite. Fenites are related to alkaline-ultramafic magmatic complexes and may replace a wide range of rocks including ultramafic rocks and carbonatites formed in the early magmatic stages, as well as acid host rocks such as granite-gneisses and more rarely sandstones. The fenitization process occurs in exocontact aureoles of nepheline syenites. Fenites are formed mainly or completely during the magmatic stage by silica-undersaturated alkaline magmatic fluids. The mass transport will modify the initial rock composition (the rocks have usually significantly less alkalis than the magmas and are silica-saturated) towards the magma composition through a series of metasomatic zones. The width of fenitization aureoles may reach several kilometres.

The zonation of fenites depends strongly on the composition of the replaced host rocks and that of the magma. In the general case the first (outermost) zone is characterized by the disappearance of quartz. The next (intermediate) zone is commonly composed of pyroxene or amphibole + orthoclase (and/or albite). The innermost zone may contain nepheline, and sometimes may even be composed of monomineralic nephelinite. This last zone contains offshoots of alkaline magma.

The SCMR definition therefore is:

Fenite: a high-temperature metasomatic rock characterized by the presence of alkali feldspar, sodic amphibole and sodic pyroxene; nepheline, calcite and biotite/phlogopite may also be present and typical accessories are titanite and apatite. Fenites occur as zoned aureoles around alkaline igneous complexes, forming in a wide range of host lithologies. They occur on the metre to kilometre scale.

Fenitization is a metasomatic process leading to the formation of fenites.

SKARN FAMILY

Skarn is a metasomatic rock formed at the contact between a silicate rock (or magmatic melt) and a carbonate rock. It consists mainly of Ca-, Mg-, Fe-, Mn-silicates, which are free from or poor in water.

Skarns (Goldschmidt, 1911) can form flat bodies along the contact (*contact skarns*) or occur as veins, pipes, etc. crossing the carbonate and/or the silicate rocks (*vein skarns*). Skarns formed from the magmatic or other silicate rocks are termed *endoskarns* and skarns formed from the carbonate rocks are termed *exoskarns*. The outermost zone of endoskarns or the nearest zone of the exoskarn to the parental magmatic body may contain not only Ca-Mg-Fe silicates but also feldspar or scapolite and/or feldspathoid. For such rocks Korzhinskii (1953) introduced the special name *near-skarn rock*.

According to their composition and genetic features skarns have been divided into two major groups, namely *magnesian skarns* developed at contacts with magnesian carbonate rocks (dolomites or magnesites), and *calc-* (or *lime-*) *skarns* formed at contacts with limestones and marbles poor in magnesium.

Skarns generally contain various metal ores. The theory of skarn genesis has been extensively studied (Korzhinskii, 1953; Zharikov, 1959, 1968, 1970, 1991; Pertsev, 1977, 1991; Einaudi *et al.*, 1981; Zaraiskii, 1989; and others).

Subfamily of magnesian skarns

Magnesian skarns may develop in both the magmatic (prograde, in contact with magmatic fluids) and postmagmatic (retrograde) stages.

Magnesian skarn is a high-temperature skarn rock containing forsterite, diopside, spinel, periclase, clinohumite, phlogopite, pargasite and formed at the contacts of magmatic and calc-magnesian or magnesian carbonate rocks. Typically magnesian skarns may host ores of iron, base metals, Cu, Au, Fe-Mg-borates and phlogopite.

Magnesian skarns related to the magmatic stage are characterized by mineral assemblages of fassaitic pyroxene, forsterite, magnesian spinel, enstatite, monticellite, åkermanite, merwinite and periclase. The typical zonation pattern (column) is: granite (or another silicate rock) → plagioclase (An_{70-95}) + diopsidic clinopyroxene → fassaitic clinopyroxene + spinel → forsterite + spinel → forsterite + spinel + calcite → dolomite (arrows show the direction of the zonal growth in the column). Orthopyroxene parageneses occur only when the chemical activity of CaO is low, for example at the contacts with magnesite. Magnesian skarn facies of the magmatic stage, which are dependent on $T - P_{CO_2}$ conditions and CaO activity, are: dolomite, periclase, magnesite, enstatite, forsterite-calcite, monticellite, monticellite-periclase, monticellite-spurrite, åkermanite, merwinite, merwinite-periclase and corundum-plagioclase (spinel occurs in each facies).

Magnesian skarns related to the postmagmatic stage are mainly developed as replacements (partly or completely) of former magnesian skarns developed during the magmatic stage. Their zonation and columns are less uniform and more complex. New characteristic minerals and parageneses appear along with the residual stable minerals of the magmatic stage. Spinel parageneses give place to phlogopite and pargasite. The main facies that are dependent on *P–T* conditions and the chemical activity of CO_2, K_2O, Na_2O, F_2, Cl_2, B_2O_3, FeO are: phlogopite-diopside, phlogopite-forsterite, magnetite-forsterite, diopside-monticellite, monticellite-brucite, monticellite-dellaite, monticellite-foshagite, pargasite-diopside, clinohumite, magnetite, ludwigite, kotoite and suanite. The temperature range in which the subfamily is formed is about 450–750 °C at pressures of 0.5–10 kbar.

Magnesian skarns of both stages are commonly replaced in varying degrees by post-magmatic calc-skarns under moderate *P–T* conditions, because of increasing CaO chemical activity with decreasing temperature and a corresponding decrease of CO_2 concentration in fluids (Pertsev, 1977). Low-temperature post-skarn alteration leads to the replacement of skarn minerals by serpentine, chlorite + actinolite, carbonates and brucite.

Common ore mineralization accompanying the formation of magnesian skarns is iron (magnetite) and borate (ludwigite, suanite and kotoite). A wide variety of ore mineralization (Cu, Mo, W, Be, Sn, Pb, Zn, Au, B and others) may be overprinted by lower-temperature hydrothermal processes. The ores are formed along with post-skarn hydrosilicates (actinolite, serpentine, talc, chlorite), carbonates and some other minerals.

A rare facies of skarn is analogous to magnesian skarns except that the magnesium in minerals is completely replaced by manganese. Typical minerals are johannsenite, tephroite, glaucochroite, sonolite and galaxite. This facies is developed at the contacts of silicate rocks with manganese carbonates. Some investigators call them *manganesian skarns*. These metasomatic rocks appear to have formed at the post-magmatic stage and at low temperatures. Ca-Mn clinopyroxenes (johannsenite) as well as Ca-Mn pyroxenoids are characteristic of the manganese-bearing facies of calc-skarns. Normally, minerals of magnesian skarns do not contain noticeable Mn.

Subfamily of calc-skarns

Calc-skarn is a high- (to medium-) temperature skarn consisting mainly of granditic garnet, salite (to ferrosalite or/and johannsenite-rich pyroxene), wollastonite or Mn-rich pyroxenoids and formed at the contacts of magmatic (or other silicate rocks) with calcium carbonate rocks. It can replace former magnesian skarns (in hypabyssal or subvolcanic conditions). Typically calc-skarns may host ores of Fe, base metals, Cu, W, Mo, Be, B, U, REE.

All (or almost all) calc-skarns belong to the postmagmatic stage. Typical minerals for the subfamily are granditic garnet, vesuvianite, clinopyroxene (salite up to hedenbergite or johannsenite), wollastonite, rhodonite and bustamite, epidote, scapolite and plagioclase. Less common are higher-temperature silicates with mole ratio $Ca/Si \geq 1.5$.

The typical facies succession, from the innermost endoskarn to the outermost exoskarn, is characterized by a bimineralic zone (where andesine or scapolite or K-feldspar is associated with salite or grossular or epidote), followed by a monomineralic (granditic garnet or calcic clinopyroxene) zone or bimineralic (garnet-clinopyroxene) zone, which may or may not be followed by a wollastonite zone. The specific mineralogies depend on both *T–P* conditions and the chemical activities of Na_2O, K_2O, F_2, Cl_2, SO_3, FeO, O_2 and CO_2. These typical calc-skarns are formed in a temperature range of about 400–650 °C at pressures of 0.5–4 kbar.

Some relatively low-temperature skarns are enriched with Fe^{2+} and Mn^{2+} (hedenbergite or johannsenite molecules in pyroxenes, spessartite in garnets, rhodonite and bustamite). Some investigators call them *iron-calc* and *mangan-calc* skarns.

High-temperature calc-skarn facies are marked by the appearance of zones containing Ca-rich silicates such as gehlenitic melilite, tilleyite, spurrite, rankinite, kilchoanite and merwinite. The temperature of their formation is 700–900 °C, at pressures of 0.5–1.5 kbar and chemical activity $X_{CO_2} < 0.05$.

Magnetite ores can be formed in coexistence with skarn garnets and clinopyroxenes.

The formation of hedenbergite and johannsenite parageneses with quartz and quartz-garnet assemblages as well as amphiboles takes place during post-skarn alterations along with or after the breakdown of skarn minerals. Different types of ores are formed at that time, namely W, Sn, Be, Mo, B (danburite and datolite), Cu, base metals, Au, Ag, U, REE.

RODINGITE FAMILY

Rodingites have mineral compositions resembling iron calc-skarns, namely grossular (or hydrogarnet), clinopyroxene of the diopside-hedenbergite series (usually enriched with Fe^{2+}, Fe^{3+} and Al), vesuvianite, epidote, scapolite, and magnetite or haematite. The rock was originally described as magmatic (Bell *et al.*, 1911) but later Grange (1927) showed that it was a metasomatic rock replacing a gabbro dyke. The physico-chemical conditions of rodingite formation were experimentally studied by Plyusnina *et al.* (1993). The formation of rodingites requires the addition of CaO to the rock. Unlike skarns, rodingites do not demonstrate distinct zonation. Rodingites are commonly situated inside serpentinite masses presumably replacing bodies of basic rocks. Many investigators believe that rodingites appear during the serpentinization of ultrabasic rocks, as a result of the alteration of basic dykes and/or inclusions. Other geologists suppose that rodingitization is an autometasomatic process connected with postmagmatic fluids released from a cooling basic magma. Possibly, there are different types of rodingite formation as they may occur not only in serpentinites but also in basalts and even as rocks replacing amphibolites in gneiss terrains. Rodingites may, in places, contain iron and gold ores.

The SCMR definition therefore is:

Rodingite is a metasomatic rock primarily composed of grossular-andradite garnet and calcic pyroxene; vesuvianite, epidote, scapolite and iron ores are characteristic accessories. Rodingite mostly replaces dykes or inclusions of basic rocks within serpentinized ultramafic bodies. It may also replace other basic rocks, such as volcanic rocks or amphibolites associated with ultramafic bodies.

Rodingitization is a metasomatic process leading to the formation of rodingites.

GREISEN FAMILY

Greisens (Lempe, 1785) are quartz-muscovite rocks commonly with topaz, fluorite, tourmaline, and locally with orthoclase or amazonite. Small amounts of lepidolite, andalusite and diaspore may also occur. They may replace granites, sandstones or micaceous schists. The greisen family of metasomatic rocks includes some facies related to varying activity ratios of F, Cl, CO_2 and B. Greisens are divided into two groups: greisens replacing Al-Si rocks (quartz, quartz-muscovite, muscovite-K-feldspar, quartz-tourmaline and quartz-topaz facies) and those replacing Al-Si undersaturated rocks, for example those with ultrabasic or carbonate compositions (fluorite-muscovite and feldspar-fluorite facies). Under specific conditions greisens replacing leucogranite can form feldspar-muscovite, fayalite-quartz, siderophyllite-quartz and other acid zones. Greisens are either the products of pervasive greisenitization in granite bodies, or form as veins, and vein borders. Rocks of this family relate to post-orogenic leucocratic granitic plutons at hypabyssal or subvolcanic depths. They may replace either the granites (*endogreisens*) or the country rocks (*exogreisens*), which are commonly metasediments. Pervasive greisenitization occurs in the apical parts of granitic plutons and is commonly accompanied with skarn and quartz-feldspar metasomatic rock formation. Greisens are typically associated with Be, W, Mo, Sn and Ta mineralization (Schwartz, 1939; Nakovnik, 1954; Rundquist *et al.*, 1971; Zaraiskii, 1989).

The SCMR definition therefore is:

Greisen is a medium-temperature metasomatic rock characterized by the presence of quartz and white mica, commonly with topaz, fluorite, tourmaline and locally with amazonite, orthoclase, andalusite and diaspore. Typically greisens may host Be, W, Mo, Sn and Ta mineralization. They are associated with high-level late-orogenic leucogranites and form as replacements either in the granite body and/or in a wide range of country rocks. Zoning may be present.

Greisenization is a metasomatic process leading to the formation of greisen.

BERESITE FAMILY

Beresite was originally a special Russian term for quartz-sericite-ankerite-pyrite metasomatic rocks and associated ore veins (Rose, 1837; Borodaevskii & Borodaevskaya, 1947; Rusinov, 1989). It originated by replacement of granite. The term was later widened to include the products of replacement of other rock types, of both igneous and sedimentary origin. These rocks are the result of acidic metasomatism under conditions of feldspar and kaolinite instability. The beresitization process in ultrabasic rocks forms *listvenite* (quartz + Fe-Mg carbonate + fuchsite + pyrite) and in the silica saturated rocks it forms beresites. Subgroups of beresite are distinguished by the following parageneses (containing pyrite): quartz-sericite-ankerite, quartz-sericite-calcite, quartz-sericite, sericite-chlorite-calcite, and hydromica-quartz. Beresites are associated with a variety of Au, Au-Ag, Ag-Pb and U ore mineralization.

The SCMR definition therefore is:

Beresite is a low-temperature metasomatic rock characterized by quartz, sericite, carbonate (ankerite) and pyrite assemblages resulting from the replacement of both igneous and sedimentary protoliths. It may be associated with a variety of Au, Au-Ag, Ag-Pb and U ore mineralizations.

Beresitization is a metasomatic process resulting in the formation of beresite or listvenite.

PROPYLITE FAMILY

Propylites (the term was introduced by Richtofen in 1868) are metasomatic greenstones, which occur in various types of ore fields replacing volcanogenic and related rocks (Rusinov, 1972, 1989). They used to be classified as 'regional' metasomatic rocks, because they form an aureole around the ore deposits and can even occupy the whole extent of an ore field. Low-temperature propylites consist of albite, chlorite, calcite, quartz and pyrite. At higher temperatures epidote, actinolite, biotite and magnetite (instead of pyrite) are formed. Propylites occur in various tectonic settings, but are commonly related to island-arc felsic or intermediate volcanism and to some post-orogenic hypabyssal granodiorite bodies. Propylite facies based on temperature variations may be distinguished, namely low-temperature calcite-chlorite-albite facies and high-temperature epidote-actinolite-biotite facies. The facies are related to the post-magmatic stage with increasing acidity of the hydrothermal solution. The metasomatizing solutions are nearly neutral. The CO_2 fugacity is higher than the stability field of pumpellyite and prehnite (prehnite can be formed only at very shallow levels, where CO_2 fugacity is low) and albite + calcite ± chlorite assemblages are formed instead of these minerals. Ore mineralization, directly related to post-propylite beresites, may overprint the propylite rocks. The ore mineralization associated with propylites varies widely, with the possibility of Cu, Pb-Zn, Au, Au-Ag, Ag-Pb-Zn, Hg and Sb minerals.

The SCMR definition therefore is:

Propylite is a low- to medium-temperature metasomatic granofels formed by the alteration of basic volcanic rocks; low-temperature varieties are principally composed of albite, calcite and chlorite; high-temperature varieties are composed of epidote, actinolite and biotite. They form at the postmagmatic stage.

Propylitization is a metasomatic process leading to the formation of propylites.

SECONDARY (OR HYDROTHERMAL) QUARTZITE FAMILY

Secondary (or hydrothermal) quartzites are metasomatic rocks consisting mainly of quartz (50–100% vol.) with high alumina minerals, replacing granites, volcanic and less commonly sedimentary rocks.[13] Some facies of the secondary quartzites are similar to products of advanced *argillisitization* (see below). The main minerals of the secondary quartzites, apart from quartz, are sericite, pyrophyllite, diaspore, andalusite, corundum, alunite, kaolinite, pyrite and haematite. Relatively common are fluorite, dumortierite, zunite, rutile, tourmaline, topaz, svanbergite and lazulite (Nakovnik, 1965; Schmidt, 1985).

Secondary quartzites relate to volcanic suites (rhyolitic or dacite-andesitic composition) and to subvolcanic granite-porphyry or monzonite-porphyry intrusions. Three groups of secondary quartzites are distinguished:

1. A higher-temperature group, closely related to intrusive bodies, containing the following parageneses: corundum + andalusite, andalusite + pyrophyllite, diaspore + pyrophyllite, quartz + sericite, zunite + sericite and dumortierite + sericite. The quartzites of this group are commonly barren of mineral ores, except for some alaskitic granites with Mo-porphyry deposits (Climax type).
2. A transitional group with quartz + sericite. This group is associated with copper-porphyry gold-bearing deposits.
3. A low-temperature sulphate group commonly associated with dacite-andesite volcanic activity, and characterized by the following parageneses: quartz + kaolinite, kaolinite + alunite, alunite + hydromica, alunite + pyrophyllite, and an exotic quartz-barite facies. Quartzites of this group are associated with deposits of Cu, Sb, Hg.

The formation of 'secondary quartzite' is commonly accompanied by the redeposition of MgO and Al_2O_3, and also some bodies of chlorite-rich rocks and sericite-rich rocks.

The SCMR definition therefore is:

'Secondary quartzite' is a medium- to low-temperature metasomatic rock mainly composed of quartz with subsidiary high-alumina minerals, such as pyrophyllite, diaspore, alunite and kaolinite. Common accessories include fluorite, dumortierite and lazulite. Secondary quartzites are associated with volcanic and subvolcanic rocks of rhyolitic to andesitic composition. Normally they form as replacements of acid igneous rocks and more rarely of sedimentary rocks. They may host mineral deposits of alunite, pyrophyllite, Au, Cu, Sb, Hg.

GUMBEITE FAMILY

Gumbeites are metasomatic rocks replacing granodiorites and associated with W- or Cu-bearing

[13] The term 'secondary (or hydrothermal) quartzite' is a translation from Russian: there is no suitable English name although such rocks are widespread. A suitable alternative might be *metasomatic quartzite*. It should be noted that 'quartzite' in this context implies that quartz is a dominant constituent of the rock but might not be ≥75% of the mode as is normally required by the SCMR recommendations.

quartz veins and veinlets. The name comes from the Gumbeika ore deposit in the Urals, Russia (Korzhinskii, 1946). The typical mineral assemblage is quartz + orthoclase + carbonate (dolomite-ankerite series). Some sulphides and scheelite also occur. The outer zones of *gumbeitization* commonly contain phlogopite or magnesian biotite.

The following temperature-based facies are distinguished within the gumbeite family, from high to low temperature: biotite-orthoclase (440–400 °C), dolomite-orthoclase (400–300 °C), and phengite-orthoclase (<300 °C). The high-temperature facies is associated with W-Mo mineralization (scheelite, Mo-scheelite, molybdenite, bismuthite and some other sulphides). The low-temperature facies (without biotite) is associated with some Au and Au-U deposits.

The SCMR definition therefore is:

Gumbeite is a medium- to low-temperature metasomatic rock mainly composed of quartz, orthoclase and carbonate. It forms as an alteration of granodiorite or syenite and is closely associated with W-Cu or Au-U-bearing veins.

Gumbeitization is a metasomatic process leading to the formation of gumbeites.

ACEITE FAMILY

Aceites are low-temperature metasomatic rocks consisting mostly of albite with subordinate carbonate and haematite; chlorite and quartz may also occur in some rocks. The name comes from the uranium mine at Ace, Canada (Omel'yanenko, 1978). The U-bearing fluorapatite is common in rocks with high CaO content, and locally in limestones it forms U-bearing apatite metasomatic rocks. The following facies, related mainly to the initial rock composition, are distinguished: quartz-albite, calcite-albite, chlorite-albite, ankerite-albite, albite-apatite and apatite-calcite.

The process of aceitization is an alkaline low-temperature metasomatism with input of Na_2O and with loss of SiO_2 (quartz is replaced by albite). The hydrothermal solutions are highly oxidized and enriched with Na-hydrocarbonate. TiO_2, P_2O_5 and ZrO_2 are mobile. Aceites occur exclusively with uranium mineralization.

The SCMR definition therefore is:

Aceite is a low-temperature alkaline metasomatic rock, mainly composed of albite with subsidiary carbonate and haematite. U-bearing apatite is a common accessory. Aceites are closely associated with U-mineralization.

ARGILLISITIZED ROCKS (ARGILLISITES) FAMILY

Argillisites are metasomatic rocks consisting of clay minerals (smectites, illite-smectites, chlorite-smectites, kaolinite group minerals, hydromicas, celadonite), zeolites, silica minerals (quartz, chalcedony, opal), carbonates (calcite, ferrous dolomite, ankerite, magnesiosiderite), alunite, jarosite and iron sulphides (pyrite, marcasite, mackinawite, greigite). They are products of hydrothermal argillisitization, which is interpreted as a low-temperature hydrothermal process resulting in the replacement of original intrusive, volcanic or sedimentary rocks by clay mineral assemblages. Lovering (1949) applied the term *argillisitized rocks* to the products of wall-rock alteration near hydrothermal ore veins. Argillisitized rocks are also widespread in areas of volcanic hydrothermal activity (*solfataric argillisites*). These two genetic types of argillisitized rocks are distinguished on the basis that the solfataric argillisites form at the Earth's surface, and *hydrothermal argillisites* form at depth and are commonly accompanied by ore mineralization (Shcherban', 1975, 1996; Rusinov, 1989).

Apart from these, two chemical groups of argillisitized rocks may be distinguished related to either strongly acidic or to neutral or slightly alkaline solutions. Acidic solutions leach base components (mainly MgO, CaO and Na_2O) and redistribute Al_2O_3, and add SiO_2. Slightly alkaline solutions produce zeolitization. Some facies of argillisites may be distinguished by their mineral assemblages, namely *acid group* (quartz-kaolinite, quartz-kaolinite-carbonate, and quartz-hydromica); *subneutral group* (smectite-zeolite). Solfataric and hydrothermal argillisites include both of the chemical rock groups.

Argillisitized rocks accompany uranium, stratabound massive sulphide Cu-Zn, and a wide range of epithermal deposits. Metasomatites closely associated with ore deposition occur around ore veins and veinlets and consist of berthierine, ferrous carbonates, adularia, mixed-layered minerals, and sometimes zeolites, fluorite and tosudite.

The SCMR definition therefore is:

Argillisite is a low-temperature metasomatic rock that is mainly composed of clay minerals, also present may be silica minerals, carbonates and iron sulphides. The rock forms from the hydrothermal alteration of both igneous and sedimentary rocks.

Argillisitization is a metasomatic process leading to the formation of argillisites.

ALKALINE METASOMATITES RELATED TO REGIONAL FAULT ZONES

Fault-related alkaline metasomatites are common in Precambrian shields. They consist primarily of albite, aegirine and riebeckite. The four most common groups of these rocks as distinguished within the Ukraine shield are: albitized granites, albitized gneisses, apojasperoids, and carbonatite-related albitites (Shcherban', 1996).

The *albitized granites and gneisses* are rather similar in their mineralogy and show the following metasomatic zonation (from initial rock to the inner zone of metasomatism): (0) initial rock (microcline + plagioclase + quartz + biotite); (1) microcline + quartz + chlorite + epidote; (2) microcline + albite + quartz + riebeckite + chlorite; (3) albite + riebeckite + aegirine. Sometimes the last zone is composed of albite + aegirine. Albitization occurs in the upper part of granite plutons. It is commonly accompanied by greisenization and mineralization of Ta-Nb and Be.

The *alkaline metasomatites* after jasperoids (*apojasperoids*) differ from albitized granites through an absence of albite, because the alumina content in the initial rocks is too low. The metasomatic zonation in these rocks is as follows: (0) initial jasperoid (quartz + magnetite + carbonate + chlorite); (1) quartz + magnetite + carbonate + riebeckite + chlorite; (2) carbonate + riebeckite + quartz + magnetite; (3) riebeckite + aegirine + quartz + magnetite. The quartz content decreases from the initial rocks into zone (3), but quartz is still present in contrast to the situation with the albitized granites.

Albitites related to *'linear' carbonatite bodies in shear zones* form rims around carbonatite lenses and veins and separate veins in the upper part of the alkaline-carbonatite rock bodies. Albitites of this type may form by the action of aqueous-CO_2 fluids replacing previously fenitized host rocks (gneisses or amphibolites).

Albitites in shear zones are commonly associated with gold mineralization. Albitite forms a network of veinlets within ore shoots and occurs as rims around quartz-gold veins or veinlets. Albitite tends to associate with tourmaline, quartz and sometimes ankerite.

CONCLUSIONS

Ten metasomatic families have been described. Others could be given but those remaining are neither typical nor widespread. Also, many different metasomatic rocks have distinct parageneses and do not form part of a wider grouping.

The SCMR recommends that in cases where the relationship of a metasomatic rock to a specific family is not clear then the rock should be named by its main mineral composition, for example brucite metasomatic rock, cordierite-sillimanite-quartz metasomatic rock, corundum-plagioclase metasomatic rock (or metasomatite). In general the SCMR recommends that new names for metasomatic rocks, facies or families should be based on their mineral compositions and not on geographical or other criteria.

2.10 Contact metamorphic and associated rocks

Ezio Callegari and Nicolaï Pertsev

2.10.1 *Introduction*

A Study Group under the leadership of E. Callegari was set up to look at definitions concerning contact metamorphism and contact metamorphic rocks. The Study Group was also asked to consider metamorphism associated with other localized heat sources, such as combustion metamorphism and lightning strikes. This section presents the report of the Study Group.

2.10.2 *Brief historical notes on contact metamorphism*

> *Les roches de ces derniers terrains (Vosges) ont souvent subi, à proximité du granite, des modifications si variées que leur nomenclature précise devient un sujet d'embarras pour le géologue.*
>
> *Daubrée, 1857.*

The first mention of contact metamorphic phenomena dates back to the end of the eighteenth century when James Hutton observed that the rocks surrounding a granitic body at Glen Tilt (Perthshire, Scotland) had suffered marked changes in either colour or structure especially in zones crossed by granitic veins (Playfair, 1822). At that time Neptunists and Plutonists were still debating the sedimentary versus igneous origin of granitic rocks, and the terms 'metamorphism' (Boué, 1820; Lyell, 1833) and 'contact metamorphism' (Delesse, 1857) had still to enter the geological vocabulary.

In the first decades of the nineteenth century, rock alterations close to contacts with granitic rocks were increasingly discovered in other localities of England and Scotland (Macculloch, 1819) as well as in other European countries. They were observed in a variety of settings (plutonic and volcanic) involving a wide spectrum of rock types. Many names appeared for this particular and localized type of metamorphism. Eventually Delesse (1857) proposed the term *contact metamorphism* which found general acceptance and is still widely used today.

Most names for contact metamorphic rocks (see Table 2.10.1) entered geological literature between the end of the eighteenth and the first two-thirds of the nineteenth century. Many of these names were coined before the study of thin sections developed and the subsequent widespread use of microscope techniques, particularly in the last quarter of the nineteenth century, did much to improve the existing definitions, removing most of the uncertainties and ambiguities. At the same time unnecessary rock names were gradually abandoned or superseded by more appropriate synonyms. The pioneering works of the early petrographers also provided a general framework for many aspects of contact metamorphism including definitions or basic concepts of *contact aureole* (von Buch, in Humboldt, 1831), *endo-* and *exomorphism* (Fournet, 1847), *contact metasomatism* (Durocher, 1846) and many other terms which are still current. Unfortunately, most of the original definitions were in languages other than English, and in some cases the original name was lost in translation.

At the onset of the twentieth century, there was, in many countries, a renewed interest in the study of contact aureoles. However, the interest of metamorphic petrologists was mainly directed towards the interpretation of the relationships between rock microstructures, mineral assemblages, rock chemistry and metamorphic conditions. It was the time when the concepts of metamorphic zones (Becke, 1903b; Grubenmann, 1904, 1907; Barrow, 1912) and metamorphic facies (Eskola, 1920) opened new horizons in the field of metamorphic petrology, and Goldschmidt (1911) discovered very simple and fixed relations between rock chemistry and mineral assemblages in the contact metamorphic rocks of the Christiania (Oslo) region, Norway. Only a restricted number of new contact metamorphic rock names appeared at this time, including the important term 'skarn' (Törnebohm, 1881a, b; Goldschmidt, 1911). Further attempts to introduce new rock names based on the mineral composition (e.g. Salomon, 1898) did not achieve much success and names such as *aviolite*, *astite*, *seebenite*, *edolite* created for different types of hornfelses were short-lived.

Metamorphic Rocks: A Classification and Glossary of Terms. Recommendations of the International Union of Geological Sciences, eds. Douglas Fettes and Jacqueline Desmons. Published by Cambridge University Press. © Cambridge University Press 2007.

In the middle part of the twentieth century there was relatively little attention paid to contact metamorphism. Indeed the question was raised as to whether the distinction between contact and regional metamorphism should be maintained or whether the two should be regarded as a part of the general system of rock metamorphism and metasomatism (Barth, 1962), even though Delesse (1857) had argued that *contact metamorphism* was the natural base for all metamorphic research because, on a limited scale, it was possible to see both the cause and the product of metamorphism. Arguments in favour and against separation of contact from regional metamorphism were brought by Read (1949), Yoder (1952) and Miyashiro (1973a); today, however, there is a general consensus on the necessity for studying contact aureoles as a tool for a better understanding of the physicochemical processes controlling regional metamorphism (Kerrick, 1991).

Following the procedures of the SCMR, the Study Group dealing with contact metamorphism examined the existing names of contact metamorphic rocks and processes in order to select a group of 'recommended names'. The Study Group also considered the metamorphic terminology associated with other localized heat sources. The reasoning and results of this process are given below.

2.10.3 *Towards a redefinition of contact metamorphism*

CONTACT OR THERMAL METAMORPHISM?

The term *thermal metamorphism* was introduced by Harker (1889) for metamorphism caused by elevated temperatures in the absence of directed stress as opposed to dynamic metamorphism. Harker argued that the rise of temperature could be due to more than one process and that contact metamorphism was therefore a type of thermal metamorphism (Harker, 1932; see also Holmes, 1920). Tyrrell (1926) regarded thermal metamorphism as the process of change where 'heat is the dominating factor', but took this to mean 'in the proximity of igneous masses' and he subdivided thermal metamorphism into various types, namely pyrometamorphism, contact metamorphism, optalic (or 'caustic') metamorphism and pneumatolytic (or 'additive') metamorphism, although previously, following Delesse (1857), all these types had been included under contact metamorphism. As Harker's division into thermal and dynamic was increasingly seen as too simplistic, the value of his distinction between thermal and contact also diminished and increasingly these two terms were taken as synonymous, 'thermal' relating to the cause of metamorphism, 'contact' to the field relationships (Turner, 1948; Barth, 1962; Spry, 1969; Yardley, 1989).

However, the term 'thermal metamorphism' conveys the idea that the metamorphism is essentially temperature dependent, the effects of directed pressure being immaterial. This is certainly not the case where the metamorphism is associated with forceful magma intrusions or where magma emplacement occurs in a region undergoing deformation, as for instance in the Donegal area of Northern Ireland (Pitcher & Read, 1963) or in the Kwoiek area of British Colombia (Hollister, 1969). Accordingly, 'distinctions between contact and regional metamorphism can only be based on spatial, genetic, kinetic or textural characters, rather than on *P–T* differences alone' (Pattison & Tracy, 1991).

Therefore to avoid any further ambiguity the SCMR proposes that the term 'thermal metamorphism' is no longer accepted as a recommended synonym of 'contact metamorphism' and should be given restricted status. The recommended definition of contact metamorphism is, therefore, as follows:

> **Contact metamorphism**: type of metamorphism of local extent that affects the country rocks around magma bodies emplaced in a variety of environments from volcanic to upper mantle depths, in both continental and oceanic settings.

The magmas are the sources of heat, mass and mechanical energy necessary for this type of metamorphism. The zone where contact metamorphism occurs is called the **contact aureole**, while the products of such metamorphism are called **contact rocks**. The thickness of the aureole ranges from the millimetre to the kilometre scale. The intensity of contact metamorphism decreases from the innermost to the outermost parts of the aureole. It is customary to separate the metamorphic effects caused by the magma on its wall rocks (**exomorphism**) from those induced by the wall rocks on the magma itself (**endomorphism**). Contact metamorphism accompanied by substantial mass transfer (change of the original rock composition) is called **contact metasomatism**.

HOW MANY TYPES OF CONTACT METAMORPHISM?

In the past two centuries several divisions have been made within the field of contact metamorphism based on geological, chemical, physicochemical, mineralogical, petrological, structural or genetic grounds. Those found in literature are given in Table 2.10.1.

PYROMETAMORPHISM REDEFINED

Unfortunately, the current usage of pyrometamorphism is at times quite different from that originally proposed, covering not only igneous-related changes but also thermal transformations due to, for example, combustion metamorphism (Cosca *et al.*, 1989), flash-heating of meteoritic material during atmospheric entry (Rietmeijer, 2004), artificial firing of carbonate-clay mixtures (Bauluz *et al.*, 2004) and even burning of rocks in prehistoric sacrificial sites (Tropper *et al.*, 2004).

The Study Group considered this position and discussed the redefinition of pyrometamorphism as follows.

The term pyrometamorphism was proposed by Brauns (1911) for a high-temperature type of metamorphism observed in ejected schist fragments found in the Laacher See tuffaceous rocks, in the Eifel region of Germany. The unusual mineral assemblage (sanidine, corundum, spinel, cordierite ± hypersthene ± glass), overprinted on both regional and contact metamorphic assemblages, suggested to him formation temperatures higher than those found in common contact metamorphism. Brauns attributed the pyrometamorphism to a steep increase in the magma temperature while it was stationary in a shallow chamber. This led to the exsolution of hot gases, which reacted with the country rocks, the latter becoming subsequently incorporated into the magma and ejected. The typical end-product of this metamorphism is a sanidine-rich rock. Eskola (1920) proposed the name *sanidinite facies* to cover the particular *P–T* conditions under which the sanidinite rocks were formed. Subsequently, Grubenmann and Niggli (1924) interpreted Brauns' pyrometamorphism as a variety of their pneumatolytic (metasomatic) contact metamorphism. Later, Tyrrell (1926) expanded the term to cover all the metamorphic products formed at very low pressure and very high magmatic temperatures whether or not associated with chemical interchange. Subsequently, Turner and Verhoogen (1960) considered pyrometamorphism as a distinct, though minor, type of metamorphism characterized by a volcanic to near-surface setting and unusual mineral assemblages typical of the sanidinite facies. The particular kinetic conditions under which pyrometamorphism occurs may favour disequilibrium melting and overstepping of equilibrium mineral reactions (Kerrick *et al.*, 1991; Grapes, 2003). The formation of quenched melts (glass or microgranophyre) either as small pods and vein fillings or as relatively thick contact zones suggests boundary conditions overlapping with ultrametamorphism (Spry & Solomon, 1964). Reverdatto (1973) offered a compilation of the 'pyrometamorphism critical minerals' for different rock systems which the Study Group has included in the definition of pyrometamorphism. Those for carbonate and calc-silicate rocks, however, should be handled cautiously because the temperatures needed for their formation are strongly dependent on the values of X_{CO_2} (Tracy & Frost, 1991).

Following the opinion of several authors (e.g. Turner, 1968; Miyashiro, 1973b; Reverdatto, 1973), the SCMR recommends that 'pyrometamorphism' is defined as a specific type of contact metamorphism, thus:

Pyrometamorphism: a very high-grade type of contact metamorphism occurring in volcanic settings or around near-surface intrusions and characterized by mineral assemblages stable at or near atmospheric pressure and very high temperatures. Critical minerals are: spurrite, tilleyite, rankinite, larnite and merwinite in silica-deficient carbonate rocks; mullite and glass in aluminous rocks; tridymite and glass in silica-oversaturated rocks.

Non-critical minerals may include monticellite, melilite, scawtite and ferruginous wollastonite in carbonate rocks; sanidine, sillimanite, cordierite, corundum, spinel and orthopyroxene in aluminous rocks. Under the *P–T* conditions of pyrometamorphism, aluminous and quartzofeldspathic rocks may show variable degrees of melting, which is indicated by the presence of glass among the component minerals or, more rarely, by the formation of glassy rocks (buchites). In the field it is found in (a) xenoliths ejected from volcanic vents or embedded in lavas, (b) narrow rims of contact rocks at the base of basaltic lava flows, and (c) small contact aureoles (centimetre to metre scale) surrounding sheets, dykes or plugs of predominantly basaltic magmas.

It should be noted that contact metamorphic effects associated with volcanic and near-surface intrusive rocks may only be defined as pyrometamorphic if at least one of the 'critical minerals' listed above is present. Otherwise contact effects, characteristic of lower temperatures, come under the general heading of contact metamorphism and may be described according to their metamorphic grade or facies.

Tyrrell (1926) proposed the term *optalic metamorphism* for the indurating, baking and fritting effects produced by lava flows or small dykes on their contact rocks: the upper boundary of 'optalic metamorphism' was characterized by incipient fusion without melting (fritting), thus marking a boundary with pyrometamorphism. However, the SCMR does not regard the process as a distinct type of contact metamorphism and considers that the term is therefore unnecessary.

2.10.4 Combustion metamorphism and other kinds of non-igneous related local metamorphism

Local metamorphic processes (excluding dislocation and impact metamorphism) unrelated to an igneous heat source may sometimes affect solid rocks and produce metamorphic aureoles with a zonal pattern similar to that of true contact aureoles. The classic occurrence of such processes is the rims or haloes of burned rocks (thermantides of Haüy, 1822) surrounding burned coal seams or other natural combustibles. However, there are other possible localized heat sources producing metamorphic effects, for example: (a) thermal shocks due to lightning; (b) flash-heating during atmospheric entry of extra-terrestrial materials; (c) the tectonic transport of hot crust or mantle slices onto cooler sedimentary or metasedimentary rock piles; (d) heat supply from 'gneiss domes' to their surrounding rocks. The resulting metamorphism is commonly accompanied by the formation of melts, which, on cooling, may generate glass if near the surface (a, b), or migmatites if at depth (c, d).

The SCMR recommends that the terms 'contact metamorphism' and 'pyrometamorphism' should not be used to describe the above processes but the following names should be used as appropriate: *combustion metamorphism*, *lightning metamorphism*, *impact metamorphism* and *hot-slab metamorphism* (Smulikowski *et al.*, Section 2.2).

The SCMR's definition of combustion metamorphism is:

Combustion metamorphism: a type of metamorphism of local extent produced by the spontaneous combustion of naturally occurring substances such as bituminous rocks, coal or oil.

In literature the process has been long considered as a distinct type of metamorphism (e.g. Delesse, 1857; Naumann, 1858; Kalkowsky, 1886; Zirkel, 1893; Arnold & Anderson, 1907; McLintock, 1932; Bentor & Kostner, 1976). The SCMR prefers the terms *combustion metamorphism* for the process and **burned** (or **burnt**) **rocks** as a collective name for the various kinds of rocks (porcellanites, buchites, paralavas) formed around burned coal seams, oil or gas fountains and bituminous shales.

Aureoles produced by combustion metamorphism usually do not exceed a few metres in width; exceptionally they may reach a few tens of metres. In some circumstances combustion metamorphism has been described improperly as pyrometamorphism (Cosca *et al.*, 1989; Clark & Peacor, 1992; Sokol *et al.*, 1998).

The SCMR's definition of lightning metamorphism is:

> **Lightning metamorphism**: a type of metamorphism of local extent that is due to a lightning strike (Smulikowski *et al.*, Section 2.2).

This is an exceptional type of metamorphism in which very high temperatures (*c.* 2000 °C) are reached in a few microseconds, causing melting, vaporization and extreme chemical reduction (with formation of metal globules) on the rock surfaces struck by lightning (Essene & Fisher, 1986). Shock wave compression structures have also been observed in unmolten mineral grains (Frenzel *et al.*, 1989). Most products of lightning-strike fusion form glassy crusts, tubules or drops to which the name *fulgurite* was first given by Arago (1821). Tube-like fulgurites are common in soils or in loose sediments, especially sands (Shrock, 1948). Their diameter does not usually exceed a centimetre, while their length may reach up to ten metres. Glassy crusts, small tubules and drops form preferentially on rock surfaces, especially on high mountains. The largest exposure of a single mass of fulgurite so far recorded has a maximum diameter of 0.3 m and a length of 5 m. It is part of a single-event system of fulgurites which extends for 30 m near Winans Lake, Livingston County, Michigan (Essene & Fisher, 1986).

The SCMR's definition of hot-slab metamorphism is:

> **Hot-slab metamorphism**: a type of metamorphism of local extent occurring beneath an emplaced hot tectonic body (Smulikowski *et al.*, Section 2.2).

This term has been proposed by the SCMR to cover the following situations: (a) the small aureoles surrounding hot slices of obducted mantle rocks (Karamata, 1968, 1985; Coleman, 1977; Okrusch *et al.*, 1978) for which Williams and Smyth (1973) proposed the term 'contact dynamo-thermal aureoles'; (b) the relatively wide aureoles of the so-called 'inverted metamorphism' observed, for example, in the High Himalayan Crystalline Sequence at the margins of hot crustal slices tectonically emplaced at upper crustal levels (Medlicott, 1864; Gansser, 1964; Le Fort, 1975; Jaupart & Provost, 1985; Swapp & Hollister, 1991; Vannay & Grasemann, 2001). The aureoles mantling typical 'gneiss domes' may be considered as similar phenomena (Allen & Chamberlain, 1989; Barton *et al.*, 1991b).

2.10.5 The rocks of contact metamorphism

Table 2.10.1 presents a list of contact metamorphic rock names found in literature. Apart from some general names these have been subdivided on the basis of their protolith (pelitic to quartzofeldspathic, carbonate to marly, mafic to ultramafic, and special rock types). Many of the names given are now obsolete. Other names are not specific to contact metamorphism, for example *marble*, *quartzite*, *calc-silicate rock*, *emery rock* are also used for regional metamorphic rocks, and *buchite*, a glassy metamorphic rock, may form by either contact or combustion metamorphism. In these cases, the SCMR recommends the use of *contact* as a prefix (e.g. contact marble, contact buchite) if it is desirable to stress the contact metamorphic origin of a particular rock type (see below).

The names recommended by the SCMR are highlighted in bold in Table 2.10.1. Some of these, although established in international literature, were initially found to have ambiguous definitions or usage, partly inherited from the original definitions, partly as a result of subsequent modifications. Redefinition of such terms was therefore necessary. The background discussion and recommended definitions are given below.

HORNFELS

Today this name is widely used for a group of compact, highly metamorphosed contact rocks, typically found in inner aureoles. It derives from an ancient term used by miners in Saxony (Germany) for hard compact rocks of various origins, characterized by their *horny appearance* and *conchoidal to subconchoidal fracture*, and commonly associated in the field with granitic rocks (Leonhard, 1823). Initially, their geological status was uncertain since at that time the igneous nature of granite was still in doubt. Boué (in Leonhard, 1823) noted a transition from hornfels to unaltered greywacke and schist, although Leonhard considered that the hornfels was in part a very fine-grained variety of granite. Cordier (1868) found the same degree of ambiguity in the corresponding French terms 'cornéenne' or 'pierre de corne'. The geological significance of the hornfels rocks emerged as the igneous character of the granitic rocks was established (Boué, 1829; Delesse, 1857; Zirkel, 1866a, b). Rosenbusch's paper (1877) on the contact rocks in the aureole of Alsace granites represents a landmark in the modern interpretation of hornfels and other contact rocks. At the end of the nineteenth century the term 'hornfels' was firmly established in geological nomenclature as a type of contact metamorphic rock (Loewinson-Lessing, 1893–4). Since then different definitions of hornfels have appeared with varying emphasis placed on compositional and structural features. For some, hornfels is an aluminous rock derived, for example, from shale; for others it also includes quartzofeldspathic protoliths such as greywackes and for some the term can be applied to any contact-metamorphic rock (including calc-silicate hornfels, quartz-hornfels, mafic and ultramafic hornfelses) provided they possess the distinctive structural characters recognized by the old Saxon miners. Another controversial point concerns the grain size of a hornfels; to some (especially those supporters of an aluminous character of the hornfels) it is very fine- to fine-grained; others regard it as very fine to coarse. Another source of uncertainty concerns whether the term hornfels should be defined in a genetic sense (for contact metamorphic rocks only) or should be used as a structural class of metamorphic rocks. Further difficulty arises from the extended use of the name hornfels by some workers to cover the partially metamorphosed rocks of the outer aureole.

To resolve these problems the Study Group decided that the name 'hornfels' should only be used for contact metamorphic rocks dominantly composed of silicate + oxide minerals in varying proportions. Contact marble and coal are considered as separate rock types. For hornfels-looking rocks produced by regional metamorphism the SCMR recommends the use of 'granofels' (e.g. calc-silicate granofels) or, alternatively, *hornfelsoid*. The Study Group further decided to base the definition mainly on structural criteria. The recommended definition of hornfels is therefore as follows:

> **Hornfels**: a hard, compact contact-metamorphic rock of any grain size, dominantly composed of silicate + oxide minerals in varying proportions, with a horny aspect and a subconchoidal to jagged fracture.

It may retain some structural features inherited from its protolith such as bedding, sedimentary laminations or metamorphic layering. Different types of hornfels may be distinguished according to structural (e.g. fine-grained, coarse-grained, spotted, layered), chemical (e.g. peraluminous) or mineralogical criteria (e.g. mafic, ultramafic, cordierite-sillimanite-spinel hornfels, diopside-wollastonite-garnet hornfels). Hornfels occurs mostly, but not exclusively, in the innermost part of contact aureoles.

The transition from unaltered schist to hornfels is marked by a gradual loss of the original fissility; the primary foliation, however, is hard to destroy and persists, as compositional differences, even in true hornfelses. Hornfelsic rocks with traces of a pre-existing schistosity have been called either 'schist hornfels' (*Schieferhornfels*) or 'schistose hornfels' (*schiefriger Hornfels*) (e.g. Rosenbusch, 1898; Salomon, 1898). The Study Group considered that neither term is satisfactory: schistose hornfels because it conveys the idea of a schistose rock, and schist hornfels because it is a compound of two different rock names. Therefore the SCMR recommends that, for such rocks, the term **hornfelsed schist** should be used. Although Harker (1932) considered that the use of hornfels as a verb 'affronted the English language', the use of 'hornfelsed' is now widespread in metamorphic literature (e.g. Spry, 1969). The term *schistose hornfels* may be retained only for the particular rock type known in the old literature as *leptynolite*, that is, a totally recrystallized rock of the innermost aureole (with the same mineralogical composition as the associated hornfelses), which is characterized by a schistose structure.

NON-HORNFELSIC CONTACT ROCKS

According to the definition of hornfels given above, not all contact metamorphic rocks are hornfelses. Among them are various rocks of the outermost aureole that commonly retain most of their original mineralogical and structural characteristics (e.g. sandstones, some limestones, schists) and can be difficult to distinguish from their unmetamorphosed equivalents, apart from a slight colour change or the development of an indistinct spotted structure. The transition from rocks unaffected by contact metamorphism to typical hornfels is marked by progressive changes in the mineral content and structure of the rocks. The classification of these transitional rocks was discussed in detail by Salomon (1898), who proposed that the rocks of the outer aureole, whose original characters are largely retained, are simply named by use of the prefix 'contact' (e.g. contact sandstone, contact schist). Unfortunately, the same prefix was used by Salomon (1891), although in hyphenated form, to mean a rock derived by contact metamorphism, thus, a contact-pyroxenite or a contact-peridotite was a rock having the composition of a pyroxenite or a peridotite and produced by contact metamorphism regardless of the nature of the protolith (e.g. a pyroxenite might be derived from a recrystallized pyroxenite, or from a marly rock).

To complicate the nomenclature further, the term 'hornfels' has been used for all the products of contact metamorphism, including rocks of the outer aureole. This usage followed the introduction of two new facies of contact metamorphism, namely hornblende-hornfels facies and albite-epidote-hornfels facies (Fyfe *et al.*, 1958).[14] Accordingly, all contact-metamorphic rocks became part of a single hornfels family. This meant that although the rocks of Eskola's (1920) original hornblende-hornfels facies were true hornfelses of the inner aureole this was not the case (or only partially the case) for rocks belonging to the albite-epidote-hornfels facies.

To resolve this situation, the SCMR recommends the following:

1. That the term 'hornfels' is used in accord with the above definition, that is, the rock is named on its structural or mineralogical characteristics, not on the facies to which it belongs.
2. That the prefix **contact** is used in front of a metamorphic rock name to distinguish that rock from another of similar composition generated by other types of metamorphism. Thus, a contact marble or contact amphibolite is a marble or an amphibolite formed by contact metamorphism and not by regional metamorphism. Also, *contact buchite* is used to distinguish a buchite formed by contact metamorphism from a *coal-fire buchite*.
3. That the prefix **contact-metamorphosed** is used in front of the name of a protolith for the rocks of the outer aureole that have been slightly modified by contact metamorphism but still retain most of their original structural and/or mineralogical characteristics; the prefix may be used in front of metamorphic protoliths, for example *contact-metamorphosed eclogite*.

Structural terms like *spotted*, *maculose* or *knotted schists* may be used to describe contact-metamorphosed schists in which the new metamorphic minerals form either spots or nodules growing on a pre-existing unaltered or slightly altered schistose matrix.

MAFIC AND ULTRAMAFIC HORNFELSES

Igneous rocks of mafic composition and/or ultramafic rocks of either oceanic or continental provenance may be present in some contact aureoles (e.g. Isle of Skye, Scotland; Sudbury layered intrusion, Ontario; Bushveld igneous complex, South Africa; Bergell pluton, Italy and Switzerland). The metabasic rocks are rather insensitive to temperature variations in this context and do not produce a great variety of contact rocks, essentially because most related chemical reactions are multivariant and do not involve the appearance or disappearance of many phases

[14] Note: these two facies are not recommended by the SCMR (see Smulikowski *et al.*, Section 2.2).

(Tracy & Frost, 1991). In the hottest part of the aureole, however, most basaltic or gabbroid rocks are converted into fine-grained hard rocks, essentially composed of plagioclase + pyroxene(s) ± olivine ± opaque minerals ± amphibole, which have been described with different names in different countries (e.g. *muscovadite*, *granulitized rocks*, *granoblast*, *sudburite*, *beerbachite*). The Study Group considers these names unnecessary and recommends the use instead of the general term **mafic hornfels** (in preference to *basic hornfels*; Spry, 1969). Further distinctions are possible on the basis of the metamorphic grade (e.g. medium-grade mafic hornfels, high-grade mafic hornfels) or the mineral assemblage (e.g. plagioclase-hornblende mafic hornfels, magnetite-olivine-augite-plagioclase mafic hornfels). In a similar manner, for the various hornfelsic rocks generated from ultramafic rocks close to igneous contacts, the Study Group recommends the use of the general term **ultramafic hornfels** (in preference to *ultrabasic hornfels*; Williams *et al.*, 1954; Spry, 1969).

CONTACT-METAMORPHIC CARBONATE ROCKS

In accord with Rosen *et al.* (Section 2.7) a contact marble is a contact rock containing more than 50% vol. of carbonate minerals (calcite and/or dolomite). Pure contact marble contains more than 95% vol. of carbonate minerals; the remainder is classified as impure contact marble. Depending on the prevailing carbonate minerals a distinction may be made between dolomitic and calcitic marble. Further differentiation can be given by reference to the metamorphic mineral assemblage (e.g. spinel-forsterite-calcite (contact) marble, diopside-wollastonite-calcite (contact) marble, scapolite (contact) marble), the microstructure (e.g. fine-grained, coarse-grained, saccharoidal) or other characteristics observable at hand-specimen scale (e.g. colour, banding, veining).

In many contact aureoles, pre-existing dolomitic rocks are largely converted to dolomitic marble; however, in the innermost part of many low-pressure aureoles, the existing T–X_{H_2O} conditions favour the formation of calcite + periclase and/or brucite from the breakdown of dolomite. As a result brucite marble is relatively common close to the igneous contact. In literature, two types of brucite marble have been distinguished by reference to their chemistry, namely **pencatite** (Roth, 1851) for a brucite marble having CaO and MgO amounts similar to those found in pure dolomite rocks (31% and 21% by weight respectively), and **predazzite** (Petzholdt, 1843) for brucite marble (derived from dolomitic limestones) having a much lower amount of MgO and a correspondingly higher value of CaO than in dolomite. The SCMR proposes that these two terms have restricted status and their use is only permissible if the rock chemistry is known. In the absence of chemical data or for general use the recommended name is **brucite marble**. In some marbles the brucite may form coarse tabular grains unrelated to the former presence of periclase (Williams *et al.*, 1954).

CONTACT-METAMORPHIC GLASSY ROCKS

Rocks heated to very high temperatures in a very short time, followed by a rapid dissipation of heat, may be converted into vitreous or semivitreous materials. Such rocks have received different names according to the cause of heating, for example tektite (by meteoritic impact), pseudotachylite (by frictional heating in fault zones), fulgurite (by a lightning strike). In volcanic to subvolcanic environments, very high-grade contact metamorphism of the immediate country rocks or of xenoliths commonly produces glassy hornfelses. Various terms have been proposed for these rocks, namely systil (Zimmermann, in Nöggerath, 1822), thermantide (Cordier, 1868), porzellanjaspis (Leonhard, 1823, 1824), basaltjaspis (Zirkel, 1866a, b), porcellanite (e.g. Cordier, 1868; Rinne, 1928) and buchite (Möhl, 1873). The last two names are still widely used in modern literature, buchite being mostly used where the protolith is a quartzofeldspathic rock, porcellanite when it is an argillisaceous rock. It is of interest to note that the two terms were originally coined for fused rocks of different origin: porcellanite for the products of combustion metamorphism (Leonhard, 1823, 1824) and buchite for the products of contact metamorphism (Möhl, 1873). The use of the terms, however, varied greatly and gave rise to many ambiguities; for example, Delesse (1857) used porcellanite to include glassy rocks formed by both combustion and contact metamorphism. Similarly, although the term buchite was originally created for vitrified sandstones formed by contact metamorphism, it was subsequently enlarged to include rocks of pelitic composition (for further details see Tomkeieff, 1940). An attempt by Tomkeieff (1940) to differentiate, on structural grounds, porcellanite (as a finely crystalline rock) from buchite (vitreous or semivitreous rock) was not particularly successful. The position is further complicated by the use of the term porcellanite by sedimentary petrologists for a special class of silica-rich rocks (e.g. Williams *et al.*, 1954; Moorhouse, 1959).

It is therefore not surprising if the terms buchite and porcellanite are used by some workers as synonyms, and by others as indicators either of the kind of metamorphism or of the protolith. Also there is a recent tendency to use buchite and/or porcellanite for any glassy material produced by any metamorphic process: from meteoritic impact to burned rocks in sacrificial sites or burned artificial mixture of clay and carbonate materials.

Against this background the Study Group agreed on the following points:

1. It is convenient to have a single name for vitreous or semivitreous rocks of contact-metamorphic origin.
2. The selected name should also be applicable to similar rocks produced by combustion metamorphism but it should not be used for vitreous rocks formed by other processes where acceptable names exist, for example fulgurite, tektite and pseudotachylite (see above).
3. The definition of names for metamorphic glassy rocks should state the minimum glass content and rock structure.

A thorough discussion of these points reached the following conclusions:

1. Vitreous rocks of contact-metamorphic origin should be named *buchites*. The name porcellanite is unnecessary and should be dropped because of its ambiguity. The name 'buchite' has no compositional implications, but the protolith lithology may be indicated by appropriate qualifiers (e.g. aluminous, marly, arenaceous).
2. Vitreous rocks produced by combustion metamorphism should also be named 'buchites'. The name porcellanite, originally proposed for such burned rocks, should be dropped because of its ambiguity.
3. The minimum glass content of buchite should be 20% by volume; contact metamorphic or burned rocks with a lower glass content should be named *fritted rocks*.

The structure of buchite ranges from massive to vesicular or slaggy. In some outcrops it may show columnar jointing. It has a glassy matrix containing abundant unmelted or partially melted mineral grains from the protolith together with new mineral grains in assemblages typical of pyrometamorphism. Fritted rocks have a structure similar to buchite, but show a wide range of colour. Sometimes they have a glazed appearance, hence their original name of 'porcellanite'. The microstructure of fritted rocks resembles that of buchite, from which they differ by their lower glass content.

The recommended definitions for buchite and fritted rock are as follows:

Buchite: a compact, vesicular or slaggy metamorphic rock of any composition containing more than 20% vol. of glass, either produced by contact metamorphism in volcanic to subvolcanic settings or generated by combustion metamorphism.

It is also used for partially melted materials obtained in laboratories by burning or heating natural rocks or artificial mineral mixtures. In hand specimen the rock is commonly characterized by a conchoidal fracture. In some outcrops buchite may show columnar jointing. In thin section the rock is composed of a glassy matrix and unmelted or partially melted mineral grains of the protolith. The glass commonly contains small grains of newly formed minerals in phase assemblages typical of pyrometamorphism. Locally the original glassy matrix is partially converted into very fine quartz-K-feldspar intergrowths resembling granophyre. Buchites formed by different metamorphic processes can be distinguished by appropriate qualifiers (e.g. contact, combustion or artificial buchite).

Fritted rock: compact, vesicular or slaggy metamorphic rock of any composition with a glass content ranging from a few per cent up to a maximum of 20% vol. and either produced by contact metamorphism in volcanic to subvolcanic settings or generated by combustion metamorphism.

It is also used for glassy materials obtained in laboratories by burning or heating natural rocks or artificial mineral mixtures. Fritted rocks commonly show changes from the protolith in either the structure or the colour even at the hand-specimen scale, commonly developing a streaky aspect due to bands of different colours. The glass is mostly interstitial between the protolith grains or is irregularly distributed in small patches or along microfractures. The glass may contain the same new minerals as in buchite, although in lesser amounts.

BURNED ROCKS

The products of combustion metamorphism have received different names in the past, for example thermantide, thermantide porcellanite, fused shale, porcellanite, porcelain jasper. However, most of these names have also been used for the products of contact metamorphism in volcanic or near-surface settings. Thus the Study Group decided to find a general name for all rocks generated by combustion metamorphism and an agreement was found for the term burned (or burnt) rock, which has been defined as follows:

Burned rock: general term for a compact, vesicular or clinkery, glassy to holocrystalline metamorphic rock of various colours, produced by the combustion metamorphism of pre-existing sedimentary rocks.

In the fused varieties the glass coexists with refractory grains and/or newly formed minerals (e.g. melilite, wollastonite, mullite, cristobalite, spinel), whose nature reflects both the very high temperature metamorphic conditions and the variable chemical composition of the original rock. The term burned rock supersedes such old names as thermantide, thermantide porcellanite, fused shale, porcellanite, porcelain jasper. The glassy or glass-bearing varieties of burned rocks are called buchite (coal-fire buchite) or fritted rock respectively. The term burned rocks also includes some typical ash deposits of refractory material remaining after the combustion of coal seams, giving rise to soft, clay-like rocks resembling volcanic cinerites, and for which the Study Group proposes the name of *coal-fire ash* to distinguish it from ash deposits of volcanic origin.

SPECIAL CONTACT METASOMATIC ROCKS: THE ADINOLE SERIES

The contact metasomatic rocks are discussed elsewhere (Zharikov *et al.*, Section 2.9). The discussion here is restricted to a particular type of contact metamorphism, well known in the Harz mountains, Germany, where it was observed in the proximity of stratiform diabase dykes injected into slates and other metasediments of Devonian to Lower Carboniferous age. Characteristic of this kind of metamorphism is the development of narrow (<2 m) aureoles with a particular zonation of the contact rocks. At the immediate contact with the diabase there is a fine-grained compact rock predominantly composed of albite and subordinate quartz, which passes rapidly into schistose or gneissic albite-rich rocks, with a characteristic spotted or banded structure, which in turn pass into slates unaffected by contact metamorphism. This rock association was interpreted by Lossen (1872) as a typical *diabase contact metasomatic aureole*, and, making use of already existing terms, he proposed the name *adinole*, for the compact albite-quartz rock, and *spilosite* and *desmosite* for the spotted and banded albite-rich schists respectively. Lossen (1872), however, also recognized the existence of other adinole rocks, with an aspect recalling the Swedish hälleflintas (the adinole of Beudant, 1824). These rocks occur as layers intercalated in a metasedimentary sequence, unrelated to diabase dykes and thus excluding a contact metamorphic origin. They were interpreted as metamorphosed silicified acid tuffs by Rosenbusch (1910). Subsequent workers (Milch, 1917; Mempel, 1935–6) tried to make a distinction between these two types of adinole-rocks as described by Lossen. For example, Mempel proposed the name 'adinolit' for the contact-type rock, and 'adinole' for the hälleflinta-looking type. However, in most textbooks and geological glossaries the term *adinole* is identified with Lossen's (1872) contact-metasomatic type.

The Study Group considered this complex position and decided the following:

1. To maintain the term adinole as a general name for all the rocks described above.
2. To maintain the distinction between the two types of adinole so far recognized, but to reject, because of their potential ambiguity, the two names proposed by Mempel.
3. To accept Milch's (1917) idea of an adinole-series, in which the various members represent different stages of an adinolization process, whose final product is the adinole rock.

The resulting proposals for the corresponding terms are the following:

Adinole: a compact, fine-grained rock with a splintery fracture, commonly with a finely banded aspect due to alternating grey, green or reddish layers of variable colour intensity; mineralogically it is essentially composed of a fine- to very fine-grained mosaic of albite and subordinate quartz, with minor amounts of other constituents (muscovite, sericite, chlorite, actinolite, epidote, rutile and/or anatase); chemically it is characterized by high (up to 10% by weight) amounts of soda, a character that easily distinguishes it from the hard siliceous shales or slates that it resembles. Hence **adinolization**, a Na-metasomatic process leading to the formation of adinole.

There exist two different varieties of adinole distinguished according to their geological settings, namely **contact adinole** found in the contact aureoles of diabase dykes and **tuffaceous adinole** found as layers within metasedimentary sequences and unrelated to the proximity of diabase dykes.

Adinole series: a group of rocks in the aureole of a diabase intrusion that show stages of contact metasomatism (adinolization) leading ultimately to the formation of adinole.

The intermediate stages of the process, still retaining part of the original schistosity, give rise to *spotted* or *banded albite-rich schist* called *spilosite* and *desmosite* respectively. Following Milch (1917) the intermediate stages could alternatively be called *adinole schist* and *adinole hornfels* according to the amount of relict schistosity.

Table 2.10.1 *Lists of selected terms for local metamorphic processes and their products.*

The terms, occurring in literature, are arranged in chronological order. Recommended terms are shown in bold font, restricted terms in regular font and unnecessary terms in italics. Not all the terms listed are given in the glossary: the full lists along with source references are given on the SCMR website. m. = metamorphism.

1. Names for contact-metamorphic processes and other types of metamorphism related to localized heat sources		
ISOCHEMICAL CONTACT METAMORPHISM		
Abnormal m.	*Diabase contact m.*	*Anaphryxis*
Caloric m.	*Pyromorphism*	*Normal contact m.*
Everse vs inverse m.	*Paroptesis*	*Thermal contact m.*
Exo- vs endomorphism	*Accidental m.*	*Thermo-contact m.*
Hydatocaustic m.	*Peripheral m.*	*Hydrothermal contact m.*
Hydatothermic m.	*Exogenic vs endogenic m.*	*M. by regeneration*
Contact m.	*Selective contact m.*	*Load-contact m.*
Special m.	*Exo- vs endomorphic contact m.*	*Alembic contact m.*
Local m.	Thermal m.	*Ultra-contact m.*
Juxtaposition m.	*Thermometamorphism*	**Isochemical contact m.**
	Piezo-contact m.	*High-pressure contact m.*
ALLOCHEMICAL/METASOMATIC CONTACT METAMORPHISM		
Metasomatism	*Pneumatolytic metasomatism*	*Additive contact m.*
Pneumatolysis	Hydrothermal metasomatism	**Metasomatic contact m.**
Perimetral m.	*Metasomatic pneumatolysis*	*Endo vs exosmotic m.*
Atmogenic m.	*Pneumatolytic contact m.*	*Pneumatolytic to*
Methylosis	*Endo- vs exopneumatolytic contact m.*	*hydrothermal contact m.*
Metachemic m.,	*Apomagmatic contact metasomatism*	*Enhanced m.*
metachemical m.	*Perimagmatic contact metasomatism*	*Pyrometasomatism*
Paramorphism	*Perimagmatic additive contact*	Pneumatolytic m.
Contact metasomatism	*metasomatism*	**Allochemical contact m.**
Additive vs subtractive metasomatism		
PYROMETAMORPHISM		
Caustic m.	*Pyromorphism*	**Pyrometamorphism**
Pyrocaustic m.	*Thermometamorphism*	*Optalic m.*
COMBUSTION METAMORPHISM		
Firing m.	*Pyromorphism*	**Combustion m.**
Burning m.	*Pyrogenic m.*	
2. Names for contact-metamorphic rocks and rocks generated by other types of metamorphism related to localized heat sources		
GENERAL NAMES		
Vitreous tube	**Fritted rock**	*Endo- vs exopneumatolytic*
Fulgurite	*Neptunopyrogenic rock*	*contact rock*
Thermantide	*Diabase contact rock*	*Contact-deposit*
Burned rock, burnt rock	*Normal contact rock*	*Contact metamorphic*
Lightning tube	*Pneumatolytic contact rock*	*ore deposit*
Contact rock		*Contactolite*
		Baked rock

Table 2.10.1 (*cont.*)

PELITIC TO QUARTZOFELDSPATHIC PROTOLITHS		
Killas	**Spotted slate/schist**	**Schistose hornfels**
Porcellanite, Porcelain-jasper	*Kornite*	*Schist-hornfels*
Cornéenne	*Cornubianite*	*Contact schist*
Tripoli	*Proteolite*	*Astite*
Basalt-jasper	**Adinole**, *adinole schist*	*Aviolite*
Hornfels	*Desmosite*	*Seebenite*
Keratite	*Spilosite*	*Edolite*
Keralite	*Leptynolite*	*Quartz hornfels*
Ebensinite	**Knotted schist**	*Injected hornfels*
Hornstone	*Hornschist*	*Paralava*
Systil	**Buchite**	*Pelitic hornfels*
Vitrified shale	*Corneite*	
MARLY AND CARBONATE PROTOLITHS		
Marble	*Barégienne*	*Tactite*
Calciphyre	*Limurite*	*Marmorite*
Predazzite	**Calc-silicate rock**	*Lime-silicate rock*
Pencatite	*Lime-silicate hornstone*	*Calc-magnesian silicate rock*
Thermocalcite	*Calc-flinta*	**Contact marble**
Plakite	*Vullinite*	*Magnesian marble*
Calc-hornfels	**Skarn**	
Calc-silicate hornfels	*Calc-iron silicate rock*	
MAFIC AND ULTRAMAFIC PROTOLITHS		
Basic hornfels	*Granoblast*	**Ultrabasic hornfels**
Mafic hornfels	*Silicoferrolite*	**Magnesian hornfels**
Beerbachite	*White trap*	
Sudburite	*Ultramafic hornfels*	
SPECIAL PROTOLITHS (COAL, LATERITE)		
Cokeite	**Coal-fire ash**	*Corundum-spinel hornfels*
Carbonite	*Corundolite*	**Emery rock**

2.11 Impactites

DIETER STÖFFLER and RICHARD GRIEVE

2.11.1 Introduction

A Study Group under the leadership of D. Stöffler (Berlin) has formulated this proposal for the classification and nomenclature of impactites. The following scientists have participated actively in the Study Group: W. von Engelhardt (Tübingen), V. I. Feldman (Moscow), R. A. F. Grieve (Ottawa), F. Hörz (Houston), and K. Keil (Honolulu). Contributions were also made by B. M. French (Washington) and W. U. Reimold (Johannesburg). After having evaluated proposals by the members of the Study Group and by scientists working with impactites, this work presents a classification and nomenclature of such rocks.

2.11.2 Classification

The term 'impactite' is a collective term for all rocks affected by one or more hypervelocity impact(s) resulting from collision(s) of planetary bodies. A classification scheme is proposed for products of single and multiple impacts (Table 2.11.1). It is applicable to terrestrial and extraterrestrial rocks, such as lunar rocks and meteorites of asteroidal, lunar and Martian provenance. The basic classification criteria are based on microstructure, degree of shock metamorphism and lithological components. Shock metamorphism is the irreversible changes in (geologic) materials resulting from the passage of a shock wave (Fig. 2.11.1). Additional criteria for a subclassification of the main types of impactites relate to the mode of occurrence with respect to the parent impact crater and to the geological or structural setting of the impactites (Fig. 2.11.2, Table 2.11.2). The proposed classification has made use of previous recommendations (Stöffler *et al.*, 1979, 1980; Stöffler & Grieve, 1994, 1996).

Impactites from a single impact are classified into three major groups (Table 2.11.1) irrespective of their geological setting which is not known for most extraterrestrial rocks such as meteorites and lunar rocks:

Table 2.11.1 *Classification of impactites formed by single and multiple impacts.*

1. **Impactites from single impacts**
   - 1.1. *Proximal impactites*
     - 1.1.1. Shocked rocks*
     - 1.1.2. Impact melt rocks[1]
       - 1.1.2.1. Clast-rich
       - 1.1.2.2. Clast-poor
       - 1.1.2.3. Clast-free
     - 1.1.3. Impact breccias
       - 1.1.3.1. Monomict breccia
       - 1.1.3.2. Lithic breccia (without melt particles)[2]
       - 1.1.3.3. Suevite (with melt particles)[2]
   - 1.2. *Distal impactites*
     - 1.2.1. Consolidated
       - 1.2.1.1. Tektite[3]
       - 1.2.1.2. Microtektite[3]
     - 1.2.2. Unconsolidated
       - 1.2.2.1. Airfall bed[4]
2. **Impactites from multiple impacts**
   - 2.1. *Unconsolidated clastic impact debris*
     - 2.1.1. Impact regolith[5]
   - 2.2. *Consolidated clastic impact debris*
     - 2.2.1. Shock-lithified impact regolith[5]
       - 2.2.1.1. Regolith breccia[5] (breccia with *in situ* formed matrix melt and melt particles)
       - 2.2.1.2. Lithic breccia[5] (breccia without matrix melt and melt particles)

* see Tables 2.11.3 to 2.11.7 for further subclassification.
1 May be subclassified into glassy, hypocrystalline, and holocrystalline varieties.
2 Generally polymict but can be monomict in a single lithology target.
3 Impact melt (generally glassy) with admixed shocked and unshocked clasts.
4 Pelitic sediment with melt spherules, shocked and unshocked clasts.
5 Generally polymict but can be monomict in a single lithology target.

Metamorphic Rocks: A Classification and Glossary of Terms. Recommendations of the International Union of Geological Sciences, eds. Douglas Fettes and Jacqueline Desmons. Published by Cambridge University Press. © Cambridge University Press 2007.

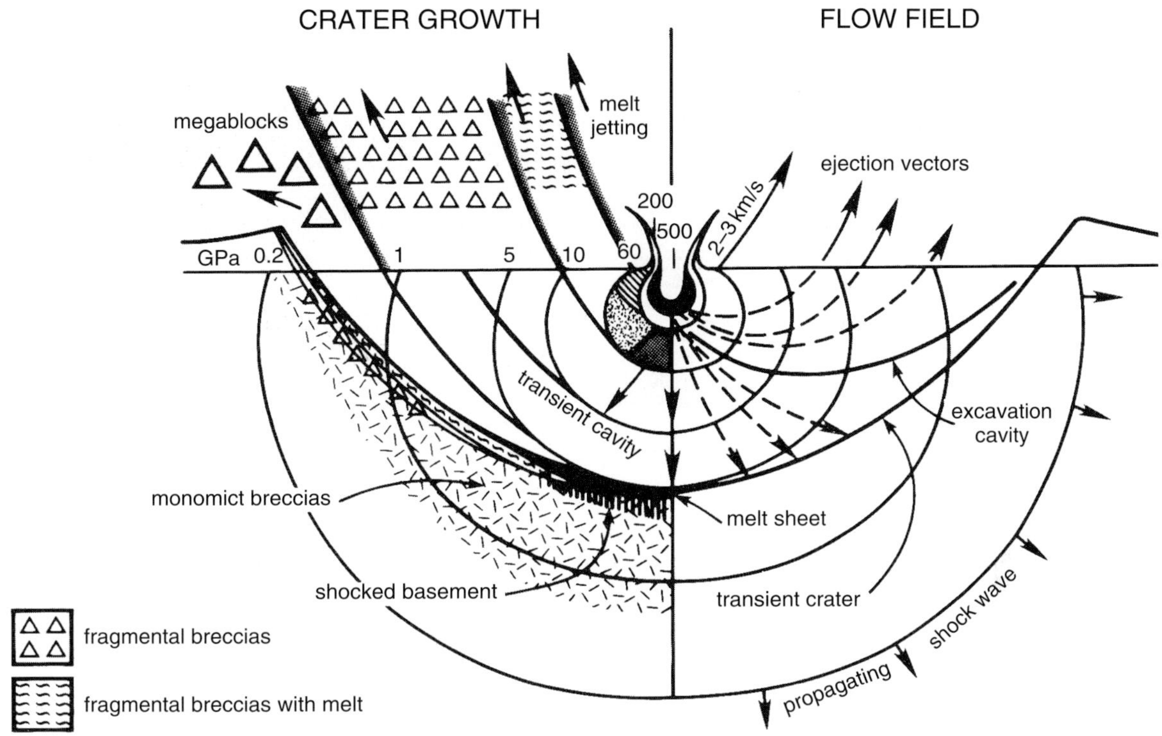

Fig. 2.11.1 Schematic representation of the shock zoning and particle motion in an impact crater based on various data and models, e.g. Dence *et al.* (1977), Grieve *et al.* (1977), Stöffler (1977), O'Keefe and Ahrens (1978), Croft (1980), Kieffer and Simonds (1980), and Orphal *et al.* (1980).

Shocked rocks are defined as non-brecciated rocks, which show unequivocal effects of shock metamorphism, exclusive of whole rock melting. They are subclassified into progressive stages of shock metamorphism (Tables 2.11.3 to 2.11.7).

Impact melt rocks are subdivided into three subgroups, according to the content of clasts. These three subtypes may be subclassified according to the degree of crystallinity into glassy, hypocrystalline and holocrystalline varieties. The first two subtypes include 'impact glass' as well as 'tektites'.

Impact breccias fall into three subgroups, according to the degree of mixing of various target lithologies and their content of melt particles. *Lithic breccias* and *suevites* are generally polymict breccias except for single lithology targets. The matrix of lithic breccias is truly clastic and consists exclusively of lithic and mineral clasts whereas the matrix of suevite additionally contains melt particles and may therefore be better called *particulate matrix*. This primary matrix of suevite may be altered by secondary (mostly hydrothermal) processes.

Impactites from multiple impacts, as known from the Moon and from meteorites, as samples of the meteorite parent bodies, are subdivided into two main groups (Table 2.11.1):

Impact regolith (unconsolidated clastic impact debris), and

Shock lithified impact regolith (consolidated clastic impact debris). This group is subclassified into *regolith breccias* (with matrix melt and melt particles) and *lithic breccias* (without matrix melt and melt particles). The term lithic breccia is synonymous with 'fragmental breccia' which has been used for lunar rocks and meteorites (Stöffler *et al.*, 1980; Bischoff & Stöffler, 1992). Note that

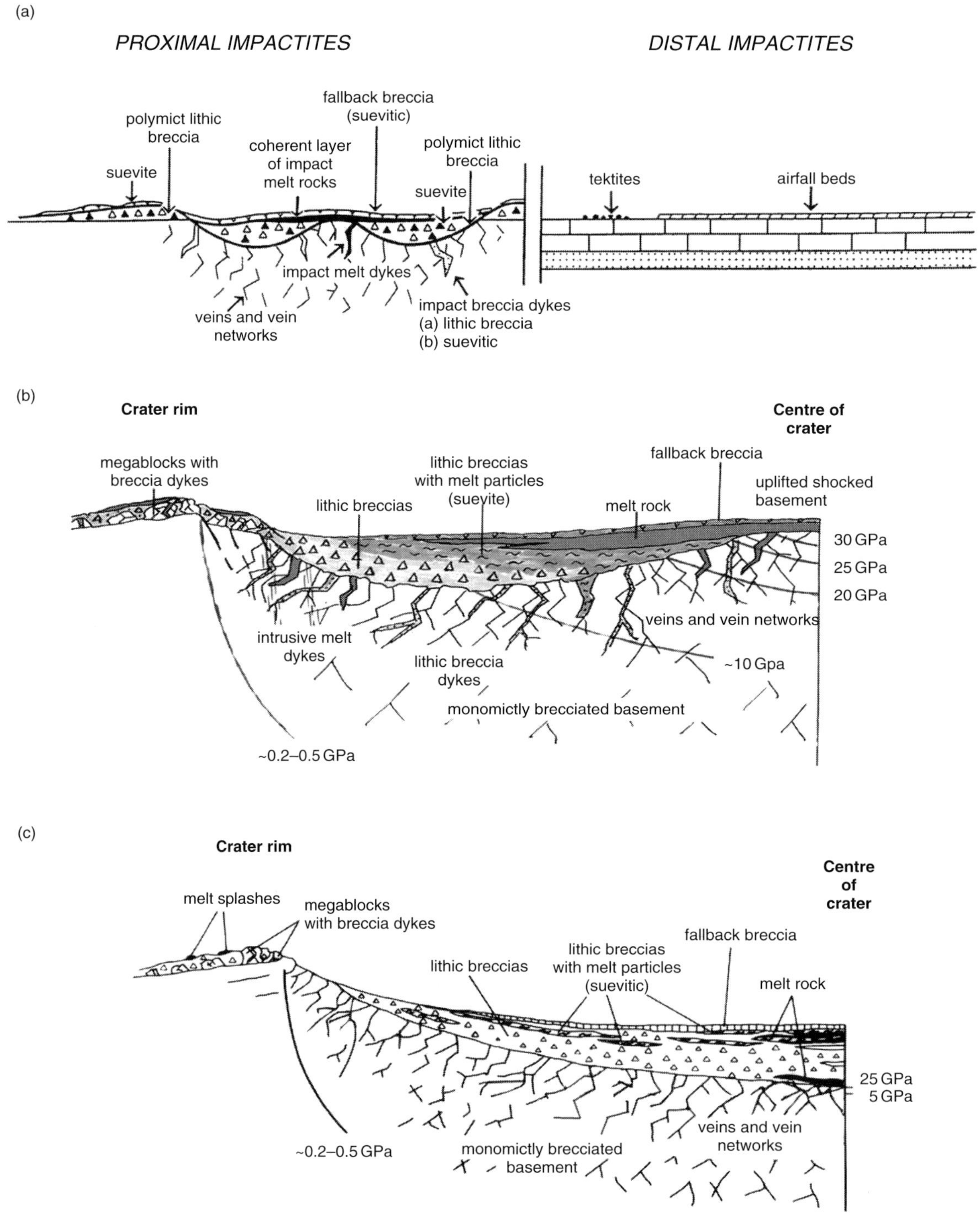

Fig. 2.11.2 Geological setting of impactites on Earth: **a**, proximal and distal impactites, **b**, proximal impactites at a simple impact crater (diameter range on Earth: ~30 m to ~2–5 km); **c**, proximal impactites at a complex impact crater with central uplift (diameter range on Earth: ~5 km to 50–60 km); shock pressure isobars are shown in the parautochthonous crater basement.

Table 2.11.2 *Classification of impactites from single impacts according to geological setting, composition and degree of shock metamorphism.*

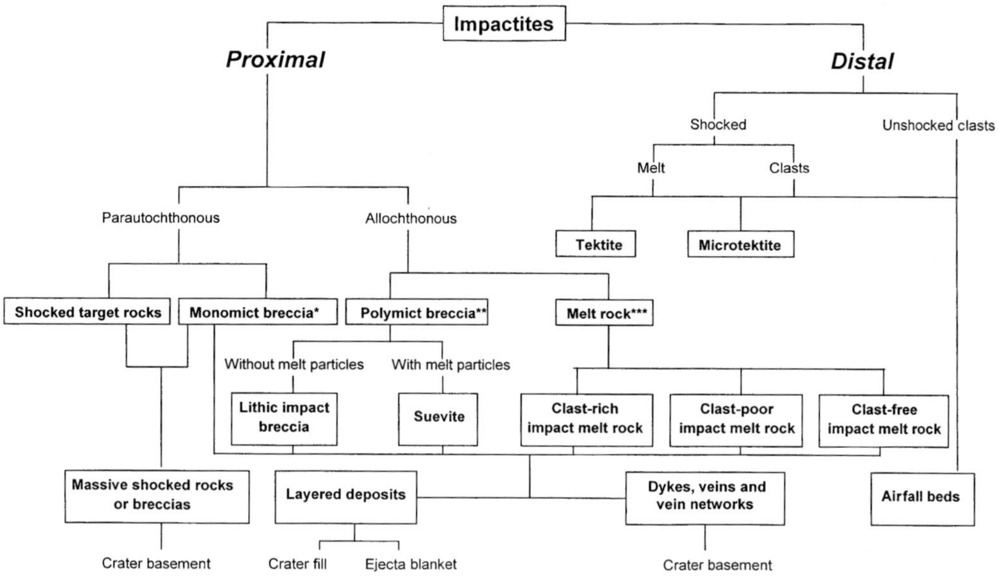

* Typically monomict.
** Generally polymict but can be monomict, e.g. in a single lithology target.
*** Includes glassy, hypocrystalline and holocrystalline varieties.

the matrix melt is formed *in situ* by intergranular melting induced by the shock lithification process (Table 2.11.7).

Irrespective of the geological setting of a specific rock type *progressive stages of shock metamorphism* (Stöffler, 1966, 1971; Chao, 1967b; Tables 2.11.3 to 2.11.7) can be identified in all target rocks affected by the shock wave. They are defined on the basis of shock effects of the constituent minerals and of the shock-induced changes of the primary rock microstructure. The definition of progressive stages of shock metamorphism depends on the mineralogical composition and on the primary microstructure (e.g. porosity) of the material shocked. Therefore, the shock classification is different for different lithologies. Since quartz, plagioclase and olivine (Chao, 1967b; Stöffler, 1972, 1974; Stöffler *et al.*, 1991; Stöffler & Langenhorst, 1994; French, 1998) are the most sensitive shock indicators, separate classification schemes have been proposed for quartzofeldspathic rocks (Table 2.11.3), basaltic-gabbroic rocks (Table 2.11.4), dunitic and chondritic rocks (Table 2.11.5), sandstone (Table 2.11.6), and particulate rock material, for example sand and regolith (Table 2.11.7). Shock metamorphism of carbonates and shales is difficult to recognize on a macroscopic and microscopic scale and reasonable classifications have not yet been established.

2.11.3 Discussion

The process that results in the formation of impactites is related to the interplanetary collisions that all planetary bodies have undergone since their formation. The term '*impact*' or more correctly '*hypervelocity impact*' is defined as the collision of two (planetary) bodies at or near cosmic velocity which causes the propagation of a shock wave in both the impactor and target body (Melosh, 1989). A shock wave is a compressional wave with material transport (whereas seismic waves are compressional waves without material transport). It can be defined as a step-like discontinuity in pressure, density, particle velocity and internal energy, which propagates in gaseous, liquid or solid matter with supersonic velocity. Shock compression is non-isentropic and results in the production of post-shock heat (waste heat), which increases with increasing pressure and eventually results in the melting or vaporization of the shocked material (Duvall & Fowles, 1963; Asay & Shahinpoor, 1993; Graham, 1993).

The material engulfed by the shock wave is affected by what is collectively called *impact metamorphism*. Impact metamorphism should be applied only for natural rocks and minerals and it includes solid state deformation, melting and vaporization of the target rock(s) and their constituent minerals. The term *shock metamorphism* is a more general term that can be used irrespective of the process which generates a shock wave: natural impacts, or artificial hypervelocity impacts, or explosions of chemical or nuclear devices (French & Short, 1968; Stöffler, 1972, 1974, 1984; Roddy *et al.*, 1977; French, 1998). Unequivocal residual shock effects in minerals of shocked rocks are generally formed above the so-called Hugoniot elastic limit (HEL), which is of the order of several gigapascals for silicate minerals. Consequently, the typical range of shock pressures resulting in remanent or residual shock effects is between 5 and 100 GPa for solid state effects and melting, and above 100 GPa for vaporization. Typical maximum pressures and temperatures at the point of impact are of the order of several 100 GPa or greater and several tens of thousands of degrees for all impacts within the inner solar system (terrestrial planets).

Impactites are formed during a complex but very short sequence of processes: shock compression of the target rocks (compression stage), decompression and material transport (excavation stage), and deposition upon ballistic transport and upon collapse of the central ejecta plume which takes place during or after the collapse of the transient crater cavity (modification stage) (Fig. 2.11.1). Consequently, shock metamorphosed material (shocked rocks and impact melts) commonly displays disequilibrium and can be mixed with unshocked lithic and mineral fragments forming polymict breccias in and around the parent crater: *layered* impact formations, such as *impact melt rocks* or *impact breccias*, and *dyke breccias*, which occur both inside the crater and as part of the continuous ejecta blanket extending some two to three crater radii (proximal impactites) and continuous *airfall beds* or discontinuous ejecta deposits, such as *tektites* (distal impactites) (Table 2.11.2). The geological setting of shocked rocks or impact melts is, therefore, variable (Fig. 2.11.2). *Impact melt lithologies* occur as (1) allochthonous coherent melt sheets, (2) inclusions in polymict impact breccias (suevite), (3) dykes and veins in the autochthonous crater basement, in displaced shocked rock fragments and in displaced (unshocked) megablocks, (4) individual melt particles on top of the ejecta blanket, glassy or crystallized spheres in global airfall beds, and (5) glassy tektites. *Shocked minerals* and *rocks* are found as allochthonous clasts within polymict impact breccias, impact melt rocks and airfall beds and as (par)autochthonous material of the crater basement. *Monomict breccias* formed during shock compression and dilatation are characteristic of the crater basement but are also common constituents of polymict breccias. Displaced megablocks within the continuous ejecta blanket are usually monomictly brecciated. *Dyke breccias* can be related to all major phases of the crater formation process and up to four generations of dykes have been observed in a single impact event (Lambert, 1981; Stöffler *et al.*, 1988; Spray, 1998). *Shock veins* and *vein networks* (previously termed 'pseudotachylites') are formed during the compression stage, since they commonly occur as clasts within later formed breccia dykes. The injection of dykes of polymict lithic breccias starts during the compression stage and continues during the excavation stage. A final generation of dykes (polymict or monomict breccias) is produced during the modification stage, while the transient crater collapses and more conventional (but still very high strain rate) faulting takes place.

The time for the formation of the final crater and of some early formed impactites (shocked rocks, melt, dykes) is of the order of seconds to minutes for craters ranging from about 1 to 100 km and the total time for the deposition of the proximal ejecta ranges from minutes to hours (Melosh, 1989; Ivanov & Artemieva, 2002). This time is very short compared with all other geological processes. Despite this, superposition contacts between layered impact formations or contacts at discordant dykes are quite common at impact craters. For example, sharp contacts of sheets of impact melt to the monomictly brecciated, unshocked or mildly shocked crater basement or contacts between the continuous ejecta deposits (polymict lithic breccias) and the overlying suevite are characteristic, as are discordant dykes intersecting displaced megablocks.

Impactites from planetary bodies with a thin or non-existent atmosphere and with very low intensity of endogenous geological activity, such as the Moon, the asteroids and, in part, Mars, show evidence of multiple impacts. This is most conspicuous for the Moon and the asteroids, where the outer zone of the crust is reworked by multiple impacts of all sizes with impactors ranging in size from hundreds of kilometres to micrometres. Because of the inverse proportionality between impactor size and impact frequency (Neukum *et al.*, 2001) the fraction of very small impactors is so large that a fine-grained *regolith* is formed in the upper few metres (5 to 15 m in the case of the Moon). This regolith rests on top of a megaregolith, which is composed of the ejecta blankets from larger impact craters superimposed on each other (Hartmann, 1973, 2003). This megaregolith was essentially formed during the so-called ‘early heavy bombardment’ of the terrestrial planets (4.5 to 3.8 Ga ago); whereas the fine-grained regolith was built up during the past 3.5 Ga when the impact rate had declined by a factor of about 1000 (Taylor, 1982; Heiken *et al.*, 1991; Neukum *et al.*, 2001; Stöffler & Ryder, 2001). Impactites from the megaregolith display all the characteristics found at single terrestrial craters (Stöffler *et al.*, 1980), whereas impactites from the *regolith* are either represented by unconsolidated

Table 2.11.3 *Classification of shocked quartzofeldspathic rocks (progressive stages of shock metamorphism).*

Modified from Stöffler (1971, 1984); post-shock temperatures are relative to an ambient temperature of 0 °C.

Shock stage	Peak pressure (GPa)	Post-shock temperature (°C)	Shock effects
0			Fractured minerals
	~5–10	~100	
Ia			Quartz with planar fractures and planar deformation features: feldspar with planar deformation features
	~20	~170	
Ib			Quartz and feldspar with planar deformation features and reduced refractive index; stishovite and minor coesite
	~35	~300	
II			Diaplectic quartz and feldspar glass; coesite and traces of stishovite; cordierite glass
	~45	~900	
III			Normal feldspar glass (vesiculated) and diaplectic quartz glass; coesite; cordierite glass
	~60	~1500	
IV			Rock glasses or crystallized melt rocks (quenched from whole rock melts)
	~80–100	>2500	
V			Rock glasses (whole rock melts condensed from silicate vapour)

clastic impact debris or by shock lithified consolidated *regolith breccias*, as sampled on the Moon during the Apollo and Luna programmes (e.g. Heiken *et al.*, 1991; see Table 2.11.1). Regolith breccias, lithic breccias, impact melt rocks and shocked rocks are represented on asteroidal meteorites in proportions that reflect the multiple cratering of asteroids and the relatively lower impact velocity in the asteroid belt. The lower impact velocity explains the scarcity of impact melt lithologies (Bischoff & Stöffler, 1992; Keil *et al.*, 1997). According to expectations, Martian meteorites are exclusively shocked rocks or monomict breccias of basaltic, gabbroic and peridotitic provenance (Nyquist *et al.*, 2001; Fritz *et al.*, 2005). In some of the Martian plutonic rocks now occurring as meteorites, more than one shock or impact event is recorded.

Table 2.11.4 *Classification of shocked basaltic-gabbroic rocks (progressive stages of shock metamorphism).*

Based on data of James, 1969; Kieffer *et al.*, 1976; Schaal and Hörz, 1977; Ostertag, 1983; Stöffler, 1984; and Stöffler *et al.*, 1986; post-shock temperatures are relative to an ambient temperature of 0 °C and in part based on Raikes and Ahrens (1979); (?) uncertain values with errors of ~±50 °C.

Shock stage	Equilibration shock pressure (GPa)	Post-shock temperature (°C)	Shock effects and textural characteristics	Accompanying disequilibrium shock effects
0			Unshocked (no unequivocal shock effects)	none
	~1–5	~0		
1			Fractured silicates; mechanical twinning on pyroxene and ilmenite; kink bands in mica; rock texture preserved	none
	~20–22	~50–150		
2a			Plagioclase with planar deformation features and partially converted to diaplectic glass	Incipient formation of localized 'mixed melt' and glassy veins
	~28–34	~200–250		
2b			Diaplectic plagioclase glass; mechanical twinning in pyroxene and ilmenite; mosaicism in olivine and other silicates	Localized 'mixed melt' and melt veins (glassy or microcrystalline)
	~42–45	~900 (?)		
3			Melted plagioclase glass with incipient flow structure and vesicles; mafic and ore minerals as in stage 2	
	~60	~1100 (?)		
4			Melted plagioclase glass with vesicles and flow structure; incipient contact melting of pyroxene; incipient recrystallization of olivine	

Table 2.11.5 *Classification of shocked chondritic meteorites and olivine-rich crystalline rocks (progressive stages of shock metamorphism) modified after Stöffler* et al. *(1991).*

Shock pressure data are based on experimental data given in Stöffler *et al.* (1991); pressures given in the final three columns indicate the upper limit of the shock stage in question; temperature data refer to the ambient temperature before shock compression.

Shock stage	Effects resulting from equilibration peak shock pressure			Effects resulting from local P–T excursions	Pressure		
	Olivine		Plagioclase		GPa* (293 K)	GPa** (293 K)	GPa** (920 K)
Unshocked S1	Sharp optical extinction Irregular fractures	Angular variation of extinction position: Low grade: <1° High grade: 1°–2°	Sharp optical extinction Irregular fractures	none	<4–5		
Very weakly shocked S2	Undulatory extinction Fractures	Angular variation of extinction position: <2°	Undulatory extinction Irregular fractures	none	5–10		
Weakly shocked S3	Planar fractures (PF) Undulatory extinction Irregular fractures	Low grade: maximum of two sets of PF High grade: three or more sets of PF	Undulatory extinction	Opaque shock veins, incipient formation of melt pockets (sometimes interconnected)	15–20	10–15	10–15
Moderately shocked S4	Mosaicism (weak)	Low grade: incipient mosaicism, PF and PDF High grade: mosaicism, PF, and PDF	Low grade: undulatory extinction High grade: partially isotropic, PDF	Melt pockets, interconnected melt veins, opaque shock veins	30–35	25–30	20–25
Strongly shocked S5	Mosaicism (strong) Planar fractures Planar deformation features (PDF)		Maskelynite (diaplectic glass)	Pervasive formation of melt pockets, veins and dykes, opaque shock veins	45–55	45–60	35–45

Table 2.11.5 (*cont.*)

Shock stage	Effects resulting from equilibration peak shock pressure		Effects resulting from local P–T excursions	Pressure GPa* (293 K)	Pressure GPa** (293 K)	Pressure GPa** (920 K)
	Restricted to local regions in or near melt zones					
Very strongly shocked S6	Recrystallization; yellow-brown staining; ringwoodite and wadsleyite; high pressure phases of pyroxene (e.g. majorite, akimotoite)	Shock melted (normal glass)	As in stage S5	75–90		45–60
Shock melted	Whole rock melting and formation of melt rocks					

* From Stöffler *et al.* (1991)
** From Schmitt (2000)

Table 2.11.6 *Classification of shocked sandstone (progressive stages of shock metamorphism).*

Modified after Kieffer (1971) and Kieffer *et al.* (1976); ranges of pressure estimates are given in parentheses; post-shock temperatures are relative to an ambient temperature of 0 °C.

Shock stage	Equilibration shock pressure (GPa)	Post-shock temperature (°C)	Shock effects
0			Undeformed sandstone
	0.2–0.9	~25	
1a			Compacted sandstone with remnant porosity
	~3.0 (2.2–4.5)	~250	
1b			Compacted sandstone compressed to zero porosity
	~5.5 (3.6–13)	~350	
2			Dense (non-porous) sandstone with 2–5% coesite, 3–10% glass and 80–95% quartz
	~13	~950	
3			Dense (non-porous) sandstone with 18–32% coesite, traces of stishovite, 0–20% glass and 45–80% quartz
	~30	>1000	
4			Dense (non-porous) sandstone with 10–30% coesite, 20–75% glass and 15–45% quartz
5			Vesicular (pumiceous) rock with 0–5% coesite, 80–100% glass (lechatelierite) and 0–15% quartz

Table 2.11.7 *Classification of unconsolidated sediments and particulate materials (progressive stages of shock metamorphism).*

Based on data from shock recovery experiments and theoretical models (e.g. Kieffer, 1975).

Equilibration shock pressure (GPa)	Particulate basalt 75035*	Lunar soils			H5 chondrite powder		L6 chondrite powder	Quartz sand 63–125 μm
		15101* 45–150 μm	Model soil	65101*	16% porosity <150 μm	<5% porosity <150 μm	ALH 85017* 125–250 μm	
	vesiculated glass	vesiculated glass					vesiculated glass (50% melt at ~65 GPa)	
40					?	?		
	lithification	lithification						vesiculated glass
30	by	by						
	glass cement	glass cement			lithification by glass cement	lithification by glass cement	intergranular glass (starting at ~25 GPa)	
20		minor intergranular glass	vesiculated glass	glass bonding	minor intergranular glass	minor intergranular glass		minor intergranular glass
10	lithification and compaction	lithification and compaction	lithification compaction	lithification	lithification and compaction	lithification and compaction	lithification and compaction (complete at 14.5 GPa)	lithification and compaction
	Schaal *et al.* (1979)	Schaal & Hörz (1980)	Kieffer (1975)	Christie *et al.* (1973)	Bischoff & Lange (1984)	Bischoff & Lange (1984)	Hörz *et al.* (2005)	Stöffler *et al.* (1975)

* Refers to lunar sample numbers and meteorite names

2.12 List of mineral abbreviations

Jaakko Siivola and Rolf Schmid

2.12.1 Introduction

This list contains abbreviations for 240 mineral species, series, subgroups and groups (Tables 2.12.1 and 2.12.2). It mainly follows the list published by Kretz (1983), critically reviewed and adapted to the current classification and nomenclature of minerals. Of the 192 abbreviations listed by Kretz, 182 have been accepted. Fifty-eight new abbreviations have been added, in part new mineral names, and in part names for groups of mineral species. Rules are given on how to expand the abbreviations to generate abbreviations for numerous other mineral names based on the same root term.

The use of prefixes and suffixes with the mineral abbreviations is also explained.

2.12.2 Importance of recommended mineral abbreviations

The use of systematic compound names as recommended by the SCMR (Schmid *et al.*, Section 2.1) requires all the mineral names to be given as prefixes to the structural root term. This may result in very long names. If the mineral names are replaced by abbreviations, the length of the terms can be reduced to less than half. For use in phase diagrams and databases, and for many other purposes, mineral abbreviations are also very helpful.

2.12.3 Basis of the list

In 1983 Kretz published a list of mineral abbreviations containing 192 entries. It has been widely accepted and its use is recommended by several journals (e.g. *American Mineralogist*). The SCMR proposes to adopt this list with the following changes.

1. Because of incompatibility with the list of mineral species in Mandarino and Back (2004), the four mineral names carnegieite, fassaite (for ferrian aluminian diopside or augite), protoenstatite and thompsonite are removed.
2. Acmite (Acm) is replaced by aegirine (Aeg).
3. Cam and Cpx which currently stand for 'Ca-clinoamphibole' and 'Ca-clinopyroxene' should simply stand for clinoamphibole and clinopyroxene.
4. Chrysotile (Ctl) becomes a group name comprising the polymorphs clino-, ortho- and parachrysotile (Cctl, Octl and Pctl).
5. The abbreviations Fac, Fed, Fts, Mkt and Mrb are replaced by Fe_2-Act, Fe_2-Ed, Fe_2-Ts, Mg-Kt and Mg-Rbk. Although longer, these abbreviations follow the general rules for using abbreviations as outlined below.

The list of names for minerals and mineral groups proposed by Mandarino and Back (2004) was examined and a number of prominent names for minerals and mineral groups were selected for use in the SCMR list (these new entries are indicated in Table 2.12.1 with the entry '*New abbreviation*' in the second last column and with no entry in the last column).

Further names for groups or subgroups that are not found in Mandarino and Back but that were thought to be useful for field-based studies have also been included in the SCMR list (these are indicated in Table 2.12.1 with the entry '*Not mentioned* ...' in the last column). These new names do not, of course, have the same character and rank as names listed by Mandarino and Back (2004). The latter are here termed *certified names*, in contrast to *uncertified names* encompassing all other names not given by Mandarino and Back.

All of these additional names were given abbreviations, mostly with three letters (three letters were used to avoid possible confusion with symbols for chemical elements, and to allow sufficient flexibility for further abbreviations in the future). Only four important mineral groups were given two-letter abbreviations, namely 'amphibole' (Am), 'pyroxene' (Px), 'carbonate mineral' (Cb) and 'opaque mineral' (Op).

2.12.4 Rules for using the abbreviations

(a) ***Mineral or chemical component? Distinction made by using capital or lower case initial letters in the mineral name or abbreviation.*** If in a paper the same mineral name or mineral

Metamorphic Rocks: A Classification and Glossary of Terms. Recommendations of the International Union of Geological Sciences, eds. Douglas Fettes and Jacqueline Desmons. Published by Cambridge University Press. © Cambridge University Press 2007.

abbreviation is used for a natural mineral as well as for a chemical component (end member), and if it is not evident which of the two connotations is meant, it is recommended that the mineral name and its abbreviation are written with a capital letter (e.g. Albite, Ab) and that lower-case is used if the reference is to a chemical component (e.g. albite, ab). In other cases where it is evident that a mineral is meant, lower-case initial letters may be used for both the mineral name and its abbreviation.

(b) ***Expansion of the abbreviations in Table 2.12.1 by specifying the chemical composition of a mineral***. The chemical composition can be specified by placing chemical element symbols with a hyphen in front of the abbreviation: e.g. Ca-Cpx (Ca-clinopyroxene). The valence of Fe (or other elements) may be indicated by Fe_2 or Fe_3, for example $CaMgFe_2$-Grt (grossular-pyrope-almandine garnet). Prefixing of chemical elements in front of mineral names or their abbreviations should follow the same rule as prefixing mineral names or mineral abbreviations in front of rock names (see below and Schmid *et al.*, Section 2.1): that is, they should be arranged in order of increasing abundance. In the case of the garnet example given above, this means that $Ca < Mg < Fe_2$ ($\mathrm{grs} < \mathrm{prp} < \mathrm{alm}$). Element symbols are not separated from each other by hyphens, but a hyphen is placed between the element symbol(s) and the mineral name or its abbreviation.

(c) ***Expansion of the abbreviations of Table 2.12.1 by prefixes***. The glossary of mineral names presented by Mandarino and Back (2004) contains many more mineral names than those present in Table 2.12.1. A large number of the Mandarino and Back entries, however, consist of compound names containing as a root term one of the names listed in Table 2.12.1, modified by one or more prefixes such as clino-, ortho-, para-, etc., or ferro-, ferri-, magnesio-, calcio-, hydroxy-, hydroxyl-, sodic-, etc. It is simple to create abbreviations for such names using the following rules. (1) Replace clino by C, ortho by O, para by P, etc., and add the letter without a hyphen directly to the root name. (2) Use the symbol for the chemical element mentioned in the mineral name according to Mandarino and Back (2004) and place it with a hyphen in front of the root name or the qualified root name (e.g. C + root name, O + root name). Ferro- is given as Fe_2-, ferri- as Fe_3-, magnesio- as Mg-, calcio- as Ca-, hydroxy- and hydroxyl- as OH-, sodic- as Na-, etc. Examples: from the root name holmquistite (Hq), clinoferroholmquistite is given as Fe_2-Chq; from the root name chrysotile (Ctl), orthochrysotile as Octl, parachrysotile (Pctl); from the root name glaucophane (Gln), ferroglaucophane as Fe_2-Gln; from the root name pyrophyllite (Prl), ferripyrophyllite as Fe_3-Prl; from the root name hastingsite (Hs), magnesiohastingsite as Mg-Hs; from the root name humite (Hu), hydroxyclinohumite as OH-Chu; from the root name gedrite (Ged), sodic-ferrogedrite as $NaFe_2$-Ged. This rule makes the abbreviations Fac, Fed, Fts, Mkt and Mrb of Kretz redundant. Some examples of expanded abbreviations following the rules (b) and (c) are already incorporated in Table 2.12.1, as a memo for this kind of expansion. They are indicated by the entry '*expanded from* ...' in the second last column of the table.

(d) ***Expansion of the abbreviations of Table 2.12.1 by suffixes***. Suffixes may be added to the abbreviations. As an example: Allanite (Aln) is a mineral of the epidote group. It may be considered as a type of mineral subgroup consisting of the mineral species Allanite-(Ce) and Allanite-(Y). It is recommended to abbreviate these names as Aln-(Ce) and Aln-(Y). The brackets around the chemical element symbols are necessary in order to distinguish the symbols from possible similar root names.

(e) ***Internal hyphenation of certified mineral names and their abbreviations***. In some cases when prefixes or suffixes are added to mineral abbreviations, it might not be clear whether these qualifiers are an integral part of a certified mineral name (as listed by Mandarino and Back, 2004), for example $NaFe_2$-Ged (sodic-ferrogedrite) or whether they are chemical modifiers additional to a name, for example $CaMgFe_2$-Grt (grossular-pyrope-almandine garnet). To remove this ambiguity it is recommended that the hyphen(s) are replaced by the underscore character (_) when the abbreviation is part of a certified mineral name, that is, $NaFe_2$_Ged (sodic_ferrogedrite), and to use the hyphen(s) in uncertified names, that is, $CaMgFe_2$-Grt (grossular-pyrope-almandine garnet). See also the following paragraph.

(f) ***Prefixing rock names with abbreviations for major minerals***. Using mineral abbreviations for specifying that a major mineral (≥5% by volume) is present in a rock: if the mineral is an essential constituent of that rock it is not mentioned. If it is not an essential mineral, then set it in front of the rock name (without a hyphen). Example: grt amphibolite, that is an amphibolite with plagioclase and amphibole in the amounts specified by its definition, and in addition garnet in an amount of ≥5%. If more than one major mineral is present, place the mineral names or the mineral abbreviations in order of increasing abundance in front of the rock name and hyphenate them (without a hyphen before the rock name). Example: am-grt pyriclasite, that is, a rock with the essential minerals pyroxene (cpx and/or opx) and plagioclase, and containing garnet and amphibole both in amounts of ≥5%. Garnet is more abundant than amphibole. **Special case**: an abbreviation within several others prefixing a rock name contains an internal hyphen. In order to make clear that this internal hyphen is part of the abbreviation and does not separate two abbreviations, the following is recommended: if the whole abbreviation with the internal hyphen is a certified mineral name, replace it by the underscore character (_) (see also paragraph (e)). If it is not a certified name: add a space in front and after the abbreviation (or mineral name) in order to separate it clearly from the adjacent abbreviations.

(g) ***Prefixing or suffixing abbreviations for minerals only sporadic in a rock series, in amounts of 0 up to ≥5%***. Example: garnet in a series of cpx amphibolites. The name of the series is: cpx amphibolites ± grt, or: cpx ± grt amphibolites (see definition of the symbol '±' in Schmid *et al.*, Section 2.1, Table 2.1.2). Another example: if either garnet or biotite occurs in a series of amphibolites, both in amounts of 0 up to ≥5%, but never together: ± grt or bt amphibolites, or: amphibolites ± grt or bt. If garnet and biotite may locally coexist: ± grt ± bt amphibolites, or: amphibolites ± grt ± bt. If garnet and biotite always occur together or are both absent: ± grt-bt amphibolites, or: amphibolites ± grt-bt.

(h) ***Prefixing abbreviations for minor minerals in front of rock names***. According to the SCMR rules for naming rocks, minor minerals (<5% by volume) need not be implicit in rock names, in contrast to major minerals. But they may be mentioned, and it is recommended to do so if they are thought to be critical (see 'critical constituent', Schmid *et al.*, Section 2.1). This should be done using the form 'mineral'-bearing. For example, bt-ttn-bearing cpx amphibolite represents a rock (an amphibolite) that contains amphibole, plagioclase and clinopyroxene as major constituents, and titanite and biotite as minor constituents.

(i) ***Prefixing or suffixing abbreviations for minerals only sporadically present in a specific rock type or rock series, in amounts <5%***. Example: rutile and/or graphite in a series of biotite-quartz-feldspar gneisses. The name of the series is: ± rt ± gr-bearing bt-qtz-fsp gneisses, or: bt-qtz-fsp gneisses ± rt ± gr-bearing. If in this series ilmenite is sometimes present as a minor mineral, but only if rutile is also present, the name of the series would be: ± ilm-rt- or ± rt- and ± gr-bearing bt-qtz-fsp gneisses, or: bt-qtz-fsp gneisses ± ilm-rt- or ± rt- and ± gr-bearing. If there is no correlation with the presence of rutile, ilmenite and garnet then the form would be ± rt ± ilm ± gr-bearing bt-qtz-fsp gneisses, or: bt-qtz-fsp gneisses ± rt ± ilm ± gr-bearing.

(j) ***Prefixing minerals to rock names***. For more information about the use of mineral prefixes and on how to deal with minor minerals (<5%) see Schmid *et al.* (Section 2.1).

2.12.5 Conclusion

The value of mineral abbreviations depends largely on their consistent and frequent use. The SCMR hopes that the list presented here may find international acceptance.

Table 2.12.1 *Abbreviations of mineral names recommended by SCMR.*

The same abbreviations can be used for chemical components of the same name. If it is not evident whether a mineral or a chemical component is meant, chemical components may be written with small letters, e.g. **ab** for pure $NaAlSi_3O_8$, whereas a natural Albite crystal may be abbreviated **Ab**, starting with a capital letter.
In this list mineral names (not their abbreviations) are arranged in alphabetical order.

		Name for:			
Mineral name	Abbreviation	Mineral group, sub-group or series	Mineral species or component	Abbreviation (name) in Kretz (1983)	Term in Mandarino and Back (2004), or Mandarino (1999)
Actinolite	Act	0	1	Identical	
Aegirine	Aeg	0	1	New abbreviation replacing Acm = Acmite not mentioned by Mandarino and Back (2004)	
Aegirine-augite	Agt	0	1	Identical	
Åkermanite (Akermanite)	Ak	0	1	Identical	
Albite	Ab	0	1	Identical	
Alkalifeldspar	Afs	1	0	New abbreviation	Not mentioned but useful for feldspar poor in an- but rich in or- and ab-component
Allanite	Aln	1	0	Identical	
Almandine	Alm	0	1	Identical	
Alumosilicate	Als	1	0	New abbreviation	Not mentioned but useful
Amphibole	Am	1	0	New abbreviation	
Analcime	Anl	0	1	Identical	
Anatase	Ant	0	1	Identical	
Andalusite	And	0	1	Identical	
Andradite	Adr	0	1	Identical	
Anhydrite	Anh	0	1	Identical	
Ankerite	Ank	0	1	Identical	
Annite	Ann	0	1	Identical	
Anorthite	An	0	1	Identical	
Anthophyllite	Ath	0	1	Identical	
Antigorite	Atg	0	1	Identical	
Apatite	Ap	1	0	Identical	

Apophyllite	Apo	1	0	Identical	
Aragonite	Arg	1	1	Identical	
Arfvedsonite	Arf	0	1	Identical	
Arsenopyrite	Apy	1	1	Identical	
Augite	Aug	0	1	Identical	
Axinite	Ax	1	0	Identical	
Barite	Brt	1	1	Identical	
Barroisite	Brs	0	1	New abbreviation	
Beryl	Brl	1	1	Identical	
Biotite	Bt	1	0	Identical	
Böhmite (Boehmite)	Bhm	0	1	Identical	
Bornite	Bn	0	1	Identical	
Brookite	Brk	0	1	Identical	
Brucite	Brc	1	1	Identical	
Bustamite	Bst	0	1	Identical	
Calcite	Cal	1	1	Identical	
Cancrinite	Ccn	1	1	Identical	
Carbonate mineral	Cb	1	0	New abbreviation	Not mentioned but useful
Carpholite	Cph	1	1	New abbreviation	
Cassiterite	Cst	0	1	Identical	
Celadonite	Cel	0	1	New abbreviation	
Celestine	Cls	0	1	Identical	
Chabazite	Cbz	1	0	Identical	
Chalcocite	Cc	0	1	Identical	
Chalcopyrite	Ccp	1	1	Identical	
Chamosite	Chm	0	1	New abbreviation	
Chlorite	Chl	1	0	Identical	
Chloritoid	Cld	0	1	Identical	
Chondrodite	Chn	0	1	Identical	
Chromite	Chr	0	1	Identical	
Chrysocolla	Ccl	0	1	Identical	
Chrysotile	Ctl	1	0	Identical	Not mentioned but useful as root name for the polymorphs clino-, ortho- and parachrysotile
Clinoamphibole	Cam	1	0	Cam = Ca clinoamphibole In this paper Ca Clinoamphibole is given as Ca-Cam (expanded from Cam)	Not mentioned but useful

Table 2.12.1 (*cont.*)

Mineral name	Abbreviation	Name for: Mineral group, sub-group or series	Name for: Mineral species or component	Abbreviation (name) in Kretz (1983)	Term in Mandarino and Back (2004), or Mandarino (1999)
Clinochlore	Clc	0	1	New abbreviation	
Clinoenstatite	Cen	0	1	Identical	
Clinoferroholmquistite	Fe_2-Chq	0	1	New abbreviation (expanded from Chq)	
Clinoferrosilite	Cfs	0	1	Identical	
Clinoholmquistite	Chq	0	1	New abbreviation	
Clinohumite	Chu	0	1	Identical	
Clinopyroxene	Cpx	1	0	Cpx = Ca clinopyroxene In this paper Ca Clinopyroxene is given as Ca-Cpx (expanded from Cpx)	Not mentioned but useful
Clinozoisite	Czo	0	1	Identical	
Coesite	Coe	0	1	New abbreviation	
Cordierite	Crd	0	1	Identical	
Corundum	Crn	0	1	Identical	
Covellite	Cv	0	1	Identical	
Cristobalite	Crs	0	1	Identical	
Cummingtonite	Cum	0	1	Identical	
Deerite	Dee	0	1	New abbreviation	
Diaspore	Dsp	0	1	Identical	
Digenite	Dg	0	1	Identical	
Diopside	Di	0	1	Identical	
Dolomite	Dol	1	1	Identical	
Dravite	Drv	0	1	Identical	
Eckermannite	Eck	0	1	Identical	
Edenite	Ed	0	1	Identical	
Elbaite	Elb	0	1	Identical	
Enstatite	En	0	1	Identical	
Epidote	Ep	1	1	Identical	
Fayalite	Fa	0	1	Identical	

Feldspar	Fsp	1	0	New abbreviation
Ferro-Actinolite	Fe_2-Act	0	1	Fac; New abbreviation (expanded from Act)
Ferro-Edenite	Fe_2-Ed	0	1	Fed; New abbreviation (expanded from Ed)
Ferrohornblende	Fe_2-Hbl	0	1	New abbreviation (expanded from Hbl)
Ferrosilite	Fs	0	1	Identical
Ferrotschermakite	Fe_2-Ts	0	1	Fts; New abbreviation (expanded from Ts)
Fluorite	Fl	0	1	Identical
Forsterite	Fo	0	1	Identical
Gadolinite	Gad	1	0	New abbreviation
Galena	Gn	1	1	Identical
Garnet	Grt	1	0	Identical
Gedrite	Ged	0	1	Identical
Gehlenite	Gh	0	1	Identical
Gibbsite	Gbs	0	1	Identical
Glauconite	Glt	0	1	Identical
Glaucophane	Gln	0	1	Identical
Goethite	Gt	0	1	Identical
Graphite	Gr	0	1	Identical
Greenalite	Gre	0	1	New abbreviation
Grossular	Grs	0	1	Identical
Grunerite	Gru	0	1	Identical
Gypsum	Gp	0	1	Identical
Haematite (Hematite)	Hem	1	1	Identical
Halite	Hl	1	1	Identical
Hastingsite	Hs	0	1	Identical
Haüyne (Haueyne)	Hyn	0	1	Identical
Hedenbergite	Hd	0	1	Identical
Hercynite	Hc	0	1	Identical
Heulandite	Hul	1	0	Identical
Högbomite (Hoegbomite)	Hgb	1	0	New abbreviation
Holmquistite	Hq	0	1	New abbreviation

Table 2.12.1 (*cont.*)

Mineral name	Abbreviation	Name for: Mineral group, sub-group or series	Name for: Mineral species or component	Abbreviation (name) in Kretz (1983)	Term in Mandarino and Back (2004), or Mandarino (1999)
Hornblende	Hbl	1	0	Identical	Not mentioned but useful as a name for the series Fe_2-Hbl – Mg-Hbl
Humite	Hu	1	1	Identical	
Illite	Ill	1	0	Identical	
Ilmenite	Ilm	1	1	Identical	
Jadeite	Jd	0	1	Identical	
Joesmithite	Joe	0	1	New abbreviation	
Johannsenite	Jh	0	1	Identical	
Kaersutite	Krs	0	1	Identical	
Kalsilite	Kls	0	1	Identical	
Kaolinite	Kln	0	1	Identical	
Kaolinite-Serpentine	Kln-Srp	1	0	New abbreviation (expanded from Kln and Srp)	
Katophorite	Ktp	0	1	Identical	
K-feldspar	Kfs	1	0	Identical	Not mentioned but useful for Alkalifeldspar containing more or- than ab-component
Kieserite	Kie	1	1	New abbreviation	
Kornerupine	Krn	0	1	Identical	
Kôzulite (Kozulite)	Koz	0	1	New abbreviation	
Kyanite	Ky	0	1	Identical	
Laumontite	Lmt	0	1	Identical	
Lawsonite	Lws	1	1	Identical	
Lazulite	Laz	1	1	New abbreviation	
Lepidolite	Lpd	1	0	Identical	
Leucite	Lct	0	1	Identical	
Limonite	Lm	1	0	Identical	Not mentioned but useful for hydrous Fe-oxides, mostly Goethite
Lizardite	Lz	0	1	Identical	

Löllingite (Loellingite)	Lo	1	1	Identical	
Maghemite	Mgh	0	1	Identical	
Magnesiohornblende	Mg-Hbl	0	1	New abbreviation (expanded from Hbl)	
Magnesiokatophorite	Mg-Ktp	0	1	Mkt; New abbreviation (expanded from Ktp)	
Magnesioriebeckite	Mg-Rbk	0	1	Mrb; New abbreviation (expanded from Rbk)	
Magnesiosadanagaite	Mg-Sdg	0	1	New abbreviation (expanded from Sdg)	
Magnesite	Mgs	0	1	Identical	
Magnetite	Mag	0	1	Identical	
Marcasite	Mrc	1	1	New abbreviation	
Margarite	Mrg	0	1	Identical	
Marialite	Mar	0	1	New abbreviation	
Meionite	Mei	0	1	New abbreviation	
Melilite	Mel	1	1	Identical	
Merwinite	Mw	0	1	New abbreviation	
Mica	Mca	1	0	New abbreviation	
Microcline	Mc	0	1	Identical	
Minnesotaite	Mns	0	1	New abbreviation	
Molybdenite	Mo	0	1	Identical	
Monazite	Mnz	1	0	Identical	
Monticellite	Mtc	0	1	Identical	
Montmorillonite	Mnt	0	1	Identical	
Mullite	Mul	0	1	Identical	
Muscovite	Ms	0	1	Identical	
Natrolite	Ntr	0	1	Identical	
Nepheline	Ne	0	1	Identical	
Norbergite	Nrb	0	1	Identical	
Nosean	Nsn	0	1	Identical	
Nyböite (Nyboeite)	Nyb	0	1	New abbreviation	
Olivine	Ol	1	0	Identical	
Omphacite	Omp	0	1	Identical	
Opaque mineral	Op	1	0	New abbreviation	Not mentioned but useful
Orthoamphibole	Oam	1	0	Identical (expanded from Am)	Not mentioned but useful
Orthoclase	Or	0	1	Identical	

Table 2.12.1 (*cont.*)

Mineral name	Abbreviation	Name for: Mineral group, sub-group or series	Name for: Mineral species or component	Abbreviation (name) in Kretz (1983)	Term in Mandarino and Back (2004), or Mandarino (1999)
Orthopyroxene	Opx	1	0	Identical (expanded from Px)	Not mentioned but useful
Osumilite	Osu	1	1	New abbreviation	
Paragonite	Pg	0	1	Identical	
Pargasite	Prg	0	1	Identical	
Pectolite	Pct	0	1	Identical	
Pentlandite	Pn	1	1	Identical	
Periclase	Per	1	1	Identical	
Perovskite	Prv	1	1	Identical	
Phengite	Phg	1	0	New abbreviation	
Phlogopite	Phl	0	1	Identical	
Pigeonite	Pgt	0	1	Identical	
Plagioclase	Pl	1	0	Identical	Not mentioned but useful for the series Ab (an = 0) – An (an = 100)
Potassic-Magnesio-sadanagaite	KMg-Sdg	0	1	New abbreviation (expanded from Sdg)	
Potassic Sadanagaite	K-Sdg	0	1	New abbreviation (expanded from Sdg)	
Prehnite	Prh	0	1	Identical	
Pumpellyite	Pmp	1	0	Identical	
Pyrite	Py	1	1	Identical	
Pyrochlore	Pcl	1	1	New abbreviation	
Pyrope	Prp	0	1	Identical	
Pyrophyllite	Prl	0	1	Identical	
Pyrophyllite-Talc	Prl-Tlc	1	0	New abbreviation (expanded from Prl and Tlc)	
Pyroxene	Px	1	0	New abbreviation	
Pyrrhotite	Po	0	1	Identical	
Quartz	Qtz	0	1	Identical	
Rhodochrosite	Rds	0	1	Identical	
Rhodonite	Rdn	0	1	Identical	

Richterite	Rit	0	1	New abbreviation	
Riebeckite	Rbk	0	1	Identical	
Rutile	Rt	1	1	Identical	
Sadanagaite	Sdg	1	0	New abbreviation	Not mentioned but useful for series Mg-Sdg – KMg-Sdg and K-Sdg – KMg-Sdg
Sanidine	Sa	0	1	Identical	
Sapphirine	Spr	0	1	Identical	
Scapolite	Scp	1	0	Identical	
Schorl	Srl	0	1	Identical	
Sepiolite	Sep	0	1	New abbreviation	
Sericite	Ser	1	0	New abbreviation	Not mentioned but useful for any fine-grained colourless mica
Serpentine	Srp	1	0	Identical	Not mentioned but useful for part of the minerals of the Kln-Srp group
Siderite	Sd	0	1	Identical	
Sillimanite	Sil	0	1	Identical	
Smectite	Sme	1	0	New abbreviation	
Sodalite	Sdl	1	1	Identical	
Spessartine	Sps	0	1	Identical	
Sphalerite	Sp	1	1	Identical	
Spinel	Spl	1	1	Identical	
Spodumene	Spd	0	1	Identical	
Staurolite	St	1	1	Identical	
Stilbite	Stb	1	0	Identical	
Stilpnomelane	Stp	0	1	Identical	
Stishovite	Stv	0	1	New abbreviation	
Strontianite	Str	0	1	Identical	
Talc	Tlc	0	1	Identical	
Taramite	Tmt	0	1	New abbreviation	
Titanite	Ttn	1	1	Identical	
Topaz	Toz	0	1	Identical	
Tourmaline	Tur	1	0	Identical	
Tremolite	Tr	0	1	Identical	
Tridymite	Trd	0	1	Identical	
Troilite	Tro	0	1	Identical	

Table 2.12.1 (*cont.*)

Mineral name	Abbreviation	Name for: Mineral group, sub-group or series	Name for: Mineral species or component	Abbreviation (name) in Kretz (1983)	Term in Mandarino and Back (2004), or Mandarino (1999)
Tschermakite	Ts	0	1	Identical	
Ulvöspinel (Ulvoespinel)	Usp	0	1	Identical	
Uvarovite	Uvt	0	1	New abbreviation	
Vermiculite	Vrm	1	0	Identical	
Vesuvianite	Ves	0	1	Identical	
Vivianite	Viv	1	1	New abbreviation	
Wairakite	Wrk	0	1	New abbreviation	
White Mica	Wmca	1	0	New abbreviation	Not mentioned but useful for any colourless mica
Winchite	Win	0	1	New abbreviation	
Witherite	Wth	0	1	Identical	
Wollastonite	Wo	1	1	Identical	
Wüstite (Wuestite)	Wus	0	1	Identical	
Xenotime	Xtm	1	0	New abbreviation	
Zeolite	Zeo	1	0	New abbreviation	
Zinnwaldite	Zwd	1	0	New abbreviation	
Zircon	Zrn	1	1	Identical	
Zoisite	Zo	0	1	Identical	

The terms ‘Carnegieite’ (Crn), ‘Protoenstatite’ (Pen), ‘Fassaite’ (Fst) and ‘Thompsonite’ (Tmp) from the list of Kretz (1983) are not included in this list, because these minerals are not mentioned in Mandarino and Back (2004).

In the column ‘Mineral name’: mineral names in brackets are written as we recommend in case the font chosen is unable to produce the special characters in that name.

In the column ‘Abbreviation (name) in Kretz (1983)’: ‘expanded from . . .’ means that the abbreviation is expanded from a root name already existing in this list according to the ‘Rules for using the abbreviations’, paragraphs b, c and d, and could therefore be removed from this list to keep it short. The abbreviations are listed as examples of correct expansions of root names. It is recommended to replace the hyphen used in such abbreviations in certain cases by the underscore character ‘_’ (see paragraph f).

Table 2.12.2 *List of mineral abbreviations arranged in alphabetical order.*

Abbreviation	Mineral name	Name for: Mineral — Group, sub-group or series	Species or component
Ab	Albite	0	1
Act	Actinolite	0	1
Adr	Andradite	0	1
Aeg	Aegirine	0	1
Afs	Alkalifeldspar	1	0
Agt	Aegirine-augite	0	1
Ak	Åkermanite (Akermanite)	0	1
Alm	Almandine	0	1
Aln	Allanite	1	0
Als	Alumosilicate	1	0
Am	Amphibole	1	0
An	Anorthite	0	1
And	Andalusite	0	1
Anh	Anhydrite	0	1
Ank	Ankerite	0	1
Anl	Analcime	0	1
Ann	Annite	0	1
Ant	Anatase	0	1
Ap	Apatite	1	0
Apo	Apophyllite	1	0
Apy	Arsenopyrite	1	1
Arf	Arfvedsonite	0	1
Arg	Aragonite	1	1
Atg	Antigorite	0	1
Ath	Anthophyllite	0	1
Aug	Augite	0	1
Ax	Axinite	1	0
Bhm	Böhmite (Boehmite)	0	1
Bn	Bornite	0	1
Brc	Brucite	1	1
Brk	Brookite	0	1
Brl	Beryl	1	1
Brs	Barroisite	0	1
Brt	Barite	1	1
Bst	Bustamite	0	1
Bt	Biotite	1	0
Cal	Calcite	1	1
Cam	Clinoamphibole	1	0
Cb	Carbonate mineral	1	0
Cbz	Chabazite	1	0
Cc	Chalcocite	0	1
Ccl	Chrysocolla	0	1
Ccn	Cancrinite	1	1
Ccp	Chalcopyrite	1	1
Cel	Celadonite	0	1

Table 2.12.2 (*cont.*)

		Name for: Mineral	
Abbreviation	Mineral name	Group, sub-group or series	Species or component
Cen	Clinoenstatite	0	1
Cfs	Clinoferrosilite	0	1
Chl	Chlorite	1	0
Chm	Chamosite	0	1
Chn	Chondrodite	0	1
Chq	Clinoholmquistite	0	1
Chr	Chromite	0	1
Chu	Clinohumite	0	1
Clc	Clinochlore	0	1
Cld	Chloritoid	0	1
Cls	Celestine	0	1
Coe	Coesite	0	1
Cph	Carpholite	1	1
Cpx	Clinopyroxene	1	0
Crd	Cordierite	0	1
Crn	Corundum	0	1
Crs	Cristobalite	0	1
Cst	Cassiterite	0	1
Ctl	Chrysotile	1	0
Cum	Cummingtonite	0	1
Cv	Covellite	0	1
Czo	Clinozoisite	0	1
Dee	Deerite	0	1
Dg	Digenite	0	1
Di	Diopside	0	1
Dol	Dolomite	1	1
Drv	Dravite	0	1
Dsp	Diaspore	0	1
Eck	Eckermannite	0	1
Ed	Edenite	0	1
Elb	Elbaite	0	1
En	Enstatite	0	1
Ep	Epidote	1	1
Fa	Fayalite	0	1
Fe_2-Act	Ferro-Actinolite	0	1
Fe_2-Chq	Clinoferroholmquistite	0	1
Fe_2-Ed	Ferro-Edenite	0	1
Fe_2-Hbl	Ferrohornblende	0	1
Fe_2-Ts	Ferrotschermakite	0	1
Fl	Fluorite	0	1
Fo	Forsterite	0	1
Fs	Ferrosilite	0	1
Fsp	Feldspar	1	0
Gad	Gadolinite	1	0
Gbs	Gibbsite	0	1

Table 2.12.2 (*cont.*)

Abbreviation	Mineral name	Name for: Mineral — Group, sub-group or series	Species or component
Ged	Gedrite	0	1
Gh	Gehlenite	0	1
Gln	Glaucophane	0	1
Glt	Glauconite	0	1
Gn	Galena	1	1
Gp	Gypsum	0	1
Gr	Graphite	0	1
Gre	Greenalite	0	1
Grs	Grossular	0	1
Grt	Garnet	1	0
Gru	Grunerite	0	1
Gt	Goethite	0	1
Hbl	Hornblende	1	0
Hc	Hercynite	0	1
Hd	Hedenbergite	0	1
Hem	Haematite (Hematite)	1	1
Hgb	Högbomite (Hoegbomite)	1	0
Hl	Halite	1	1
Hq	Holmquistite	0	1
Hs	Hastingsite	0	1
Hu	Humite	1	1
Hul	Heulandite	1	0
Hyn	Haüyne (Haueyne)	0	1
Ill	Illite	1	0
Ilm	Ilmenite	1	1
Jd	Jadeite	0	1
Jh	Johannsenite	0	1
Joe	Joesmithite	0	1
Kfs	K-feldspar	1	0
Kie	Kieserite	1	1
Kln	Kaolinite	0	1
Kln-Srp	Kaolinite-Serpentine	1	0
Kls	Kalsilite	0	1
KMg-Sdg	Potassic-Magnesiosadanagaite	0	1
Koz	Kôzulite (Kozulite)	0	1
Krn	Kornerupine	0	1
Krs	Kaersutite	0	1
K-Sdg	Potassic Sadanagaite	0	1
Ktp	Katophorite	0	1
Ky	Kyanite	0	1
Laz	Lazulite	1	1
Lct	Leucite	0	1
Lm	Limonite	1	0
Lmt	Laumontite	0	1

Table 2.12.2 (*cont.*)

Abbreviation	Mineral name	Name for: Mineral Group, sub-group or series	Name for: Mineral Species or component
Lo	Löllingite (Loellingite)	1	1
Lpd	Lepidolite	1	0
Lws	Lawsonite	1	1
Lz	Lizardite	0	1
Mag	Magnetite	0	1
Mar	Marialite	0	1
Mc	Microcline	0	1
Mca	Mica	1	0
Mei	Meionite	0	1
Mel	Melilite	1	1
Mgh	Maghemite	0	1
Mg-Hbl	Magnesiohornblende	0	1
Mg-Ktp	Magnesiokatophorite	0	1
Mg-Rbk	Magnesioriebeckite	0	1
Mgs	Magnesite	0	1
Mg-Sdg	Magnesiosadanagaite	0	1
Mns	Minnesotaite	0	1
Mnt	Montmorillonite	0	1
Mnz	Monazite	1	0
Mo	Molybdenite	0	1
Mrc	Marcasite	1	1
Mrg	Margarite	0	1
Ms	Muscovite	0	1
Mtc	Monticellite	0	1
Mul	Mullite	0	1
Mw	Merwinite	0	1
Ne	Nepheline	0	1
Nrb	Norbergite	0	1
Nsn	Nosean	0	1
Ntr	Natrolite	0	1
Nyb	Nyböite (Nyboeite)	0	1
Oam	Orthoamphibole	1	0
Ol	Olivine	1	0
Omp	Omphacite	0	1
Op	Opaque mineral	1	0
Opx	Orthopyroxene	1	0
Or	Orthoclase	0	1
Osu	Osumilite	1	1
Pcl	Pyrochlore	1	1
Pct	Pectolite	0	1
Per	Periclase	1	1
Pg	Paragonite	0	1
Pgt	Pigeonite	0	1
Phg	Phengite	1	0
Phl	Phlogopite	0	1

Table 2.12.2 (*cont.*)

Abbreviation	Mineral name	Name for: Mineral — Group, sub-group or series	Species or component
Pl	Plagioclase	1	0
Pmp	Pumpellyite	1	0
Pn	Pentlandite	1	1
Po	Pyrrhotite	0	1
Prg	Pargasite	0	1
Prh	Prehnite	0	1
Prl	Pyrophyllite	0	1
Prl-Tlc	Pyrophyllite-Talc	1	0
Prp	Pyrope	0	1
Prv	Perovskite	1	1
Px	Pyroxene	1	0
Py	Pyrite	1	1
Qtz	Quartz	0	1
Rbk	Riebeckite	0	1
Rdn	Rhodonite	0	1
Rds	Rhodochrosite	0	1
Rit	Richterite	0	1
Rt	Rutile	1	1
Sa	Sanidine	0	1
Scp	Scapolite	1	0
Sd	Siderite	0	1
Sdg	Sadanagaite	1	0
Sdl	Sodalite	1	1
Sep	Sepiolite	0	1
Ser	Sericite	1	0
Sil	Sillimanite	0	1
Sme	Smectite	1	0
Sp	Sphalerite	1	1
Spd	Spodumene	0	1
Spl	Spinel	1	1
Spr	Sapphirine	0	1
Sps	Spessartine	0	1
Srl	Schorl	0	1
Srp	Serpentine	1	0
St	Staurolite	1	1
Stb	Stilbite	1	0
Stp	Stilpnomelane	0	1
Str	Strontianite	0	1
Stv	Stishovite	0	1
Tlc	Talc	0	1
Tmt	Taramite	0	1
Toz	Topaz	0	1
Tr	Tremolite	0	1
Trd	Tridymite	0	1
Tro	Troilite	0	1
Ts	Tschermakite	0	1

Table 2.12.2 (*cont.*)

		Name for: Mineral	
Abbreviation	Mineral name	Group, sub-group or series	Species or component
Ttn	Titanite	1	1
Tur	Tourmaline	1	0
Usp	Ulvöspinel (Ulvoespinel)	0	1
Uvt	Uvarovite	0	1
Ves	Vesuvianite	0	1
Viv	Vivianite	1	1
Vrm	Vermiculite	1	0
Win	Winchite	0	1
Wmca	White Mica	1	0
Wo	Wollastonite	1	1
Wrk	Wairakite	0	1
Wth	Witherite	0	1
Wus	Wüstite (Wuestite)	0	1
Xtm	Xenotime	1	0
Zeo	Zeolite	1	0
Zo	Zoisite	0	1
Zrn	Zircon	1	1
Zwd	Zinnwaldite	1	0

3 Glossary

3.1 Introduction

The nomenclature scheme devised by the SCMR aims to rationalize the current situation of many contradictory and ambiguous terms. Recommendations concerning the systematics of the main terms have been given in the specialist sections forming the first part of this book.

The glossary contains, in alphabetical order, definitions and literature references for a comprehensive list of terms related to metamorphic rocks and their structures, including fault rocks and the whole field of impactites. In addition to the recommended terms and those of restricted usage the glossary includes numerous terms classified as unnecessary (for the categories of terms, see Schmid *et al.*, Section 2.1). The inclusion of unnecessary terms is considered important both for those researching older literature and in order, even for relatively modern terms, to give better or preferred synonyms.

The number of entries is 1127, among which 544 are for terms of recommended (443) or restricted (101) status; the total of defined terms is about 1220. There are two lists of the recommended and restricted terms, which are intended to aid research: the first is an alphabetical list and the second is a list where the terms are classified according to subject (e.g. structure, migmatites, impactites). There is also a table that shows the relationships between the main types of metamorphism and selected subsidiary terms.

Many metamorphic terms were first used in a language other than English. These terms are defined in the glossary in those cases where they are or have been used, or are recommended for use, in English, either translated (e.g. knotted schist, a term that has approximately the same meaning as the German Knotenschiefer and the obsolete French terms macline, schiste maclifère) or in their non-English form (e.g. augen, hälleflinta). Most un-translated terms that have been used for a very short time, commonly by their proposer, have not been included in the glossary. Nor have those terms that constitute the exact equivalents of the English terms in other languages (e.g. schiste ardoisier, the French term for slate in its structural meaning; Blitzrohr, Blitzglas and Blitzstein, German terms used for the present-day fulgurite).

3.1.1 Glossary entries

Each entry in the glossary includes the following items.

1. The ***term*** to be defined, which is printed in a style indicating the category:
   - bold and upper-case for **RECOMMENDED** terms,
   - bold and lower-case for terms of **restricted** category,
   - small capitals for terms classified as UNNECESSARY.

   In a few cases a synonym or an alternative spelling is given after the name being defined (e.g. metamorphic assemblage, metamorphic association; schindolith, skindolith). An erroneous spelling that can be found in the literature may be included between square brackets (e.g. olenite [ollenite]).

   An arrow symbol (⇒) after the entry refers the reader to the relevant paper or papers (in this vol.) for extended discussion or additional information.
2. The ***etymology*** of the term and the ***locality*** where the term has first been used: these are printed in small characters and between brackets. The source term, a non-English or dead-language term, a person's name or a locality, is italicized. The etymology helps to clarify the meaning of the term and may act as a guide for further use.

   The etymology is not repeated for terms that derive from the same source and whose entries follow each other in the glossary: it is given for only one of these terms, preferably a

Metamorphic Rocks: A Classification and Glossary of Terms. Recommendations of the International Union of Geological Sciences, eds. Douglas Fettes and Jacqueline Desmons. Published by Cambridge University Press. © Cambridge University Press 2007.

recommended one and the first or second in the sequence.

The source terms are:

- common English language terms; the etymology of these terms has normally not been given, as it can be found in English language dictionaries;
- miners' or quarrymen's terms (e.g. the old-Saxonian 'gneiss');
- or they have been especially created to designate specific rocks or processes. For this purpose the learned scientists of the eighteenth to twentieth centuries, educated in the classics, generally turned to Greek roots; workers on migmatitic rocks were particularly inventive in this respect.

The original spelling is given. Greek terms have been transliterated with e standing for ε (epsilon), ê for η (eta), o for ο (omicron) and ô for ω (omega).

3. The ***definition***. For terms of recommended or restricted status the essential, necessary, part of the definition is contained in one sentence and ends with a full stop; the wording accords with that in the papers. *Complementary information* may be given in other sentences. For unnecessary terms, the original meaning and context are given as far as possible; the recommended synonym (rec.syn.) is also given if appropriate. Only recommended terms are normally used in defining terms; terms of a lower category are not used for defining terms of a higher category. Cross-references to other entries, generally of the same category, are mentioned as synonyms (syn., equivalent to), antonyms (ant.) and related terms (see, see also, cf.).

   All terms used in the definition that themselves constitute entries in the glossary are *italicized*, regardless of the category (including the derived adjectives and verbs). Exceptions are the following common terms (where used alone): metamorphism, metamorphic, metamorphosed, metamorphic rock, structure, schist, schistose, gneiss, gneissose, granofels, granofelsic, grain size, acid, basic, intermediate, ultrabasic, grade, layer, layered, layering, contact, altered and the prefix meta– (but types of metamorphism and types of structure are italicized, e.g. contact metamorphism, nematoblastic structure).

   Terms related to an entry, that may be defined with respect to the entry and do not merit their own entries are ***italicized and set in bold*** within the definition: examples would be a rock and the process leading to the formation of that rock (gumbeitization defined under gumbeite), the structure related to a type of mineral grain (crystalloblastic structure defined under crystalloblast), or subdivisions (different types of lineation defined under lineation).

4. ***References*** are given in brackets at the end of the entry; they are printed in italics and give the author(s), the year and the page of the definition or discussion of the term. The references mostly appear in chronological order, so that the first mentioned is the original reference, i.e. the first written trace found: that of the creator of the term or of the author who first defined it or described the designated rock or process. If no original reference is known, the letters OU stand for 'origin unknown/uncertain'. If not otherwise specified, papers and books mentioned in the references have been consulted by an SCMR member. In thoses cases where the original paper was unavailable to an SCMR member, the reference is given as 'X in Y, year, page'. Tomkeieff (the 1983 edition of the *Dictionary of Petrology*), Bates and Jackson (various editions from 1979 to 1997, and Neuendorff *et al.*, 2005) and other glossaries are mentioned in the references when they have been the only available source of information. Only the first page is given for several running pages of long descriptions or discussions. Missing page numbers may reflect a lack of information or the lack of pagination of the consulted copy.

The two oldest references mentioned are to Herodotos, the Greek historian of the fifth century BC, and to Theophrastos, the Greek philosopher and scholar of the fourth and third centuries BC: these are the first written traces found for 'marble'. The third oldest is Pliny the Elder, for schist, ophite, coticule and also for marble. Traces of older, but unwritten, etymologies are found in present-day languages, such as the pre-Indo-European sound 'cal' or 'c', meaning stone or rock, preserved through Greek, Gallic and Latin, into modern European languages: for example the radical calc-, the French terms caillou (pebble, boulder), chaux (lime) and stony localities (e.g. Karst, Causses, Queyras, Crau).

The next oldest references are of the sixteenth century, namely Georgius Agricola (Georg Bauer) who mentioned gneiss, and Vasari who mentioned cipollino.

In the eighteenth and nineteenth centuries numerous scientists enthusiastically explored the new young sciences of geology and mineralogy-petrography, commonly with an amazing perspicacity. After the rock catalogues of Cronstedt (second edition in 1758) and Werner (1780, 1786, 1791–2), translations and other catalogues in various languages were published at the very end of the eighteenth and in the nineteenth century.[1]

Glossaries of the twentieth and twenty-first centuries are useful as sources for recent meanings of the terms.

The number of references mentioned for each term vary widely, reflecting the long or short history of the term, its various meanings, its evolutions in different countries prior to the SCMR systematic work, and also the variable success achieved by the SCMR members in tracking down old books or rare journals.

3.1.2 *Reference list*

In order to avoid repetitions the consolidated ***reference list*** at the end of the book contains all the references cited in both the specialist sections and the glossary. The total number is *c*. 1050. In contrast to Le Maitre (1989, 2002), the terms for which the paper or book has been consulted are not listed under the reference, because of the very considerable additional length it would have involved.

Different editions of the same book may have been used by different contributors. All the editions are referenced, as the same term generally appears on different pages in the different editions. Titles in Russian and Japanese have been translated into English.

The glossary and the reference list are the result of the joint efforts of many contributors, all members of the SC. As far as possible all references were checked at first hand by the SCMR members. Nevertheless, in spite of these efforts, a few mistakes and omissions are likely to remain. Readers who might detect these are requested to inform the editors (c/o Cambridge University Press). As written in the first century of our era, 'Let experience [of the readers] make up for what is missing in the present work'.[2]

[1] In the eighteenth century: Karsten (1789), Kirwan (1794), Estner (1795–7), Klaproth (1795–7), Napione (1797), Brochant (1798, 1800), Emmerling (1799). In the first half of the nineteenth century: Brongniart (1813, 1825, 1827), Haüy (1822), Beudant (1830), Breithaupt (1830, 1847), Omalius d'Halloy (1831, 1868), Dana (1850, 1892). In the second half of the nineteenth century: Coquand (1857), Cordier (1868); and many others.

[2] Quintilianus Marcus Fabius, in Emmerling (1799), reported by A. Dudek: '*Quae presenti opusculo desunt, suppleat aetas*'.

3.2 Alphabetical list of recommended and restricted terms

Recommended terms in regular font; restricted terms in italics
f. – facies, fmt. – formation, gn. – gneiss, ln. – lineation, m. – metamorphism, ms. – metasomatism, str. – structure

Aceite
Acicular
Acid
Adinole (adinolization, contact adinole, tuffaceous adinole)
Adinole series
Agmatite
Alkremite
Allochemical contact m.
Allochemical m.
Allochthonous impact breccia
Allofacial
Allogenic
Alteration
Amphibolite
Amphibolite f.
Ana-
Anakit-type contact m.
Anatexis
Anatexite
Anchimetamorphism
Anchizone
Andalusite-sillimanite series
Anhedral
Antitaxial
Aphanitic
Argillisite (argillisitization)
Armoured relict
Arterite
Assemblage
Atoll str.
Augen
Augen gneiss
Augen mylonite
Augen str.
Aureole
Authigenesis
Authigenic
Autochthonous (parautochthonous) impact breccia
Autometamorphism
Autometasomatism
Axial plane foliation
Baked rock (baking)
Band (banding, banded)
Banded gneiss
Banded str.
Baric types of m.
Barrovian-type facies series
Barrow's zones (Barrovian zones)
Basic
Bathograd
Bathozone
Beerbachite
Beresite (beresitization)
Bimetasomatism
Bimineralic
Blast, blasto-, blastic
Blastesis
Blastomylonite
Blueschist
Blueschist f. (epidote-blueschist facies)
Bosost-type contact m.
Boudin
Boudinage
Boundary ms.
Bow-tie structure
Breccia
Breccia dyke
Buchite
Bunte breccia
Burial m.
Burned rock, burnt rock
Burning m.
Calciphyre
Calc-schist
Calc-silicate mineral
Calc-silicate rock
Californite
Carbonate-silicate rock
Cata-
Cataclasite (cataclastic structure, cataclasis)
Central uplift
Charnockite (charnockitic series)
Chemographical diagram
Cipolin
Cleavage (domainal, M-domain, Q-domain)
Coal-fire ash
Cokeite
Combustion m.
Complex impact crater
Composition phase diagram
Comrie-type contact m.
Contact
Contact aureole (inner aureole, outer aureole)
Contact aureole systematics
Contact marble
Contact m.
Contact ms. (endo-, exocontact zone)
Contact rock
Contact zone
Continuous cleavage
Continuous reaction
Core-and-mantle str.
Corona (corona str.)
Crenulation
Crenulation cleavage/schistosity
Critical mineral assemblage
Crystalline limestone
Crystalloblast (crystalloblastic str.)
Crystalloblastesis
C/S fabric
Dactylitic str.
Decorated planar deformation features
Decussate str.
Deformation lamellae/bands
Degranitization
Dehydration (reaction)

Depositional temperature
Depth zones
Diablastic str.
Diagenesis (shallow diagenesis, deep diagenesis)
Diagenetic zone
Diaphthoresis
Diaphthorite
Diaplectic glass (solid state glass)
Diastathermal m.
Diatexis
Diatexite, diatectite
Dictyonite
Differentiation
Diffusional ms.
Discontinuous reaction
Disequilibrium assemblage
Disjunctive cleavage
Dislocation m.
Dissolution creep
Distal impactite
Domain
Dyke (impact) breccia
Eclogite
Eclogite f.
Ejecta
Ejecta blanket
Ejecta plume
Emery rock
Enderbite
Endocontact zone
Endometamorphism
Endomorphism
Endoskarn
Epi-
Epidote-amphibolite f.
Epigenetic
Epitaxis (epitaxy, epitaxial growth)
Epithermal
Epizone (epimetamorphic zone)
Equant
Equigranular
Equilibration shock pressure
Euhedral
Exocontact zone
Exometamorphism
Exomorphism
Exoskarn
Fabric (mega, meso, microfabric, linear f., planar f., random f.)
Facies
Facies series (metamorphic facies series)
Fallback (fallback fmt.)
Fallout (fallout fmt.)
Fascicular (fascicular str.)
Faserkiesel
Fault
Fault breccia
Fault gouge
Fault rock
Fault zone
Feldspathization
Felsic granulite
Felsic minerals (felsic rock)
Fenite (fenitization)
Flaser gabbro
Flaser str.
Foliated str.
Foliation (s-surface)
Fracture
Fracture cleavage
Fritted rock (fritting)
Fulgurite
Garbenschiefer
Geobarometer
Geotherm
Geothermal gradient
Geothermometer
Ghost stratigraphy (ghost str.)
Glaucophane-schist f.
Glaucophanite
Glomeroblastic str.
Gneiss (lineated gn.)
Gneissose str.
Gondite
Gorotubite
Gouge
Grain size
Granitization
Granoblastic str.
Granoblastic-polygonal str.
Granofels
Granofelsic str.
Granulite
Granulite f.
Granulitic
Granulitic (impact) breccia
Greenschist
Greenschist f.
Greenstone
Greenstone belt
Greisen (greisenization)
Griquaite
Grospydite
Gumbeite (gumbeitization)
Hartschiefer
Helicitic str.
Heteroblastic str.
Homeoblastic str.
Hornblendite
Hornfels
Hornfelsed schist
Hot-slab m.
Hugoniot curve
Hugoniot elastic limit
Hugoniot equation-of-state (Rankine–Hugoniot equation)
Hybridization (hybrid)
Hydrothermal m.
Hydrothermal ms.
Hypidioblast (hypidioblastic str.)
Hypothermal
Idioblast (idioblastic str.)
Impact (hypervelocity impact)
Impact breccia
Impact crater
Impact ejecta
Impact formation (inner imp. fmt., outer imp. fmt.)
Impact glass
Impactite
Impact melt
Impact melt rock
Impact m.
Impactoclast
Impactoclastic airfall bed
Impactoclastic deposit
Impactor
Impact pseudotachylite
Impact regolith
Impact structure

Inclusion (inclusion trails, internal foliation (S_i), external foliation (S_e))
Index mineral
Inequant
Infiltration ms.
Injection
Intermediate
Inverted metamorphic zones
Isochemical contact m.
Isochemical m.
Isochemical series
Isofacial
Isograd
Isograde
Itabirite
Itacolumite
-ite
Joint (joint set, jointing)
Kata-
Katazone
Kelyphite (kelyphitic str.)
Khondalite
Kinzigite
Knotted schist
Kodurite
Kyanite-sillimanite series
Laminated str.
Layer (layered, layering)
Lepidoblastic str.
Leucocratic
Leucophyllite
Leucosome
Lightning m.
Lineation (mineral ln., crenulation ln., intersection ln.)
Listvenite
Lithic (impact) breccia
Lit-par-lit injection (lit-par-lit gneiss)
Local m.
Mafic granulite
Mafic hornfels
Mafic minerals (mafic rock)
Magnesian hornfels
Main Donegal-type contact m.
Mantled porphyroclast
Marble (pure marble, impure marble)
Maskelynite
Megabreccia
Megaregolith
Megastructure
Melanocratic
Melanosome
Melt pocket
Melt vein
Meso-
Mesocataclasite
Mesomylonite
Mesosome
Mesostructure
Mesothermal
Mesozone
Meta-
Metablastesis
Metamorphic assemblage, metamorphic association
Metamorphic differentiation
Metamorphic event
Metamorphic facies
Metamorphic facies series
Metamorphic geotherm
Metamorphic grade
Metamorphic phase
Metamorphic rock
Metamorphic subfacies
Metamorphic zone
Metamorphism
Metasomatic column
Metasomatic contact m.
Metasomatic facies
Metasomatic family
Metasomatic rock
Metasomatic zone
Metasomatism
Metatect
Metatexis
Metatexite
Meteorite crater
Mica-fish
Mica schist
Microbreccia
Microcrystite
Microfabric
Microlithon
Microstructure
Microtektite
Migmatite (migmatization)
Millipede str.
Mimetic growth/crystallization
Mineral assemblage/association
Mineral facies
Mineral zone
Monometamorphism
Monomict impact breccia
Monomineralic
Monophase m.
Mortar str.
Mosaic str.
Mosaicism
Mullion str.
Multi-ring (impact) basin
Multi-ring (impact) crater
Mylonite
Near-vein ms.
Nebulite
Nematoblastic str.
Neoblast
Neomineralization
Neosome
Ocean-floor m.
-oid
Ophicalcite
Ophicarbonate (ophimagnesite)
Orogenic m.
Ortho-
Paired metamorphic belts
Palaeoblast
Palaeosome
Palimpsest str.
Palingenesis (palingenite)
Para-
Paragenesis (paragenetic sequence)
Parent rock
Particle velocity
Pelite
Pencatite
Petrogenetic grid
Phacoid (phacoidal str.)
Phaneritic
Phlebite
Phyllite
Phyllonite

Piezo-thermic array
Pinch-and-swell str.
Pinitization
Planar deformation features (PDF)
Planar deformation str.
Planar fractures
Planar microstructure
Plurifacial m.
Poikiloblast (poikilitic str.)
Polygonal str.
Polymetamorphism
Polymict impact breccia
Polymineralic
Polymorphic transformation (polymorphism)
Polyphase m.
Porphyroblast (porphyroblastic str.)
Porphyroclast (porphyroclastic str.)
Porphyroclast system
Porphyroid
Post-shock temperature (progressive shock m.)
Prasinite
Predazzite
Preferred orientation
Prehnite-actinolite f.
Prehnite-pumpellyite f.
Pressure shadow
Pressure solution (press. sol. striping/cleavage, solution transfer, dissolution creep)
Prograde m. (progressive m.)
Prograde shock m. (progressive shock m.)
Projectile
Propylite (propylitization)
Protocataclasite
Protolith
Protomylonite
Proximal impactite
Psammite
Pseudomorph
Pseudotachylite
P–T–t path
Ptygmatic folding
Pumpellyite-actinolite f.
Pyrometamorphism
Pyroxene-hornfels f.
Pyroxenite
Reaction rim (reaction border)
Recrystallization
Regional m.
Regional ms.
Regolith
Regolith breccia
Relict str.
Resister
Restite
Retrograde m. (retrogressive m.)
Ribbon str. (ribbon quartz)
Rodingite (rodingitization)
Sanidinite f.
Saussuritization
S-C fabric
Schist
Schistoid
Schistose hornfels
Schistose str.
Schistosity
Schlieren
Schollen
Secondary
Secondary quartzite
Semipelite
Sericitization
Serpentinite
Shatter cone
Sheaf str.
Shock deformation
Shocked rocks
Shock effect (residual shock effect)
Shock front
Shock impedance
Shock melting
Shock m.
Shock stage
Shock state
Shock temperature
Shock vaporization
Shock vein
Shock wave
Shock wave velocity
Shock zone
Sieve str.
Simple impact crater
Skarn (magnesian sk., calc-sk.)
Slate
Slaty cleavage
Snowball str.
Solution transfer
Solvus
Spaced cleavage
Spaced schistosity
Spilite
Spotted schist/slate
Stromatite (stromatitic str.)
Stronalite
Structure
Subfabric
Subfacies
Subgrain
Subgrain boundary
Subgreenschist f.
Subhedral
Subidioblast (subidioblastic str.)
Suevite, suevite breccia
Sutured boundary
Symplectite (symplectic intergrowth)
Synantetic
Syngenetic
Syntaxial (syntaxial vein, syntaxial growth)
Tabular
Tagamite
Tanohata-type contact m.
Target
Target lithology
Target rock
Tectonite (L-tect., S-tect., L-S tect.)
Tektite
Texture
Thermal m.
Thermochron
Topochemical change
Topotaxis (topotaxy, topotaxial growth)
Transient crater (transient cavity)
Trimineralic
Types of metamorphism
Typomorphic minerals

Ultrabasic
Ultracataclasite
Ultrahigh-pressure m.
Ultramafic hornfels
Ultrametamorphism
Ultramylonite
Undulose extinction
Uralitization
Vapour plume
Venite
Verdite
Verdolite
Vert antique
Vogtland-type contact m.
Whiteschist
Xenoblast (xenoblastic str.)
Zeolite f.

3.3 Recommended and restricted terms by subject

Recommended terms in regular font; restricted terms in italics
f – facies, fmt. – formation, gn. – gneiss, ln. – lineation, m. – metamorphism, ms. – metasomatism, str. – structure

SPECIFIC ROCK NAMES BASED ON MINERAL CONTENT

Alkremite
Amphibolite
Blueschist
Buchite
Calciphyre
Calc-schist
Calc-silicate rock
Californite
Carbonate-silicate rock
Charnockite (charnockitic series)
Cipolin
Crystalline limestone
Eclogite
Emery rock
Enderbite
Felsic granulite
Glaucophanite
Gondite
Gorotubite
Granulite
Greenschist
Greenstone
Griquaite
Grospydite
Hornblendite
Itabirite
Itacolumite
Khondalite
Kinzigite
Kodurite
Mafic granulite
Marble (pure marble, impure marble)
Mica schist
Ophicalcite
Ophicarbonate (ophimagnesite)
Pelite
Pencatite
Porphyroid
Prasinite
Predazzite
Psammite
Pyroxenite
Rodingite
Semipelite
Serpentinite
Skarn
Spilite
Stronalite
Verdite
Verdolite
Vert antique
Whiteschist

SPECIFIC ROCK NAMES BASED ON STRUCTURE

Augen gneiss
Augen mylonite
Banded gneiss
Blastomylonite
Cataclasite (cataclastic structure, cataclasis)
Fault breccia
Fault gouge
Fault rock
Flaser gabbro
Gneiss (lineated gn.)
Granofels
Hartschiefer
Knotted schist
Leucophyllite
Mesocataclasite
Mesomylonite
Mylonite
Phyllite
Phyllonite
Protocataclasite
Protomylonite
Pseudotachylite
Schist
Slate
Spotted schist/slate
Ultracataclasite
Ultramylonite

TYPES OF METAMORPHISM

Allochemical contact m.
Allochemical m.
Anakit-type contact m.
Anchimetamorphism
Autometamorphism
Bosost-type contact m.
Burial m.
Burning m.
Combustion m.
Comrie-type contact m.
Contact m.
Diastathermal m.
Dislocation m.
Endometamorphism
Endomorphism
Exometamorphism
Exomorphism
Hot-slab m.
Hydrothermal m.
Impact m.
Isochemical contact m.
Isochemical m.
Lightning m.
Local m.
Main Donegal-type contact m.
Metamorphism
Metasomatism

Monometamorphism
Monophase m.
Ocean-floor m.
Orogenic m.
Plurifacial m.
Polymetamorphism
Polyphase m.
Prograde m. (progressive m.)
Pyrometamorphism
Regional m.
Regional ms.
Retrograde m. (retrogressive m.)
Shock m.
Tanohata-type contact m.
Thermal m.
Types of metamorphism
Ultrahigh-pressure m.
Ultrametamorphism
Vogtland-type contact m.

STRUCTURE

Acicular
Anhedral
Antitaxial
Aphanitic
Armoured relict
Atoll str.
Augen
Augen str.
Axial plane foliation
Band (banding, banded)
Banded gneiss
Banded str.
Blast, blasto-, blastic
Boudin
Boudinage
Bow-tie structure
Cleavage (domainal, M-domain, Q-domain)
Continuous cleavage
Core-and-mantle str.
Corona (corona str.)
Crenulation
Crenulation cleavage/schistosity
Crystalloblast (crystalloblastic structure)
C/S fabric
Dactylitic str.
Decussate str.
Deformation lamellae/bands
Diablastic str.
Disjunctive cleavage
Dissolution creep
Epitaxis (epitaxy, epitaxial growth)
Equant
Equigranular
Euhedral
Fabric (mega, meso, microfabric, linear f., planar f., random f.)
Fascicular (fascicular str.)
Faserkiesel
Fault
Fault breccia
Fault gouge
Fault rock
Fault zone
Flaser str.
Foliated str.
Foliation (s-surface)
Fracture
Fracture cleavage
Garbenschiefer
Glomeroblastic str.
Gneissose str.
Gouge
Granoblastic str.
Granoblastic-polygonal str.
Granofelsic str.
Granulitic
Helicitic str.
Heteroblastic str.
Homeoblastic str.
Hypidioblast (hypidioblastic str.)
Idioblast (idioblastic str.)
Inclusion (inclusion trails, internal foliation (S_i), external foliation (S_e))
Inequant
Joint (joint set, jointing)
Kelyphite (kelyphitic str.)
Laminated str.
Layer (layered, layering)
Lepidoblastic str.
Lineation (mineral ln., crenulation ln., intersection ln.)
Mantled porphyroclast
Megastructure
Mesostructure
Mica-fish
Microfabric
Microlithon
Microstructure
Millipede str.
Mimetic growth/crystallization
Mortar str.
Mosaic str.
Mullion str.
Nematoblastic str.
Neoblast
Palaeoblast
Phacoid (phacoidal str.)
Pinch-and-swell str.
Poikiloblast (poikilitic str.)
Polygonal str.
Porphyroblast (porphyroblastic str.)
Porphyroclast (porphyroclastic str.)
Porphyroclast system
Preferred orientation
Pressure shadow
Pressure solution (press. soln striping/cleavage, solution transfer, dissolution creep)
Pseudomorph
Ptygmatic folding
Reaction rim (reaction border)
Relict str.
Ribbon str. (ribbon quartz)

S-C fabric
Schistoid
Schistose str.
Schistosity
Sheaf str.
Sieve str.
Slaty cleavage
Snowball str.
Solution transfer
Spaced cleavage/schistosity
Structure
Subfabric
Subgrain
Subgrain boundary
Subhedral
Subidioblast (subidioblastic str.)
Sutured boundary
Symplectite (symplectic intergrowth)
Synantetic
Syntaxial (syntaxial vein, syntaxial growth)
Tabular
Tectonite (L-tect., S-tect., L-S tect.)
Texture
Topotaxis (topotaxy, topotaxial growth)
Undulose extinction
Xenoblast (xenoblastic str.)

MIGMATITE

Agmatite
Allogenic
Anatexis
Anatexite
Arterite
Degranitization
Diatexis
Diatexite, diatectite
Dictyonite
Feldspathization
Ghost stratigraphy (ghost str.)
Granitization
Hybridization (hybrid)
Injection
Leucosome
Lit-par-lit injection (lit-par-lit gneiss)
Melanosome
Mesosome
Metablastesis
Metamorphic differentiation
Metatect
Metatexis
Metatexite
Migmatite (migmatization)
Nebulite
Neosome
Palaeosome
Palimpsest str.
Palingenesis (palingenite)
Phlebite
Ptygmatic fold
Resister
Restite
Schlieren
Schollen
Stromatite (stromatitic str.)
Venite

METASOMATISM

Aceite
Argillisite (argillisitization)
Autometasomatism
Beresite (beresitization)
Bimetasomatism
Boundary ms.
Contact ms.
Diffusional ms.
Endoskarn
Epigenetic
Epithermal
Exoskarn
Feldspathization
Fenite (fenitization)
Greisen (greisenization)
Gumbeite (gumbeitization)
Hydrothermal ms.
Hypothermal
Infiltration ms.
Listvenite
Mesothermal
Metasomatic column
Metasomatic contact ms.
Metasomatic facies
Metasomatic family
Metasomatic rock
Metasomatic zone
Metasomatism
Near-vein ms.
Propylite (propylitization)
Regional ms.
Rodingite (rodingitization)
Secondary quartzite
Skarn (magnesian sk., calc sk.)
Syngenetic

CONTACT METAMORPHISM

Adinole (adinolization, contact adinole, tuffaceous adinole)
Adinole series
Aureole
Baked rock (baking)
Beerbachite
Buchite
Burned rock, burnt rock
Burning m.
Coal-fire ash
Cokeite
Combustion m.
Contact
Contact aureole (inner aur., outer aur.)
Contact aureole systematics
Contact marble
Contact m.

Contact ms.
Contact rock
Contact zone
Emery rock
Endocontact zone
Endometamorphism
Endomorphism
Endoskarn
Exocontact zone
Exometamorphism
Exomorphism
Exoskarn
Fritted rock (fritting)
Fulgurite
Hornfels
Hornfelsed schist
Isochemical contact m.
Lightning m.
Mafic hornfels
Magnesian hornfels
Pencatite
Predazzite
Pyrometamorphism
Schistose hornfels

METAMORPHIC FACIES

Allofacial
Amphibolite f.
Andalusite-sillimanite series
Baric types of m.
Barrovian-type facies series
Barrow's zones (Barrovian zones)
Bathozone
Blueschist f. (epidote-blueschist f.)
Depth zones
Diagenetic zone
Eclogite f.
Epidote-amphibolite f.
Epizone (epimetamorphic zone)
Facies
Facies series (metamorphic facies series)
Glaucophane-schist f.
Granulite f.
Greenschist f.
Index mineral
Inverted metamorphic zones
Isofacial
Isograd
Isograde
Katazone
Kyanite-sillimanite series
Mesozone
Metamorphic facies
Metamorphic facies series
Metamorphic subfacies
Metamorphic zone
Mineral facies
Mineral zone
Paired metamorphic belts
Plurifacial m.
Prehnite-actinolite f.
Prehnite-pumpellyite f.
Pumpellyite-actinolite f.
Pyroxene-hornfels f.
Sanidinite f.
Subfacies
Subgreenschist f.
Zeolite f.

METAMORPHIC AND RELATED PROCESSES

Alteration
Authigenesis
Blastesis
Continuous reaction
Crystalloblastesis
Degranitization
Dehydration (reaction)
Diagenesis (shallow diag., deep diag.)
Diaphthoresis
Differentiation
Discontinuous reaction
Feldspathization
Granitization
Hybridization
Metablastesis
Metamorphic differentiation
Neomineralization
Pinitization
Polymorphic transformation (polymorphism)
Recrystallization
Saussuritization
Sericitization
Uralitization

IMPACT METAMORPHISM

Allochthonous impact breccia
Autochthonous (parautochthonous) impact breccia
Breccia
Breccia dyke
Bunte breccia
Central uplift
Complex impact crater
Decorated planar deformation features
Deformation lamellae
Depositional temperature
Diaplectic glass (solid state glass)
Distal impactite
Dyke (impact) breccia
Ejecta
Ejecta blanket
Ejecta plume
Equilibration shock pressure
Fallback (fallback fmt.)
Fallout (fallout fmt.)
Granulitic (impact) breccia
Hugoniot curve
Hugoniot elastic limit
Hugoniot equation-of-state (Rankine–Hugoniot equation)

Impact (hypervelocity impact)
Impact breccia
Impact crater
Impact ejecta
Impact formation (inner imp. fmt., outer imp. fmt.)
Impact glass
Impactite
Impact melt
Impact melt rock
Impact m.
Impactoclast
Impactoclastic airfall bed
Impactoclastic deposit
Impactor
Impact pseudotachylite
Impact regolith
Impact structure
Lithic (impact) breccia
Maskelynite
Megabreccia
Megaregolith
Melt pocket
Melt vein
Meteorite crater
Microbreccia
Microcrystite
Microtektite
Monomict impact breccia
Mosaicism
Multi-ring (impact) basin
Multi-ring (impact) crater
Particle velocity
Planar deformation features (PDF)
Planar deformation str.
Planar fractures
Planar microstructure
Polymict impact breccia
Post-shock temperature
Prograde shock m. (progressive shock m.)
Projectile
Proximal impactite
Regolith
Regolith breccia
Shatter cone
Shock deformation
Shocked rocks
Shock effect (residual shock effect)
Shock front
Shock impedance
Shock melting
Shock m.
Shock stage
Shock state
Shock temperature
Shock vaporization
Shock vein
Shock wave
Shock wave velocity
Shock zone
Simple impact crater
Suevite, suevite breccia
Tagamite
Target
Target lithology
Target rock
Tektite
Transient crater (transient cavity)
Vapour plume

DESCRIPTIVE TERMS

Acid
Aphanitic
Basic
Bimineralic
Felsic minerals (felsic rock)
Grain size
Intermediate
Leucocratic
Mafic minerals (mafic rock)
Melanocratic
Monomineralic
Ortho-
Para-
Phaneritic
Polymineralic
Trimineralic
Ultrabasic

OTHER TERMS

Allofacial
Ana-
Anchizone
Assemblage
Authigenic
Bathograd
Calc-silicate mineral
Cata-
Chemographical diagram
Composition phase diagram
Critical mineral assemblage
Diaphthorite
Disequilibrium assemblage
Domain
Epi-
Epigenetic
Geobarometer
Geotherm
Geothermal gradient
Geothermometer
Ghost stratigraphy
Greenstone belt
Injection
Isochemical series
Isograde
-ite
Kata-
Meso-
Meta-
Metamorphic assemblage
Metamorphic event
Metamorphic facies
Metamorphic facies series
Metamorphic geotherm
Metamorphic grade
Metamorphic phase
Metamorphic rock
Metamorphic subfacies
Metamorphic zone

Mineral assemblage/association
Mineral facies
Mineral zone
Ortho-
Para-
Paragenesis (paragenetic sequence)
Parent rock
Pelite
Petrogenetic grid
Piezo-thermic array
Protolith
Psammite
P–T–t path
Secondary
Semipelite
Solvus
Syngenetic
Thermochron
Topochemical change
Typomorphic minerals

3.4 Main types of metamorphism with selected synonyms

(m. = metamorphism, ms. = metasomatism)

Main terms	Related terms	
	Recommended or restricted status	Unnecessary or obsolete status
1. *Classified according to the extent and the cause/setting of the metamorphism: mainly* ***isochemical m.***		
REGIONAL M.		Free m., General m., Normal m.
Burial	*Diastathermal m.*	Geothermal m. (in part), Load m., Static m., Statohydral m.
Ocean-floor m.		Sub-sea floor m., Geothermal m. (in part)
Orogenic m.		Dynamo-thermal m., Thermodynamic m.
LOCAL M.		Abnormal m., Special m.
Combustion m.	*Burning m.*	Pyrogenic m., Pyromorphism (in part)
Contact m.		Caloric m., Juxtaposition m., Normal contact m., Peripheral m., Thermal contact m.
	Autometamorphism	
	Thermal m.	Thermometamorphism
	Endo(meta)morphism	Endogenic m., Inverse m.
	Exo(meta)morphism	Exogenic m., Everse m.
	Pyrometamorphism	Optalic m., (Pyro)caustic m., Pyromorphism (in part), Hydatocaustic m. (in part)
	Hydrothermal m. (in part)	Alembic contact m., Atmogenic m., Hydatocaustic m. (in part), Hydrothermal contact m., Pneumatolytic (contact) m.
Hot-slab m.		Dynamo-static m., Upside-down m.
Hydrothermal m.		Hydatothermic m., Pneumatolytic m.
Impact m.	Shock m.	
Lightning m.		
2. *Allochemical m.*		
Metasomatism (also commonly present in Ocean-floor m. and Hydrothermal m.)	Autometasomatism, Boundary ms., Contact ms., Diffusional ms., *Hydrothermal ms.*, Infiltrational ms., *Near-vein ms.*	Additive m., Additive/subtractive ms., Chemical m., Hydrometasomatism, Hydrochemical m., Pneumatolytic ms., Pyrometasomatism
3. *Classified according to grade and/or pressure*		
Very low- to low-grade m.	Anchimetamorphism	Archometamorphism, Cryptic m., Eometamorphism, Epimetamorphism,
Medium-grade m.		Mesometamorphism
High-grade m.		
Ultrahigh-pressure m.	Ultrametamorphism	
4. *Other terms*		
Monometamorphism		Monocyclic m.
Polymetamorphism		Polycyclic m.
Prograde/progressive m.		
Retrograde/retrogressive m.		Retrometamorphism

3.5 Main glossary of terms

ABNORMAL METAMORPHISM. First used to designate what is now called *contact metamorphism*, the term was later used to contrast with *normal* (or *regional*) *metamorphism* and subdivided into the equivalents of *combustion*, *contact*, *hydrothermal* and *pneumatolytic metamorphism*. *(Élie de Beaumont 1833, in Touret & Nijland 2002, p. 115; Dufrénoy & Élie de Beaumont 1841, p. 42; Naumann 1858, p. 718)*

ABUKUMA(-TYPE) FACIES SERIES (*Abukuma* Plateau in Japan). Type of *facies series* characterized by low *P*/*T* values and the development of andalusite and sillimanite. Term now superseded by low *P*/*T* type or *andalusite-sillimanite series*. *(Miyashiro 1961, p. 279; Winkler 1974, p. 91; Kornprobst 2002, p. 104)*

Aceite ⇒ Section 2.9, Fig. 2.9.1. (*Ace* uranium mine, Goldfields, Saskatchewan, Canada). Low-temperature alkaline *metasomatic* rock, mainly composed of albite with subsidiary carbonate and haematite. U-bearing apatite is a common accessory. Aceites are closely associated with U-mineralization. *(Omel'yanenko 1978, p. 171)*

ACICULAR (Latin *acicula*, small needle). Said of a crystal that is needle-shaped. *(OU)*

ACID, INTERMEDIATE, BASIC, ULTRABASIC ⇒ Section 2.1, Table 2.1.2. Terms defining the chemical composition of rocks, as defined by Le Maitre. *(Le Maitre 1989; 2002)*

ADDITIVE METAMORPHISM, ADDITIVE CONTACT METAMORPHISM. Terms proposed for a type of *contact metamorphism* characterized by a mass transfer of material from the igneous body into its host rocks. Cf. *contact metasomatism*, *additive metasomatism*, *pneumatolytic contact metamorphism*, *allochemical metamorphism*. *(Spurr* et al. *1912, p. 455; Tilley 1920, p. 492, 499; Tyrrell 1926, p. 254)*

ADDITIVE METASOMATISM. Type of *contact metasomatism* characterized by a mass transfer of material into a rock body. Ant. *subtractive metasomatism*. *(Barrell 1907, p. 117)*

ADINOLE ⇒ Section 2.10 (Greek *adinos*, compact; Salberg, Sala, Sweden). Compact, fine-grained rock with a splintery fracture, commonly with a finely *banded* aspect due to alternating grey, green or reddish layers of variable colour intensity; mineralogically it is essentially composed of a fine- to very fine-grained mosaic of albite and subordinate quartz, with minor amounts of other constituents (muscovite, sericite, chlorite, actinolite, epidote, rutile and/or anatase); chemically it is characterized by high (up to 10% by weight) amounts of soda, a character that easily distinguishes it from the hard siliceous shales or *slates* that it resembles. Hence ***adinolization***, a Na-*metasomatic* process leading to the formation of adinole. There exist two different varieties of adinole distinguished according to their geological settings, namely ***contact adinole*** found in the *contact aureoles* of diabase dykes and ***tuffaceous adinole*** found as layers within metasedimentary sequences and unrelated to the proximity of diabase dykes. *(Beudant 1824; Hausmann 1828, p. 654; Beudant 1832, p. 126; Lossen 1872, p. 739; Rosenbusch 1910, p. 626; Milch 1917, p. 361; Mügge, in Rosenbusch 1923, p. 616; Mempel 1935–6, p. 17; Zellmer 1997, p. 460, 464)*

ADINOLE SERIES ⇒ Section 2.10. Group of rocks in the *aureole* of a diabase intrusion that show stages of *contact metasomatism* (*adinolization*), leading ultimately to the formation of *adinole*. The intermediate stages of the process, still retaining part of the original schistosity, give rise to *spotted* or *banded* albite-rich *schist* (formerly called *spilosite* and *desmosite* respectively). Following Milch the intermediate stage had been alternatively called adinole schist and adinole hornfels according to the amount of relict schistosity. *(Milch 1917, p. 357)*

AETHOBALISM, AETHOBALLISM (Greek *aethô*, to light up, burn out, and *ballô*, to throw). Proposed synonym for *contact metamorphism*. Hence ***aethobal(l)ic rocks***. Cf. *symphrattism*. *(Grabau 1904, p. 236)*

AGENTS MINÉRALISATEURS (French). Mineralizing solutions. *(Sainte-Claire Deville 1861, p. 1264)*

AGGRADATION RECRYSTALLIZATION. Type of *recrystallization* which results in a relative increase in the size of the crystal grains. Ant. *degradation recrystallization*. Cf. *aggrading neomorphism*. *(OU; Bates & Jackson)*

AGGRADING NEOMORPHISM, AGGRADATION NEOMORPHISM. Type of *neomorphism* that results in a relative increase in the size of the crystal grains. See *aggradation recrystallization*. Ant. *degradation neomorphism*. *(Folk 1965, p. 20)*

AGMATITE (Greek *agma*, fragment). *Migmatite* with breccia-like structure. *(Sederholm 1923, p. 117; Dietrich & Mehnert 1960, p. 59; Mehnert 1968, p. 354)*

AKYRIOSOME [akyrosome] (Greek *a*, not, *kyrios*, master, and *sôma*, body). Subordinate part of a *chorismite*. *(Huber 1943, p. 90)*

ALBITE-EPIDOTE-HORNFELS FACIES. *Facies* name proposed for *contact rocks* of relatively low grade. Cf. *hornblende-hornfels facies*. *(Fyfe* et al. *1958, p. 203)*

ALEMBIC CONTACT METAMORPHISM. Term proposed for a rare type of *contact metamorphism* interpreted as due to pure vapour-driven processes. *(Daly 1917, p. 410)*

Alkremite (mnemonic name after the main element contents: *al*uminium, silicon, in Russian '*kre*mnii', and *m*agnesium; Udachnaia Pipe, Siberia). Ultramafic rock composed of spinel and pyrope-rich garnet that together form about 75% of the rock and both of which are present in large proportion. It is found in kimberlite pipes and may contain some secondary minerals such as phlogopite, diopside or plagioclase. Syn. spinel-pyrope rock/gneiss/granofels. *(Ponomarenko 1975, p. 157, 931; Carswell 1990a, p. 18, 320)*

ALLALINITE (*Allalin* glacier, Saas Valley, Switzerland). Ophiolitic metagabbro with Cr-bearing omphacite ('smaragdite') and whose original structure is still preserved despite the total *pseudomorphic alteration* of the igneous *assemblage*. It is an *eclogite-facies* ophiolitic metagabbro. Cf. *flaser*-metagabbro. *(Rosenbusch 1896, p. 328; Meyer 1983)*

ALLOCHEMICAL CONTACT METAMORPHISM (Greek *allos*, other, different). Synonymous with *contact metasomatism*, *metasomatic contact metamorphism*. Ant. *isochemical contact metamorphism*. *(Reverdatto* et al. *1974, p. 287)*

ALLOCHEMICAL METAMORPHISM. Type of metamorphism characterized by a pronounced change of the original bulk composition by mass transfer processes, either addition or subtraction of material or both processes. Syn. *metasomatism*. Ant. *isochemical metamorphism*. *(Scheumann 1936a, p. 402; Eskola 1939, p. 264; Turner 1948, p. 3)*

ALLOCHTHONOUS IMPACT BRECCIA. *Impact breccia* in which the component materials have been displaced from their point of origin. It includes *polymict breccias* (*lithic breccias*, *suevite*) and *(impact) melt rocks* of the proximal *impact formations*. Ant. *autochthonous impact breccia*. *(OU; Dence 1964, p. 249; Pohl* et al. *1977, p. 352; Grieve 1998, Fig. 7)*

ALLOFACIAL (Greek *allos*, other, different, and Latin *facies*, form, aspect). Said of a group of rocks belonging to different *metamorphic facies*. Ant. *isofacial*.

ALLOGENIC (Greek *allos*, other, different, and *genesis*, origin). Said of a substance or mineral that has been introduced into a system from outside, for example injected material in a *migmatite*. Ant. *authigenic*.

ALLOGENIC IMPACT BRECCIA. Synonymous with *allochthonous impact breccia*. *(OU; Cassidy 1968, p. 120)*

ALLOMETAMORPHISM. Type of metamorphism caused by external forces that are later than and unrelated to the formation of the rock. *(Grubenmann & Niggli 1924, p. 180)*

ALLOMIGMATITE. Type of *migmatite* containing introduced material and originating from injection and assimilation. *(Polkanov 1935, p. 290)*

ALLOPHASE METAMORPHISM. Type of metamorphism that results in new mineral phases replacing the old ones. Ant. *isophase metamorphism*. *(Eskola 1939, p. 264)*

ALLOSKARN. Type of *skarn* formed by the replacement of the host carbonate rock along or near the contact with an intrusive rock. Equivalent to *exoskarn*. Cf. *autoskarn*. *(Abdullaev 1947, p. 245)*

ALLOTHIMORPHIC (Greek *allothi*, of other, and *morphê*, form; Vorderrheintal, Switzerland). Obsolete adjective applied to constituents in a metamorphic rock (e.g. crystals) that have preserved their original shape or outline from the *parent rock*. An ***allothimorph*** is any constituent of a metamorphic rock (e.g. crystal) that in the new rock still possesses its original outline. Ant. *authimorphic*. *(Milch 1894, p. 107; Johannsen 1939, p. 161)*

ALLOTRIOBLAST (Greek *allotrios*, foreign, and *blastos*, bud, sprout). Little-used synonym for *xenoblast*. *(OU)*

ALPINE-TYPE FACIES SERIES. Type of *facies series* of high *P/T* style from the Alpine region, that is characterized by the stability of pumpellyite at very low *metamorphic grades* and by *glaucophane-schist/blueschist facies assemblages* at higher grades. It is the highest-pressure type *facies series* of Hietanen. *(Hietanen 1967, p. 203)*

ALPINITE (Swiss Alps). Local name for a fine-grained quartz-plagioclase-mica gneiss (most probably an orthogneiss). *(Simler 1862, p. 20)*

ALTERATION (Latin *alter*, other). General term for any change in the mineralogical or chemical composition of a rock. The term is normally restricted to *secondary* effects such as weathering or the action of *hydrothermal* solutions. *(OU; Brongniart 1825, p. 151)*

AMAUSITE (French *émail*, pl. *émaux*, enamel). Obsolete term for *granulite*, *leptynite* or *Weissstein*. *(Gerhard 1814–15, in Tomkeieff)*

AMPHIBOLITE ⇒ Section 2.8, Fig. 2.8.2. Gneissose or granofelsic metamorphic rock mainly consisting of green, brown or black amphibole and plagioclase (including albite), which combined form ≥75% of the rock and both of which are present as major constituents; the amphibole constitutes ≥50% of the total mafic constituents and is present in an amount of ≥30%. Other common minerals include quartz, clinopyroxene, garnet, epidote-group minerals, biotite, titanite and scapolite. Amphibolites are mostly present in *amphibolite-facies* metamorphic complexes, or in the highest *subfacies* of the *greenschist facies* (quartz-albite-epidote-almandine *subfacies*). *(Brongniart 1813, p. 40; Aubuisson de Voisins 1819, p. 1045; Leonhard 1823–4, p. 244; Cordier 1868, p. 142; Loewinson-Lessing 1893–4, p. 11; Rosenbusch 1898)*

AMPHIBOLITE FACIES ⇒ Section 2.2, Fig. 2.2.4, Table 2.2.1. *Metamorphic facies* representing moderate pressure and temperature conditions. It lies between the *epidote-amphibolite* and *granulite facies*. It is characterized by hornblende-plagioclase ($>An_{17}$) *assemblages* in rocks of basaltic composition. *(Eskola 1920, p. 155, 165)*

AMPHICLASITE (mnemonic after the mineral content: *amphi*bole + plagio*clase*). Obsolete term for a metamorphic rock composed of amphibole and plagioclase and derived from a member of the *charnockite series* by replacement of hypersthene by hornblende. Rec.syn. *amphibolite. (OU; Ryka & Maliszewska 1982, p. 30; Dostal* et al. *1996, p. 621)*

ANA- (Greek *ana*, up, anew). Prefix referring to processes occurring in the deeper zones of the crust, e.g. *anamorphism*, *anatexis*. *(Becke 1892, p. 298; Van Hise 1904, p. 43)*

Anakit-type contact metamorphism (*Anakit*, a tributary of the lower Tunguska River, Siberia). Type of *isochemical contact metamorphism* mostly developed around *mafic* subvolcanic intrusive *rocks* and in which the *mineral assemblages* show either the *sanidinite facies* or the *pyroxene-hornfels facies*, rarely both. The aureole thickness does not exceed a few tens of metres. See *contact aureole systematics*. *(Reverdatto* et al. *1970, p. 311)*

ANAMIGMATIZATION (Greek *ana*, up, anew). Production of a melt in a pre-existing rock under high-temperature and high-pressure conditions. *(OU; Bates & Jackson)*

ANAMORPHIC ZONE, ZONE OF ANAMORPHISM (Greek *ana*, up, anew, and *morphê*, form). Zone in the Earth's crust where *anamorphism* takes place. Ant. *katamorphic zone*. *(Van Hise 1904, p. 43, 167)*

ANAMORPHISM. Process acting in the deep crust, whereby high-temperature and high-pressure conditions lead to rock flow and where the metamorphism results in the production of complex compounds from simpler ones. The important reactions are silicification (formation of silicates), *dehydration*, decarbonation and de-oxidation. Cf. *anamorphic zone*. Ant. *katamorphism*. *(Van Hise 1904, p. 43, 167)*

ANAPHRYXIS (Greek *ana*, up, anew, and *phryxis*, roasting). Unnecessary name for *contact metamorphism*. *(Gürich 1905, p. 250)*

ANATECTITE. Synonymous with *anatexite*.

ANATECTONICS (contraction of *ana*texis and *tectonics*; Argentera massif, Italian Western Alps). Structures whose production involves *anatexis* rather than just simple deformational mechanisms. *(Malaroda 1993, p. 259)*

ANATEXIS (Greek *ana*, up, anew, and *têkô*, to melt). Melting of a rock. The term is used irrespective of the proportion of melt formed, which may be indicated by adjectives such as intergranular, initial, advanced, partial, differential, complete, etc. Cf. *palingenesis*. *(Grubenmann 1904, p. 72; Sederholm 1907, p. 49, 102; Dietrich & Mehnert 1960, p. 59; Mehnert 1968, p. 353)*

ANATEXITE ⇒ Section 2.6. Rock still showing the evidence of *in situ* formation by *anatexis*. *(Jung & Roques 1936, p. 26; Loewinson-Lessing 1949, p. 451; Dietrich & Mehnert 1960, p. 59; Mehnert 1968, p. 353)*

ANCHIMETAMORPHISM (Greek *anchi*, approximate). Type of metamorphism found in the *anchizone*. Originally defined as the changes in the mineral content of rocks under the pressure and temperature conditions prevailing between the Earth's surface and the zone of 'true' metamorphism. It rougly corresponds to very low-grade metamorphism. Syn. *anchizonal* metamorphism. *(Harrassowitz 1927, p. 9)*

ANCHIZONE ⇒ Section 2.5, Fig. 2.5.1 (Greek *anchi*, approximate). *Metamorphic zone*, transitional between the *diagenetic zone* and the *epizone*, which roughly corresponds to very low-grade metamorphism and which is characterized by illite Kübler index (KI) mean values between 0.42 and $0.25°\Delta 2\theta CuK_{\alpha}$. Metamorphism in this zone is called *anchimetamorphism*, which roughly corresponds to very low-grade metamorphism. Syn. *anchimetamorphic* zone. *(Harrassowitz 1927, p. 9; Kübler 1967, p. 111; Kübler 1968, p. 393; Kübler 1984, p. 578)*

ANDALUSITE-SILLIMANITE SERIES. See *baric types of metamorphism*.

ANHEDRAL (Greek *an*, without, and *edra*, side, face). Said of a crystal that is not bounded by any of its own crystal faces. Term originally used to describe igneous rocks but now more widely applied. The specific recommended term for metamorphic rocks is *xenoblastic*. Ant. *euhedral*. *(Cross* et al. *1906, p. 698; Passchier & Trouw 1996, p. 255; Barker 1998, p. 229; Vernon 2004, p. 476)*

ANOGENIC METAMORPHISM, ANOGENE METAMORPHISM (Greek *anô*, upper, and *gennô*, to give birth to). Ambiguous term for metamorphism (in the original paper *dynamometamorphism*) of the upper zones of the Earth's crust as established by Becke after his studies in Hrubý Jeseník (Hohes Gesenke). Van Hise used a similar term for the changes in the deep zones of the crust. Ant. *katogenic metamorphism*. *(Becke 1892, p. 298; Van Hise 1904, p. 162)*

ANTI-STRESS MINERALS. Minerals such as andalusite, cordierite and forsterite, formed in metamorphic rocks and which were thought to develop where shearing stress was absent. As such these minerals were regarded as typical products of *thermal* or *contact metamorphism*. Ant. *stress minerals*. *(Harker 1918, p. 78)*

ANTITAXIAL (Greek *anti*, opposed to, and *taxis*, arrangement, ordering). Said of a vein-infilling in which fibres grow from the centre towards the walls. Hence ***antitaxial vein***, ***antitaxial growth***. Ant. *syntaxial*. *(Ramsay & Huber 1983, p. 242; Shelley 1993, p. 290)*

AORITE (*Greek aôros*, immature). Term originally given to a *migmatite* considered as an early stage in the process leading to a plutonic end member and from which the composition of the end member may be determined, for example diorite aorite, syenite aorite, granite aorite, etc. Hence ***aoritic stage***. *(Erdmannsdoerffer 1946, p. 96)*

APHANITE, APHANITE-SCHIST. Obsolete name given to various fine-grained rocks or schists. *(Cotta 1862, p. 96; Omalius d'Halloy, in Cordier 1868, p. 101)*

APHANITIC (Greek *a*, not, and *phaenô*, to be visible). Said of a rock in which the individual grains are not visible with the unaided eye (*c*. <0.1 mm). Ant. *phaneritic*.

APO- (Greek *apo*, from). Prefix used for metamorphic or *metasomatic* rocks in which either the primary structure (e.g. aporhyolite, a devitrified rhyolite; apogranite, an albitized granite) or primary minerals (e.g. apodolomite) are preserved. *(Bascom 1893, p. 828; Van Hise 1904, p. 777; Shabynin 1973, p. 199; Beus & Scherbakova 1993, p. 221)*

APOLLONIAN METAMORPHIC ROCKS (*Apollo* mission to the Moon, itself from the Greek god). Pre-Imbrian lunar highland rocks, generally of noritic-anorthositic composition, with *granoblastic* to *poikiloblastic microstructure* (lunar *granulitic* rocks). These rocks are ubiquitous in the lunar highlands and have been found at all lunar landing sites. However, it is not clear whether this kind of *thermal metamorphism* is a more local effect possibly induced by proximity to cooling *impact melt* sheets within the lunar *megaregolith* or whether it represents a global process that affected the whole lunar crust at some depth. The rocks are not *granulites* in the terrestrial sense because they result from very low-pressure–high-temperature metamorphism. Consequently, the process and the heat sources are not well defined although the rocks are clearly *recrystallized* crustal rocks with the average composition of a noritic anorthosite. *(Stewart 1975, p. 774)*

APOLLONIAN METAMORPHISM. *Thermal metamorphism* of the pre-Imbrian lunar crust leading to the formation of lunar *granulitic* rocks. *(Stewart 1975, p. 774)*

APOMAGMATIC CONTACT METASOMATISM (Greek *apo*, from). Obsolete term for *contact metasomatism* giving *epigenetic* ore deposits in the more distal parts of a *contact aureole* and for which there is no clear genetic connection with a plutonic mass (Bergeat). Redefined by Grubenmann & Niggli as *meso-* to *epithermal* deposits found outside the *contact aureole* in regions of igneous activity. Cf. *perimagmatic (additive) contact metasomatism*. Note: the prefix '*apo*' is used here with a meaning different from that defined above. *(Bergeat 1912, p. 11; Grubenmann & Niggli 1924, p. 316)*

ARCHOMETAMORPHISM (Greek *archaios*, ancient). Obsolete term for a kind of metamorphism dominated by dynamic effects and characteristic of rocks situated above the *epizone*. Rec.syn. very low-grade metamorphism, *anchimetamorphism*. *(Rinne 1928, p. 6; Rinne 1928–40, p. 323, 401)*

Argillisite ⇒ Section 2.9, Fig. 2.9.1. Low-temperature *metasomatic* rock that is mainly composed of clay minerals; also present may be silica minerals, carbonates and iron sulphides. The rock forms from the *hydrothermal alteration* of both igneous and sedimentary rocks. Hence ***argillisitization***, *metasomatic* process leading to the formation of argillisite. *(Lovering 1941, p. 236; Lovering 1949, p. 25; Shcherban' 1996, p. 10)*

ARKÉSINE (Greek *archaios*, ancient; Mt Blanc, French–Italian Alps). Obsolete term for a meta-granitic rock composed of quartz, feldspar, hornblende, white mica, chlorite and titanite. Cf. *protogine*. *(Jurine 1806, p. 373)*

ARMOURED RELICT. Relict mineral(s) surrounded by a rim of *alteration* products and protected from further reaction by the rim. *(Eskola 1915, p. 120; Joplin 1968, p. 5)*

ARTERITE (Latin *arteria*, vein). Originally defined as a term for a veined *migmatite* irrespective of the genesis of the *leucosome* (veins), but later taken to mean a type of *migmatite* where the darker parts are injected by veins of lighter material (*leucosome*) introduced from outside. Cf. *venite*, *phlebite*. *(Sederholm 1897–9, p. 134, 241; Scheumann 1936a, p. 298; Dietrich & Mehnert 1960, p. 59)*

ASSEMBLAGE. See *metamorphic assemblage*.

ASTITE (Maso waterstream, Cima d'*Asta*, Trentino, Italy). Obsolete term for a mica-andalusite *hornfels*. *(Salomon 1898, p. 150)*

ASTRIDITE (in honour of Queen *Astrid* of Belgium; New Guinea). Local name for a dark-green, veined, sometimes precious ornamental rock from New Guinea, consisting mainly of chromian jadeite and chromian hercynite (spinel) which are intergrown with quartz, opal and limonite. It is reported to have been derived from an olivine rock. Rec.syn. jadeite granofels. *(Willems 1934, p. 120; Webster 1975, p. 232)*

ASTROBLEME (Greek *astrô*, star, and *blêma*, wound). Unnecessary synonym for *impact crater*. *(OU; Dietz 1961, p. 2)*

ATATSCHITE, ATACHITE (*Atacha* Mt in the Ural). Type of metasomatized trachyte found near the contact with a magmatic body. The rock is mainly composed of sanidine, magnetite, tourmaline, sericite and apatite, with subordinate quartz, plagioclase, epidote, hornblende, andalusite, leucoxene. *(Morozevich 1901, in Tomkeieff; Zavaritskii 1922, in Tomkeieff)*

ATEXITE, ATECTITE (Greek *a*, not, and *têkô*, to melt). Basic material that remained unchanged during *anatexis*. *(Smulikowski 1947, p. 267; Dietrich & Mehnert 1960, p. 59)*

ATMOGENIC METAMORPHISM. Type of metamorphism due to the activity of volcanic gases. Syn. *pneumatolytic metamorphism*. *(Kalkowsky 1886, p. 35; Hatch 1888, p. 424)*

ATOLL STRUCTURE. Type of structure characterized by a ring or shell of one mineral around a core of another mineral or other minerals, resembling an atoll. The ring may be complete or not. The core may consist of a single crystal or an aggregate of crystals. The core may be older or younger than the rim. See also *corona structure*. *(Edwards 1954, p. 118; Joplin 1968, p. 28; Spry 1969, p. 306; Passchier & Trouw 1996, p. 255)*

AUGEN (German for eyes). Lensoid grains or grain aggregates usually on the mesoscopic scale. Augen generally occur in gneissose or schistose rocks and the schistose planes bend around the augen. Augen are usually the products of deformation (e.g. relict clasts in a *mylonite*). Hence *augen structure*. *(OU; Joplin 1968, p. 5; Vernon 2004, p. 476)*

AUGEN-BLAST (Greek *blastos*, bud, sprout). Newly crystallized *augen* or *porphyroblasts* lying in a *mylonitic* matrix. Cf. *augen-clast*. *(Bayly 1968, p. 233)*

AUGEN-CLAST (Greek *klastos*, broken). Relict *augen* or clast present in a *cataclastic* or *mylonitic* matrix. Cf. *augen-blast*. *(Bayly 1968, p. 233)*

AUGEN GNEISS, AUGENGNEISS. Type of gneiss characterized by the presence of *augen*. *(Cotta 1855, p. 127; Holmes 1920, p. 39; Joplin 1968, p. 19)*

AUGENHÄLLEFLINTA-GRANULITE, AUGENHÄLLEFLINT (Granulitgebirge in Saxony, Germany). Obsolete term for a *banded mylonite* with feldspar *augen*. Rec.syn. *augen mylonite*. *(Scheumann 1936c, p. 419)*

AUGEN MIGMATITE. Type of *migmatite* with relict *augen* arranged parallel to the *foliation*. The *augen* are usually large plates of orthoclase. *(Cheng 1943, p. 129)*

AUGEN MYLONITE. Type of *mylonite* containing distinctive large crystals or lithic fragments around which the *foliated* fine-grained matrix is wrapped, often forming symmetric or asymmetric trails. *(Scheumann 1936c, p. 462)*

AUGEN SCHIST. Type of *mylonite* with relict *augen*-shaped *porphyroclasts* and a significantly *recrystallized* matrix. Rec.syn. *augen mylonite*. *(Lapworth 1885, p. 559; Higgins 1971, p. 72)*

AUGEN STRUCTURE. Type of structure characterized by the presence of *augen*. *(Lapworth 1885, p. 559; Grubenmann 1904, p. 83)*

AUGITE-AMPHIBOLITE. Obsolete term for an intermediate rock between an augite-plagioclase gneiss and an *amphibolite*. The term is listed by Grubenmann and Niggli as one of the rock types of the *katazone* whereas *amphibolite* is given as belonging to the *mesozone*. *(Grubenmann & Niggli 1924, p. 501)*

AURA GRANITICA (Latin *aura*, breeze). Gaseous emanations (*granitic fumaroles*) from a granitic body that may give rise to ore and/or mineral deposits in the margins of the igneous body or in its surrounding rocks. Cf. *pneumatolytic metasomatism*. *(Élie de Beaumont 1847, p. 1295, 1314)*

AUREOLE. See *contact aureole*.

AUTHIGENESIS (Greek *authi*, here, in this place, and *genesis*, origin). Reactions between the constituents of a sediment or sedimentary rock, that form new minerals or result in the enlargement of existing minerals (see *diagenesis*). *(OU; Pettijohn 1957, p. 650, 661)*

AUTHIGENIC. Said of a substance or mineral that has formed in place. Ant. *allogenic*. *(Tester & Atwater 1934, p. 25)*

AUTHIGENIC IMPACT BRECCIA. Synonymous with *autochthonous impact breccia*. *(OU; Cassidy 1968, p. 121)*

AUTHIMORPHIC (Greek *authi*, here, in this place, and *morphê*, form; Vorderrheintal, Switzerland). Obsolete adjective applied to those constituents (e.g. crystals) of a metamorphic rock that have changed their original shape or outline during metamorphism. Hence ***authimorph***, such a constituent. *(Milch 1894, p. 107; Johannsen 1939, p. 163)*

AUTOCHTHONOUS (PARAUTOCHTHONOUS) IMPACT BRECCIA. *Cataclastic (monomict) impact breccia* in which the component materials have not been displaced any significant distance from their point of origin. Ant. *allochthonous impact breccia*. *(OU; Dence 1964, p. 249; Pohl* et al. *1977, p. 352; Grieve 1998, Fig. 7)*

AUTOCLASTIC STRUCTURE (Greek *autos*, self, and *klastos*, broken). Type of structure in which the rock has a broken or fragmented appearance due to tectonic processes acting *in situ*, for example *fault breccia*. *(Smythe 1891, p. 331)*

AUTOMETAMORPHISM (Greek *autos*, self). Type of metamorphism in igneous rocks where the process of change is driven by inherent features of the igneous rock rather than by external conditions, for example reaction with its own volatile phases, *recrystallization* due to the residual heat of the rock. See *autometasomatism*. *(Sederholm 1916, p. 4; Sargent 1917, p. 59; Grubenmann & Niggli 1924, p. 180; Turner 1948, p. 5)*

AUTOMETASOMATISM. Type of *metasomatism* that occurs at the top of magmatic bodies during the early postmagmatic stage. It is produced by reaction of the igneous rock with its own volatile phases. Typical autometasomatic processes, for example, are albitization in granitic plutons and serpentinization in ultramafic rocks. *(Korzhinskii 1953, p. 378; Bowes 1989a, p. 362)*

AUTOMIGMATITE. Term originally given to a *migmatite* formed through *autometamorphic* processes. *(Polkanov 1935, p. 290)*

AUTOSKARN. Type of *skarn* formed by the replacement of a magmatic rock due to *skarn*-producing fluids generated by the magmatic rock. Equivalent to *endoskarn*. Cf. *alloskarn*. *(Abdullaev 1947, p. 245)*

AVIOLITE (Mt *Aviolo*, Adamello batholith, upper Val Camonica, northern Italy). Obsolete term for a cordierite-mica *hornfels*. *(Salomon 1898, p. 150)*

AXIAL PLANE FOLIATION. Type of *foliation* developed parallel to the axial plane of a fold. Hence ***axial plane schistosity***, ***axial plane cleavage***. *(Fairbairn 1935, p. 592)*

BAKED ROCK. Rock slightly affected by heat at the contact with a lava flow, a dyke or burning combustibles. The *alteration* includes bleaching of carbonaceous rocks, reddening of iron-rich rocks, elimination of water and other volatile constituents, indurating effects and a small degree of *fritting*. It is the first stage of the sequence baked rock, *fritted rock* and glassy rock (*buchite*) reflecting increasing heat conditions. Hence ***baking***, the formation of baked rock. Term also rarely used as a synonym of *burned rock*. *(Zirkel 1866a, p. 443; Arnold & Anderson 1907, p. 753; Tyrrell 1926, p. 301)*

BAND, **BANDING**, **BANDED**. Synonymous with layer, layering, layered. *(OU)*

BANDED GNEISS. Type of gneiss characterized by alternating *bands*, on the mesoscopic scale, of markedly different composition and/or structure. *(Teall 1887, p. 484; Dietrich 1960, p. 36; Joplin 1968, p. 19)*

BANDED STRUCTURE. In metamorphic rocks, type of structure characterized by alternating parallel *bands*, on the mesoscopic scale, of markedly different composition and/or structure. Equivalent to *banding* and *layering*. *(Harker 1950, p. 205)*

BARAMITE (*Baramia* mine, SE Arabia desert, Upper Egypt). Altered ultramafic plutonic rock composed of serpentine, abundant magnesite, and opal. *(Hume* et al. *1935, p. 28; Tröger 1938, p. 83)*

BARÉGIENNE [barrégienne] (*Barèges*, Hautes-Pyrénées, France). Local name given to a highly folded, varicoloured *calc-silicate rock* composed of finely-alternating limestone and *slate*, later shown to be of *contact-metamorphic* origin. *(Ramond 1801; Magnan 1877, p. 33; Lacroix 1899; Bresson 1903, p. 104; Jung 1963, p. 164; Aubouin* et al. *1968, p. 525; Foucault & Raoult 1984, p. 36, 82)*

BARIC TYPES OF METAMORPHISM ⇒ Section 2.2. Term used instead of *facies series* in a classification proposed by Miyashiro: (i) a low-P/T type (also referred to as the *andalusite-sillimanite series*), (ii) a medium-P/T type (also referred to as the *kyanite-sillimanite series* or *Barrovian type*), and (iii) a high-P/T type (also referred to as *glaucophanic metamorphism*); the *eclogite facies* is considered as part of this type. *(Miyashiro 1973a, p. 73; Yardley 1989, p. 188; Miyashiro 1994, p. 200, Fig. 8.3)*

BARROVIAN-TYPE FACIES SERIES ⇒ Section 2.2 (in honour of G. *Barrow*, 1853–1932, who first described and mapped series in 1893 and 1912, using mineral zones; Grampian Highlands of Scotland, Dalradian Supergroup, UK). Type of *facies series* characterized by the progressive development of garnet and kyanite from lower to higher grades; andalusite, cordierite and glaucophane are absent. The facies represents medium P/T conditions and is common in *regional metamorphism*. It is equivalent to the medium-P/T type of *regional metamorphism* of Miyashiro (see *baric types of metamorphism*). See also *Barrow's zones*. *(Read 1952, p. 278; Fyfe* et al. *1958, p. 228; Miyashiro 1961, p. 278; Hietanen 1967, p. 195; Turner 1981, p. 375)*

Barrow's zones, **Barrovian zones**. Sequence of *metamorphic zones* characteristic of *Barrovian-type facies series* metamorphism. The zones, in order of increasing grade, are defined by the presence of chlorite, biotite, garnet, staurolite, kyanite and sillimanite. *(Barrow 1893, p. 343; Hietanen 1967, p. 195)*

BASAL DEFORMATION LAMELLAE. See *deformation lamellae*.

BASALT JASPER (German *Basaltjaspis*). Obsolete term for a metasedimentary rock of jaspery appearance that has been affected by *pyrometamorphism* at the contact with basaltic rocks; it is a *buchite* or *fritted rock* according to the glass contents. (Not to be confused with either *porcellanite* or *porcellanjasper* which are typical products of *combustion metamorphism*.) Cf. *systil*. *(Werner, in Haüy 1822, p. 582; Leonhard 1823, p. 106; Leonhard 1824, p. 535; Zirkel 1866a,b, p. 619)*

BASIC ⇒ Section 2.1, Table 2.1.2. See *acid*, *intermediate*, *basic*, *ultrabasic*.

BASIC BEHIND. Term proposed for residual basic masses produced during the process of *granitization*. It was considered that the more acid components had been removed from the mass leaving it relatively enriched in basic material. Equivalent to *restite*. Ant. *basic front*. *(Read 1951, p. 11)*

BASIC FRONT. Term proposed for a zone of enrichment in basic material formed ahead of a front of *granitization*. It was considered that the zone was formed by the introduction of basic elements such as calcium, magnesium and iron driven out of the rocks being *granitized*. Ant. *basic behind*. *(Holmes & Reynolds 1947, p. 58)*

BASIC HORNFELS ⇒ Section 2.10. Collective name mostly used for *hornfelses* formed by *contact metamorphism* of pre-existing rocks of basic composition, initially also including marly rocks (Williams *et al.*), subsequently restricted to *hornfelsed* basic igneous rocks (Turner, Spry). Because the most obvious feature of these *hornfelses* is their *mafic mineral* content as opposed to their basic chemistry, the SCMR prefers to refer to them as *mafic hornfels*. *(MacGregor 1931, p. 508; Williams* et al. *1954, p. 186, 195; Turner 1968, p. 4; Spry 1969, p. 198)*

BASIFICATION. Enrichment of a rock in basic elements such as calcium, magnesium and iron. *(Reynolds 1946, p. 432)*

Bathograd (Greek *bathos*, depth, and Latin *gradus*, step, degree). Mappable line, based on an invariant model reaction, that separates occurrences of a higher-*P assemblage* from occurrences of a lower-*P assemblage*. Bathograds do not need to be isochronous on the regional scale. *(Carmichael 1974, p. 680; Carmichael 1978, p. 771; Bucher & Frey 1994, p. 110)*

Bathozone. Mappable area comprising *metamorphic mineral assemblages* indicating a similar *P*-range, and bounded by *bathograds*. Six bathozones have been distinguished. *(Carmichael 1978, p. 770; Bucher & Frey 1994, p. 110)*

BEARDED STRUCTURE. See *pressure shadow*.

Beerbachite (Nieder *Beerbach*, Odenwald Mt, Hessen, Germany). High-grade, fine-grained *granoblastic contact-metamorphic* rock containing pyroxene(s), plagioclase, iron oxide, with minor olivine and/or amphibole and/or biotite. Originally interpreted as gabbro-aplite dyke, then as a sedimentary xenolith; ultimately identified as a basaltic dyke rock metamorphosed by contact with gabbroic rock. Rec.syn. *mafic hornfels*. Cf. *sudburite*, *muscovadite*, *granoblast*. *(Chelius 1892; Klemm 1926; MacGregor 1931)*

BELOCHERITE (*Beloretsk*, central Ural). Local name for a variety of Precambrian quartzite from the central and southern Ural. *(OU; Loewinson-Lessing & Struwie 1937, p. 46; Loewinson-Lessing & Struwie 1963, p. 41)*

Beresite ⇒ Section 2.9, Fig. 2.9.1 (*Beresovsk* gold deposit, Ural, Russia). Low-temperature *metasomatic* rock characterized by quartz, sericite, carbonate (ankerite) and pyrite *assemblages*, resulting from the replacement of both igneous and sedimentary *protoliths*. It may be associated with a variety of Au-, Au-Ag-, Ag-Pb-, or U-ore mineralizations. Hence ***beresitization***, *metasomatic* process leading to the formation of *beresite* or *listvenite*. *(Rose 1837, p. 186; Johannsen 1932, p. 23; Kouznetsov 1924, p. 23; Korzhinskii 1953, p. 443; Marignac* et al. *1996, p. 785)*

BERGMANITE [bergmannite] (in honour of T. *Bergman*; Norway). Obsolete term for a variety of *serpentinite*. *(Pinkerton 1811b, p. 53; Humble 1860, p. 52; Tomkeieff)*

BIBLIOLITE (Greek *biblion*, book, and *lithos*, stone). Laminated schistose rock. Syn. *bookstone*. *(OU; Fay 1920, p. 77)*

Bimetasomatism. Variety of *boundary metasomatism* that causes replacement in both rocks in contact because of two-way diffusion of different components across the contact. *(OU; Korzhinskii 1970, p. 134)*

BIMINERALIC. Said of a rock in which ≥95% of the modal content is composed by two minerals, as opposed to *polymineralic*, *monomineralic*, *trimineralic*.

BIOTITE-CROSS-SCHIST (German *Biotitquerschiefer*). Obsolete term for a *mica schist* with biotite *porphyroblasts* orientated at an angle to the *schistosity*. *(Grubenmann & Niggli 1924, p. 505)*

BIRBIRITE (River *Birbir*, Ethiopia). Brownish rock composed of chalcedony, limonite, chromite and sperrylite, derived from the hydration and silicification of dunite in the sequence dunite – *serpentinite* – birbirite. *(Duparc* et al. *1927, p. 139)*

BIRD'S EYE SLATE (English Lake District). Type of *cleaved* tuff with accretionary lapilli. The lapilli have an elliptical shape and are zoned with an outer light-coloured rim and a dark-green centre, thus resembling a bird's eye. *(Marr 1916, p. 25, 87)*

BISTAGITE (*Bis-Tag* Range, Yenisey province, Siberia, Russia). Obsolete, local name for a variety of clinopyroxenite composed of diopside, partly serpentinized, and K-feldspar. *(Yachevskii 1909, p. 46; Ryka & Maliszewska 1982, p. 48)*

BLACOLITE (in honour of the chemist Joseph *Black*, 1728–99). Obsolete term for a variety of *serpentinite* containing grains of bastite (a coarse-platey variety of serpentine formed by *alteration* of pyroxene). *(Pinkerton 1811b, p. 53; Loewinson-Lessing & Struwie 1937, p. 50)*

BLAST, **BLASTO-**, **-BLASTIC** (Greek *blastos*, bud, sprout). Suffix or prefix used to describe the structures produced during metamorphism as distinct from those produced in igneous rocks. When used as a prefix (blasto-) the term refers to *relict structures* or features of the rock (see *blastoporphyritic*, *blastomylonite*, *blastolaminar*, etc.). When used as a suffix (-blast, -blastic) the term refers to new structures or features of the rock (see *idioblastic*, *xenoblastic*, *granoblastic*, *granuloblastic*, *lepidoblastic*, *nematoblastic*, *poikiloblastic*, *porphyroblastic*, *glomeroblastic*, etc.). *(Becke 1903a, p. 570; Becke 1903b, p. 48; Joplin 1968, p. 28)*

BLASTESIS. See *metablastesis*.

BLASTOGRANITIC STRUCTURE. Type of structure in a metamorphic rock that is dominated by the *relict* granitic *structure* of the *parent rock*. *(Becke 1903b, p. 48)*

BLASTOGRANULAR STRUCTURE. Type of *porphyroclastic structure* characterized by relatively small strains as evidenced by the weak development of a *foliation* in the rock. Ant. *blastolaminar structure*. *(OU)*

BLASTOHYPIDIOMORPHIC (Greek *blastos*, bud, sprout, *hypo*, under, *idios*, own, particular, and *morphê*, form). Type of structure in a metamorphic rock, that is dominated by relict hypidiomorphic crystal grains of the *parent* igneous *rock*. *(Eskola 1939, p. 275)*

BLASTOLAMINAR STRUCTURE (Greek *blastos*, bud, sprout, and Latin *lamina*, lamina). Type of *porphyroclastic structure* characterized by relatively large strains as evidenced by the strong development of a *foliation* in the rock. Ant. *blastogranular structure*. *(OU)*

BLASTOMYLONITE. *Mylonite* that displays a significant degree of grain growth related to or following deformation. *(Sander 1912, p. 250; Knopf 1931, p. 13; Waters & Campbell 1935, p. 477; Spry 1969, p. 230; Higgins 1971, p. 72; Passchier & Trouw 1996, p. 255)*

BLASTOPELITIC STRUCTURE (Greek *pêlos*, mud). Type of structure that is dominated by the *relict structure* of the *parent rock* and is characteristic of metamorphosed mudstones. *(Grubenmann 1904, p. 100)*

BLASTOPHILIC [blastophyllic, blastophylic] (Greek *philos*, friend). Obsolete term for minerals that can be formed only in metamorphic rocks. Ant. *blastophobic*. *(Loewinson-Lessing 1933, p. 55; Tomkeieff)*

BLASTOPHITIC STRUCTURE (Greek *ophis*, snake). Type of structure in a metamorphic rock that is dominated by the *relict ophitic structure* of the *parent rock*. *(Becke 1903b, p. 48)*

BLASTOPHOBIC (Greek *phobos*, fear). Obsolete term for minerals that do not form in metamorphic rocks. Ant. *blastophilic*. *(Loewinson-Lessing 1933, p. 77; Tomkeieff)*

BLASTOPORPHYRITIC STRUCTURE. Type of structure in a metamorphic rock that is dominated by the *relict* porphyritic *structure* of the *parent rock*. *(Becke 1903b, p. 48)*

BLASTOPSAMMITIC STRUCTURE (Greek *psammos*, sand). Type of structure characteristic of metamorphosed sandstones, that is dominated by the *relict structure* of the *parent rock*. *(Becke 1903b, p. 48)*

BLASTOPSEPHITIC STRUCTURE (*psephis*, pebble, gravel). Type of structure characteristic of metamorphosed conglomerates or breccias and that is dominated by the *relict structure* of the *parent rock*. *(Becke 1903b, p. 48)*

BLAVIÉRITE (in honour of the French geologist *Blavier*; Mayenne, France). Obsolete term for a schistose *contact-metamorphosed* porphyry, mainly composed of fine-grained white mica, quartz ± sericitized feldspars. *(Munier-Chalmas, in Oehlert 1862, p. 134; Jacquet & Michel-Lévy 1886, p. 523; Bergeron 1888, p. 58)*

BLEPHARITIC STRUCTURE (Greek *blepharon*, eyelid). Type of structure in a metamorphic rock in which the *foliation* planes curve around *porphyroblasts*; the *foliation* is characteristically formed by phyllosilicate-rich layers. *(Ussher* et al. *1913, p. 65)*

Blueschist ⇒ Section 2.4. Schist whose bluish colour is due to the presence of sodic amphibole. More precise terms should be used wherever possible (e.g. glaucophane schist). The term is also used as a *facies* name and if used in this sense the *facies* context should be made clear by saying *blueschist-facies* rock (or schist, or gneiss, or granofels). *(Bailey 1962, p. 4; Bailey* et al. *1964, p. 91; Fyfe 1967, p. 36)*

BLUESCHIST FACIES ⇒ Sections 2.2 & 2.4, Fig. 2.2.4, Table 2.2.1. See *glaucophane-schist facies*. Glaucophane-bearing mineral associations lacking in lawsonite but containing epidote have been separated by Evans as forming an ***epidote-blueschist facies***. *(Bailey 1962, p. 4; Bailey* et al. *1964, p. 91; Brothers & Blake 1987, p. 773; Evans & Brown 1987, p. 774; Evans 1990, p. 3)*

BOETONITE (*Boeton* Island, SE Celebes, Indonesia). Vein rock composed mainly of opal and chalcedony with some chromite and marcasite and locally magnesite. It occurs predominantly in ultrabasic rocks. *(Hetzel 1938, p. 150; Tomkeieff)*

BOOKSTONE. See *bibliolite*.

BOOK STRUCTURE. Type of structure characterized by the alternation of parallel sheets of schist/*slate* and quartz veins or in-fillings. *(Lindgren 1928, p. 193)*

BORZOLITE (*Borzoli*, Genoa, Italy). Hornblende-bearing rock with carbonate amygdales, associated with *serpentinite*. *(Issel 1880, p. 187; Capacci 1881, p. 301; Tomkeieff)*

Bosost-type contact metamorphism (*Bosost* area, central Pyrenees, Spain). Type of *isochemical contact metamorphism* of *regional* extent, associated with synmetamorphic granitic melts and in which the rocks formed range from typical biotite-sillimanite gneiss to muscovite-staurolite schist down to *greenschist*. Typical *granulite-facies assemblages* are absent. The *aureole* thickness may reach 7–10 km. See *contact aureole systematics*. *(Reverdatto* et al. *1970, p. 312)*

BOUDIN (French *boudin*, blood sausage; Bastogne, Ardennes, SE Belgium). Fragment of a disrupted competent bed, barrel-shaped in cross-section and separated from the next fragment either by a commonly fibrous in-filling, consisting of quartz or calcite ± other minerals, or by surrounding incompetent layers which have been pinched in. It results from the failure of the competent bed under compressional stress, whereas the incompetent layer yields by shearing and/or flowing. ***Pinch-and-swell structures*** result from pinching-in and thinning of the competent layer but without failure of the layer. *(Lohest* et al. *1908, p. B.371; Lohest 1909, p. B.278; Wilson 1961, p. 496; Price & Cosgrove 1990, p. 405–433; Jongmans & Cosgrove 1993, p. 131)*

BOUDINAGE. (1) Fragmentation of a competent rock layer into *boudins*; (2) fragmented state of a competent rock layer as *boudins*. *(Price & Cosgrove 1990, p. 1)*

BOUNDARY METASOMATISM ⇒ Section 2.9. Type of *metasomatism* that occurs at the contact between two rock types. *(OU)*

BOWENITE (in honour of G. T. *Bowen* who first analysed it; Smithfield, Rhode Island, USA for the mineral, New Zealand for the rock). Originally defined as a mineral close to nephrite, but differing from it by a large water content. Now used as a local name for a variety of *serpentinite* resembling nephrite in colour. Syn. *tangiwai*. *(Bowen 1822, p. 346; Dana 1850, p. 154, 265)*

Bow-tie structure. Synonymous with *garbenschiefer* structure. *(OU; Barker 1998, p. 230)*

BRECCIA (old German *brecha*, fracture, and Ligurian-Italian *breccia*, broken stone). See *impact breccia*.

BRECCIA DYKE. Dyke formed in the (par) autochthonous basement or in displaced megablocks of *impact craters* consisting of *impact breccia* (*polymict breccias* such as *impact melt rock*, *suevite*, *lithic breccia* or more rarely *monomict breccia*). *(OU; Dence 1971, p. 5555; Lambert 1981, p. 61, Table 1; Stöffler* et al. *1988, p. 287, Table 2)*

BROWNSTONE FACIES ⇒ Section 2.5. Obsolete term indicating a low-temperature *mineral facies* encompassing ocean-floor weathering and low-temperature *hydrothermal alteration* of the ocean floor. *(Cann 1979, p. 230; Bowes 1989a, p. 69)*

BUCHAN-TYPE FACIES SERIES, BUCHAN-TYPE METAMORPHISM (Dalradian Supergroup, Ythan Valley, Aberdeenshire, Scotland, UK). Type of *facies series* characterized by the development of cordierite and andalusite at lower grades and sillimanite at higher grades; kyanite is absent. The *facies* represents low P/T conditions (see Smulikowski *et al.*, Section 2.2) and is common in *regional metamorphism*. Equivalent to the low-pressure *facies series* of Miyashiro. *(Read 1952, p. 278; Fyfe* et al. *1958, p. 206; Miyashiro 1961, p. 281; Hietanen 1967, p. 193; Turner 1981, p. 368)*

BUCHITE ⇒ Section 2.10 (in honour of the German geologist L. von *Buch*, 1774–1853; Vogel Mts., Germany). Compact, vesicular or slaggy metamorphic rock of any composition containing more than 20% vol. of glass, either produced by *contact metamorphism* in volcanic to subvolcanic settings or generated by *combustion metamorphism*. It is also used for partially melted materials obtained in laboratories by burning or heating natural rocks or artificial mineral mixtures. In hand specimen the rock is commonly characterized by a conchoidal fracture. In some outcrops buchite may show columnar jointing. In thin section the rock is composed of a glassy matrix and unmelted or partially melted mineral grains of the *protolith*. The glass commonly contains small grains of newly formed minerals in phase *assemblages* typical of *pyrometamorphism*. Locally the original glassy matrix is partially converted into very fine quartz-K-feldspar intergrowths resembling granophyre. Buchites formed by different metamorphic processes can be distinguished by appropriate qualifiers (e.g. *contact*, *combustion* or artificial buchite). See *fritted rock*. *(Möhl 1873, p. 83; Flett 1908, p. 129; Thomas 1922, p. 240; Tomkeieff 1940, p. 54; Spry & Solomon 1964, p. 535)*

Bunte breccia (German for vari-coloured; Ries impact crater, Germany). Local term for *polymict lithic breccia* forming a *continuous ejecta blanket*. First described as 'Kalkbreccie'. *(Gümbel 1870, p. 182; Bentz 1925, p. 97; Engelhardt 1971, p. 5567; Pohl* et al. *1977, p. 352)*

BURIAL METAMORPHISM ⇒ Section 2.2, Fig. 2.2.1. Type of *metamorphism*, mostly of *regional* extent, which affects rocks deeply buried under a sedimentary-volcanic pile, and is typically not associated with deformation or magmatism. The resultant rocks are partially or completely *recrystallized* and generally lack *schistosity*. It commonly involves from very low to medium metamorphic temperatures and low to medium P/T ratios. *(Coombs 1961, p. 214)*

BURNED ROCK, BURNT ROCK ⇒ Section 2.10. General term for a compact, vesicular or clinkery, glassy to holocrystalline metamorphic rock of various colours, produced by the *combustion metamorphism* of pre-existing sedimentary rocks. In the fused varieties the glass coexists with refractory grains and/or newly formed minerals (melilite, wollastonite, mullite, cristobalite, spinel, etc.) whose nature reflects the very high-temperature metamorphic conditions. The term burned rock supersedes such names as *thermantide*, *thermantide porcellanite*, *porcellanite*, *porcelain jasper*, fused shale. The glassy or glass-bearing varieties of burned rocks are called *buchite* (coal-fire *buchite*) or *fritted rock* respectively. See also *coal-fire ash*. *(Leonhard 1824, p. 500;*

Arnold & Anderson 1907, p. 755; Reverdatto 1973, p. 87)

Burning metamorphism. Term proposed by the SCMR as an acceptable alternative to the recommended term *combustion metamorphism*.

CALC-AUGEN PHYLLITE (Löwenzeiler Mühle bei Stromberg, Hunsrück, Germany). Type of *phyllite* containing roundish calcite crystals, 0.2–1.0 cm in size, set in a muscovite-chlorite groundmass. Alternative terms such as carbonate-bearing muscovite-chlorite schist (or *phyllite*) should be used instead. *(Beyenberg 1930, p. 318)*

CALC-FLINTA (Bodmin and Camelford areas, Cornwall, UK). Obsolete term for a very fine-grained rock of flinty appearance formed by *contact metamorphism* of a calcareous mudstone, possibly with some accompanying *pneumatolytic* action. The new minerals include feldspars, tremolite-actinolite and *calc-silicate minerals*, the latter being less abundant than in calc-silicate *hornfels*. *(Barrow & Thomas 1908, p. 113; Harker 1932, p. 89; Harker 1954, p. 262)*

Calciphyre (Latin *calx*, lime, chalk, and -phyre as in the French *porphyre*, porphyry). Metacarbonate rock containing a conspicuous amount of calcium-silicate and/or magnesium-silicate minerals. The term *carbonate-silicate rock* should be used in preference if the non-carbonate mineral content is higher than 50% vol. and the term *impure marble* should be used in preference if the non-carbonate mineral content is lower than 50% vol. *(Brongniart 1813, p. 38; Zirkel 1894b, p. 454; Artini & Melzi 1900; Tacconi 1911, p. 80; Brongniart 1927, p. 97; Holmes 1928, p. 51; Hieke 1945, p. 22; Smolin 1959, p. 43; Dobretsov* et al. *1992, p. 69)*

CALC-MICA-SCHIST (German *Kalkglimmerschiefer*). Metacarbonate rock with a *schistose structure* and composed of calcite, oriented mica and quartz. The compound term 'quartz-mica-carbonate schist' should be used in preference. *(Cotta 1855, p. 141; Zirkel 1894b, p. 292; Harker 1932, p. 206)*

Calc-schist. Metamorphosed argillaceous limestone containing calcite as a substantial component and with a *schistose structure* produced by parallelism of platy minerals. The term 'carbonate-silicate schist' should be used in preference if the non-carbonate mineral content is >50% vol. and the term 'schistose *impure marble*' if this content is <50% vol. *(Brongniart 1813, p. 36; Holmes 1928, p. 52)*

CALC-SILICATE MARBLE. Type of *marble* in which Ca- and/or Mg-silicate minerals are conspicuous. The term *impure marble* should be used in preference.

CALC-SILICATE MINERAL ⇒ Section 2.7, Table 2.7.1. Calcium-rich silicate mineral. The SCMR noted that there was no precise definition of calc-silicate minerals; attempts to define them solely on the CaO content of the minerals proved difficult. The SCMR, therefore, proposes the following list of the main calc-silicate minerals: calcic garnet (ugrandite), calcic plagioclase, calcic scapolite, diopside-hedenbergite, epidote group minerals, hydrogrossular, johannsenite, prehnite, pumpellyite, titanite, vesuvianite, wollastonite. *(OU; Shand 1931, p. 188)*

CALC-SILICATE ROCK ⇒ Section 2.7, Fig. 2.7.1. Metamorphic rock mainly composed of *calc-silicate minerals* and containing less than 5% vol. of carbonate minerals (calcite and/or aragonite and/or dolomite). Cf. *lime-silicate rock*. *(Harker 1902, p. 397)*

Californite (Fresno, Siskiyou and Tulare Counties in *California*, USA). Local term for a dark-green to grass-green jade-looking rock composed essentially of vesuvianite. The term is also used locally for a white variety of grossular garnet from Fresno County, California. Syn. vesuvianitite, vesuvianite rock, vesuvianite granofels. *(Webster 1975, p. 232)*

CALORIC METAMORPHISM (German *kalorische Metamorphose*). Obsolete term for *thermal metamorphism*, now superseded by *isochemical contact metamorphism*. Cf. *normal contact metamorphism*. *(Durocher 1846, p. 576, 643; Milch 1922, p. 288)*

CANGA. See *tapanhoacanga*.

CARBONATE-SILICATE ROCK ⇒ Section 2.7, Fig. 2.7.1. Metamorphic rock mainly composed of silicate minerals (including *calc-silicate minerals*) and containing between 5 and 50% vol. of carbonate minerals (calcite and/or aragonite and/or dolomite). More precise systematic (mineral-structural root) names should be used wherever possible. *(Rosen* et al. *Section 2.7)*

CARBONITE. As a metamorphic rock name, term used for a coal altered by an igneous intrusion. Also used with other meanings: (a) a fossil coal, and (b) a very brittle black variety of bitumen. *(Heinrich 1875, in Tomkeieff 1954, p. 35)*

CARVOEIRA, CARVOÏRA (old Brazilian *carvâo*, storage place, e.g. for coal; Minas Geraes, Brazil). Obsolete local name for a black fine-grained quartz-tourmaline rock. Cf. *tourmalite*. *(Eschwege 1832, p. 178)*

Cata- (Greek *kata-*, downwards, against, completely). See *kata-*.

CATABLASTIC STRUCTURE (KATABLASTIC STRUCTURE) (Greek *kata-*, downwards, against, completely, and *blastos*, bud, sprout). Originally proposed as a term for the structure in metamorphic rocks caused by *recrystallization* in the solid state. Rec.syn. *crystalloblastic structure*. *(Loewinson-Lessing 1905, p. 8)*

CATACLASITE ⇒ Section 2.3, Fig. 2.3.1 (Greek *kata-*, downwards, against, completely, and *klasis*, crushing). *Fault rock* that is cohesive with a poorly developed or absent *schistosity*, or that is incohesive, characterized by generally angular *porphyroclasts* and lithic fragments in a finer-grained matrix of similar composition. Hence, ***cataclastic structure***, the structure of a cataclasite; ***cataclasis***, the process leading to the production of a cataclasite. Cataclasite may be subdivided according to the relative proportion of finer-grained matrix into *protocataclasite*, *mesocataclasite* and *ultracataclasite*. *(Daubrée 1862, in Tomkeieff; Loewinson-Lessing 1905, p. 549; Grubenmann & Niggli 1924, p. 219; Tyrrell 1926, p. 285; Waters & Campbell 1935, p. 477; Gidon 1987, p. 80)*

CATACLASTIC BRECCIA. See *fault breccia*.

CATAGENESIS, KATAGENESIS. Obsolete term used, especially by Russian authors, to indicate changes occurring in (an already lithified) sedimentary rock buried under a distinct covering layer, characterized by *P–T* conditions that are significantly different from those of both deposition and metamorphism. The term is synonymous with *epigenesis*; both terms are equivalent to *deep diagenesis*. *(Fersman 1922, p. 30; Strakhov 1967, p. 90)*

CATAWBARITE [catawberite, catabirite] (*Catawba* Indians living in the region; South Carolina, USA, many quarries in York, Union, etc., districts). Weakly schistose metamorphic rock composed of talc and magnetite. *(Lieber 1860; Zirkel 1866a, p. 351)*

CAUSTIC METAMORPHISM (Greek *kaustikos*, burning). Type of dry *contact metamorphism* restricted to the vicinity of basaltic lava flows or dykes, thus not found under plutonic conditions, and characterized by *baking*, *fritting* or melting processes, and by the formation of glassy or glass-bearing rocks. In part superseded by *pyrometamorphism*. Syn. *optalic metamorphism*. *(Leonhard 1832, p. 288; Naumann 1858, p. 744; Zirkel 1893, p. 584; Milch 1922, p. 288; Tyrrell 1926, p. 301)*

CENTRAL UPLIFT. Structurally uplifted central volume, which can be manifested as a central peak (commonly with an irregular circular shape in plan view), in *complex impact craters* of intermediate size, formed by the dynamic collapse of the *transient crater cavity*. *(OU; Dence 1964, p. 249; Grieve 1987, p. 248; Melosh 1989, p. 18; Dence 2004, p. 277, Fig. 10)*

CHARNOCKITE (from the tombstone of J. *Charnock*, the founder of Calcutta). Orthopyroxene-bearing rock of granitoid composition. A member of the ***charnockitic series*** of rocks. Definition adopted from Le Maitre (1989, 2002, section B.9), which see for further description. Cf. *enderbite*, *khondalite*.

CHEMICAL METAMORPHISM. Type of metamorphism in which chemical interchange and reaction is a significant feature. Equivalent to *metasomatism*. *(Van Hise 1904, p. 39)*

Chemographic(al) diagram (Arabic *alkimya*, itself from the Greek *chêmeia*, chemistry, and Greek *graphô*, to write). Isothermal-isobaric diagram representing the chemical composition of the minerals that constitute a metamorphic rock. Minerals in a binary system are depicted as points along a straight line, in a ternary system as points on or within a triangle. Equivalent to ***composition phase diagram***. *(OU; Bucher & Frey 1994, p. 28)*

CHEMOMETAMORPHISM. Metamorphism of igneous rocks resulting from the influence of high temperature and the presence of solvent solutions. Equivalent to *metasomatism*. *(OU; Tomkeieff)*

CHLORITOLITE (Greek *chloros*, green, and *lithos*, stone). *Monomineralic* chlorite rock (chloritite) occurring as veins and resulting from *hydrothermal metasomatism*. The rock may contain inclusions of the altered or host rock. It is associated with polymetallic ores. *(Bessmertnaya & Gorzhevsky 1958, p. 21)*

CHLOROGRISONITE SCHISTS (Greek *chloros*, green, and *Grisons*, Rhaeto-Romanic name of the Graubünden region in eastern Switzerland). Obsolete term for a group of *greenschists* within the Bündnerschiefer complex, characterized by variable contents of plagioclase, actinolite, epidote, chlorite and magnetite, and subdivided into *valrheinite*, *gadriolite*, *cucalite*, *paradiorite* and *hypholite*. Equivalent to *prasinite*. *(Rolle 1879, p. 40)*

CHORISMITE (Greek *chôrisma*, partition). Composite rock consisting of at least two parts at the megascopic scale, that are petrographically distinct and may be of different or uncertain origin; for example, a *migmatite*. See *exochorismite*, *endochorismite*. *(Huber 1943, p. 89; Mehnert 1968, p. 198, 353)*

CHYMOGENETIC, CHYMOGENIC (Greek *chymos*, juice, and *genesis*, from *gennô*, to bear, give birth to). Related to the portion of a composite rock such as a *migmatite*, that crystallized from a fluid phase (magma, *hydrothermal* solution or vapour). Ant. *stereogenetic*. *(Huber 1943, p. 90; Dietrich & Mehnert 1960, p. 60; Mehnert 1968, p. 353)*

Cipolin (Italian *cipollino*, itself from *cipolla*, onion). Metacarbonate rock rich in chlorite and other phyllosilicates, and displaying a saccharoidal structure. The term was also used in France for any *crystalline limestone* but this use is now obsolete. Terms such as 'carbonate-chlorite schist' should be used in preference if chlorite and other phyllosilicates constitute more than 50% vol. of the rock, and the term '*impure marble*' should be used if chlorite and other phyllosilicates constitute less than 50% vol. *(Brongniart 1813, p. 38; Cordier 1868, p. 287; Jung 1963, p. 176; Foucault & Raoult 1984)*

CIPOLLINO. Commercial term used in the Italian *marble* industry for a variety of *impure marble* containing micaceous layers. Micaceous *impure marble* should be used in preference. *(Vasari 1568, p. 47; Ferber 1776, p. 328; Cotta 1855, p. 175; Zirkel 1894b, p. 52)*

CLASTOAMPHIBOLE-SLATE (Greek *klastos*, broken). Obsolete term for an actinolite schist with a *cataclastic structure*. *(Koto 1889, in Tomkeieff)*

CLAY SLATE. See *slate*.

CLEAVAGE ⇒ Section 2.3 (old English *cleofan* or Dutch *klieven*, to split; first used for diamonds). Property of a rock to split along a regular set of parallel or sub-parallel closely spaced surfaces. A cleavage is said to be ***domainal*** when it is not continuous or completely penetrative in a rock, that is a *spaced cleavage*. The cleavage domain, or ***M-domain*** (after mica-rich), is that area occupied by the cleavage plane as opposed to the area in between, which is termed a *microlithon* or ***Q-domain*** (after quartz-rich). *(OU; Sedgewick 1835, p. 469; Harker 1885, p. 836; Leith 1905, p. 11; Wilson 1961, p. 457; Powell 1979, p. 21; Barker 1998, p. 231)*

COAL-FIRE ASH ⇒ Section 2.10. Soft, clay-like ash deposit remaining after the combustion of a coal seam. It is included in the *burned rocks*. *(Delesse 1857, p. 105; Naumann 1858, p. 736; McLintock 1932, p. 211; Suk 1983, p. 20)*

COCKADE GNEISS (French *cocarde*, a cocade or rosette borne on military head-gear, itself from *crête de coq*, cockscomb; Los Pozos, Sierra de Famatina, Provincia la Rioja, Argentina). Obsolete term for an *augen gneiss* containing a fine-grained hornblende groundmass with numerous large crystals of quartz and plagioclase surrounded by rims of hornblende and biotite. *(Stelzner 1885, p. 22)*

COFACIAL. Synonymous with *isofacial*.

COKEITE (English *coke*, the volatile-poor, distilled coal). Natural coke of metamorphic origin formed either by *contact metamorphism* of coal seams, or by subterranean fires in coal mines. *(Coquand 1857, p. 342; Delesse 1857, p. 106; Zirkel 1893, p. 601; Lacroix 1910, p. 648)*

COLLECTIVE CRYSTALLIZATION (German *Sammelkrystallisation*). Type of *recrystallization* in a metamorphic rock, by which smaller grains become unstable and larger grains grow at the expense of the smaller ones. *(Rinne 1908, p. 167, 300; Rinne & Boeke 1908, p. 393)*

COLLOBRIÉRITE (*Collobrières*, near Toulon, France). Layered metamorphic rock, probably derived from an oolithic iron ore and composed of fayalite, garnet, grunerite, ilmenite, magnetite and apatite. *(Lacroix 1917, p. 67; Rosenbusch 1923, p. 779; Gueirard 1957, p. 2339)*

COLMITE. See *kolmite*.

COLONNES FILTRANTES (French *filterable columns*). Obsolete French term for vertical flows or emanations in the Earth's crust, considered as an agent of *regional metamorphism*. *(Termier 1903; Termier 1912, p. 592)*

COMBUSTION METAMORPHISM ⇒ Sections 2.2 & 2.10, Fig. 2.2.1. Type of *metamorphism* of *local* extent produced by the spontaneous combustion of naturally occurring substances such as bituminous rocks, coal or oil. The very high temperatures reached during the combustion may either fuse the rocks or convert them into holocrystalline aggregates of high-temperature metamorphic minerals (see *burned rock*, *fritted rock*, *buchite*). The residual products of coal combustion may give rise to local ash deposits (*coal-fire ash*). The term combustion metamorphism subsumes a variety of old names. The SCMR recognized it as the main term for this process although it noted that *burning metamorphism* might prove an acceptable alternative. *(Breislak 1822, p. 4; Delesse 1857, p. 104; Naumann 1858, p. 736; Zirkel 1893, p. 602; Arnold & Anderson 1907, p. 750; McLintock 1932, p. 208; Bentor* et al. *1963, p. 478; Kolodny & Gross 1974, p. 489; Bentor & Kostner 1976, p. 486)*

COMPLEX IMPACT CRATER. *Impact crater* with relatively low depth/diameter ratio and with *central uplift*, annular trough, and down-*faulted*, terraced rim structure. *Central uplift* can be expressed topographically as a peak or a peak ring. In very large craters the *central uplift* is replaced by two or more concentric topographic rings (*multi-ring crater*). A complex crater forms by the collapse of a deep, bowl-shaped *transient crater cavity*. Ant. *simple impact crater*. *(OU; Dence 1964, p. 259; Dence 1968, p. 170, 180; Grieve 1987, p. 248; Melosh 1989, p. 18; Dence 2004, p. 276)*

COMPOSITE GNEISS. Type of gneiss produced by the interlayering or mixture of two distinct rock types, one generally having an igneous aspect. Cf. *lit-par-lit gneiss*, *migmatite*. *(Cole 1902, p. 204; Cole 1915, p. 183)*

Composition phase diagram. See *chemographical diagram*.

Comrie-type contact metamorphism (*Comrie* area, Scotland, UK). Type of *isochemical contact metamorphism* mostly associated with mafic intrusive bodies (thick dykes and sills, igneous stocks) and, subordinately, with granitoids or alkaline intrusive rocks, in which the *mineral assemblages* were ascribed to the *pyroxene-hornfels facies*, *hornblende-hornfels facies* and *albite-epidote-hornfels facies*. The *aureole* thickness ranges from hundreds of metres to 1–2 km. See *contact aureole systematics*. *(Reverdatto* et al. *1970, p. 312)*

CONSTRUCTIVE METAMORPHISM. Type of metamorphism, which may be *thermal* or *hydrothermal*, leading to the formation of *authigenic* minerals, as opposed to *dynamic* (*dislocation*) *metamorphism* which is purely mechanical producing new structures but not producing *authigenic* minerals. *(Cunningham-Craig 1904, p. 11)*

CONTACT ⇒ Section 2.10. Widely used qualifier for processes, geologic units, rocks or minerals related to *contact metamorphism* (e.g. *contact metasomatism*, *contact aureole*, *contact zone*, *contact ore deposits*, *contact rock*, contact mineral). The SCMR recommends the use of contact as a qualifier to a rock name to indicate that the rock was generated by *contact metamorphism* (e.g. *contact marble*) and of the qualifier *contact-metamorphosed* to the *protolith* of a rock subjected to *contact metamorphism* (e.g. *contact-metamorphosed* limestone). *(Bunsen 1851, p. 239; Lossen 1869, p. 287; Salomon 1890, p. 486, 495; 1891, p. 482; 1898, p. 143)*

CONTACT AUREOLE (French *aureole*, itself from the Latin *aureus*, golden). Zone around a magma body, plutonic, near-surface or volcanic, in which *contact metamorphism* due to that body can be recognized. The thickness ranges from centimetres to kilometres, depending on the dimensions and composition of the magma body, the physico-chemical properties of the surrounding rocks and the depth of the magma emplacement. The metamorphic effects are more pronounced close to the igneous mass, i.e. in the ***inner aureole***, and tend to diminish away to the ***outer aureole***. Many contact aureoles can be subdivided into *metamorphic zones* (see *contact zone*); anomalous zonal patterns may reveal complications due to *metasomatizing* fluids. See *contact aureole systematics. (Buch, in Humboldt 1831; Zirkel 1866a,b, p. 518; Lossen 1869, p. 281, 294, 327; Rosenbusch 1877, p. 56; Rosenbusch 1887, p. 39; Salomon 1890, p. 491, 494; Beck 1892)*

CONTACT AUREOLE SYSTEMATICS. Classification of *contact aureoles* based on the type and succession of the *contact-metamorphic zones* in relation to the size, composition and emplacement level of the igneous body, and the geotectonic setting. Examples in the case of *isochemical contact metamorphism*: *Anakit-*, *Bosost-*, *Comrie-*, *Main Donegal-*, *Tanohata-* and *Vogtland-type contact metamorphism. (Turner 1968, p. 16, 193; Reverdatto* et al. *1970, p. 330; Barton* et al. *1991b, p. 725, 730; Pattison & Tracy 1991, p. 107)*

CONTACT DYNAMO-THERMAL AUREOLE. Unnecessary term for the zone affected by what is now called *hot-slab metamorphism. (Williams & Smyth 1973, p. 615)*

CONTACT MARBLE ⇒ Sections 2.7 & 2.10. *Marble* resulting from *contact metamorphism*.

CONTACT METAMORPHISM ⇒ Sections 2.2 & 2.10, Fig. 2.2.1. Type of *metamorphism* of *local* extent that affects the country rocks around magma bodies emplaced in a variety of environments from volcanic to upper mantle depths, in both continental and oceanic settings. It is essentially caused by the heat transfer from the intruded magma body into the country rocks. The range of metamorphic temperatures may be very wide. It may or may not be accompanied by significant deformation depending on the dynamics of the intrusion. The zone where contact metamorphism occurs is called the *contact aureole*, whereas the products of such metamorphism are called the *contact rocks*. See *contact metasomatism*; cf. *endomorphism*, *exomorphism. (Delesse 1857, p. 90; Omalius d'Halloy 1868, p. 469; Bonney 1886, p. 61; Barrow 1893, p. 336; Kerrick 1991, p. 1)*

CONTACT METASOMATISM. Type of *metasomatism* that occurs at or near the contact between a magmatic body and another rock. It may occur at various stages in the magmatic evolution. ***Endocontact zones*** develop by replacement of the magmatic rocks and ***exocontact zones*** are formed by replacement of the host rocks. Contact metasomatism was originally divided into *pneumatolytic* and *hydrothermal* types. The former represented a hotter gas-dominated stage, the latter a cooler hydrous-dominated stage. In modern usage pneumatolytic has been largely subsumed by hydrothermal, the range and associations of which have greatly increased. See *pneumatolytic* and *hydrothermal metasomatism. (Durocher 1846, p. 606; Barrell 1907, p. 117; Lindgren 1933, p. 115; Eskola 1939, p. 372; Turner 1948, p. 5)*

CONTACTOLITE (Greek *lithos*, stone). Obsolete term synonymous with *contact rock. (Loewinson-Lessing 1925, p. 444)*

CONTACT ROCK. General name for all rocks produced by *contact metamorphism*. See *contact*.

CONTACT ZONE (Barr-Andlau granite aureole, Vosges, France). One of the spatial subdivisions of a *contact aureole* based on mineralogical, structural or *facies* criteria. In the classical example there are three zones, from the *outer* to the *inner aureole*, namely *spotted slate*, *knotted schist* and *hornfels*. The term has sometimes been used incorrectly as a synonym of *contact aureole*. See *endocontact zone*, *exocontact zone. (Zirkel 1866a, p. 518; Rosenbusch 1877, p. 79; Rosenbusch 1887, p. 38; Turner 1968, p. 6, 197)*

CONTINUOUS CLEAVAGE. Type of *cleavage* characterized by the preferred orientation of all the *inequant* mineral constituents of a rock, and in which the *cleavage* planes are developed at the grain-size scale (cf. *spaced cleavage*). Originally introduced as a non-genetic alternative to *flow cleavage*. Equivalent to *slaty cleavage* and well-developed *schistosity. (Chidester 1962, p. 22; Powell 1979, p. 33; Davis & Reynolds 1996, p. 429; Barker 1998, p. 45)*

CONTINUOUS EJECTA BLANKET. See *ejecta blanket*.

CONTINUOUS REACTION. Metamorphic reaction that continuously achieves equilibrium over a range of temperature at constant pressure (or vice versa) owing to compositional variability among the minerals, for example, in the Fe/Mg value. A *paragenetic* diagram will show a continuous variation in the line orientations but no topological change. Achievement of equilibrium is normally facilitated by the presence of an intergranular fluid with a constant composition. *(Bowen 1928, p. 54; Thompson 1957, p. 857; Miyashiro 1994, p. 347)*

CORE-AND-MANTLE STRUCTURE. Type of structure where a deformed crystalline core, usually a single crystal, is surrounded by a mantle of finer grains of the same mineral. This is thought to occur by the preferential dynamic *recrystallization* of the outer edge of a large single crystal and is most common in feldspar in quartzofeldspathic rocks. Supersedes *mortar structure*, a term that erroneously implies *cataclasis*. *(OU; White 1976, p. 73; Passchier & Trouw 1996, p. 37; Vernon 2004, p. 478)*

CORNEAN (French *cornéenne*; Trap Formation of Pembrokeshire). Obsolete term for a fine-grained compact feldspar-quartz-hornblende rock similar to the French *cornéenne*. *(De la Beche 1826, p. 3)*

CORNÉENNE (French *corne*, horn). Term used in French initially for a heterogeneous group of compact rocks of horny aspect, and subsequently equivalent to *hornfels*. *(Delesse 1857, p. 757; Zirkel 1866a,b, p. 517; Dolomieu, in Cordier 1868, p. 101, 465; Lapparent 1893, p. 1403; Foucault & Raoult 1984, p. 75)*

CORNEITE (French *corne*, horn). Biotite *hornfels*. *(Gosselet 1888, p. 767; Harker 1889, p. 17)*

CORNUBIANITE (Latin *Cornubia*, Cornwall; tin and copper deposits). Term first used for a *banded*, fine-grained, compact to schistose rock essentially composed of quartz, feldspar and micas, generated by *contact metamorphism* of a gneissose to schistose *protolith*. Later variably used for (a) schistose andalusite *hornfels* (cf. *proteolite*, *leptynolite*), and (b) tourmaline-quartz-mica *hornfels*. In Italy, largely used as equivalent to *hornfels* s.l. *(Boase 1832, p. 390, 394; Lapparent 1882, p. 618; Bonney 1886, p. 104; Rosenbusch 1887, p. 42; Artini 1952, p. 312)*

CORONA (Latin *corona*, crown). Zone or zones of minerals arranged concentrically around a core mineral. The minerals usually exhibit a radial arrangement. The zone or zones may be primary or result from *secondary* reactions. The term was originally only used in the description of igneous rocks, but is now also applied to metamorphic rocks. The term includes *reaction rim* and *kelyphite*.[3] Hence ***corona structure***. *(Törnebohm 1877, p. 37; Lacroix 1889, p. 230; Sederholm 1916, p. 9–41; Spry 1969, p. 104; Vernon 2004, p. 478)*

CORONITE. Term for any rock containing *coronas*. *(Sederholm 1916, p. 34; Buddington 1939, p. 294; Shand 1945, p. 251)*

CORUNDOLITE. *Contact-metamorphic* rock predominantly composed of corundum. Rec.syn. *emery rock*. *(Loewinson-Lessing 1925)*

COTICULE (Latin *cos*, *cotis*, any hard stone; Stavelot massif, Belgium). Low-grade, very fine-grained metamorphic rock consisting of spessartine garnet in a matrix of white mica ± submicroscopic quartz. Sometimes used, incorrectly, for all spessartine quartzites. Cf. *novaculite*, *gondite*. *(Plinius 80 Lib. 36–47, p. 526; Omalius d'Halloy 1828, p. 110; Renard 1878, p. 12; Kramm 1976, p. 135; Kennan 1986, p. 141)*

CRENULATION (French *crénulation* and *créneau*, themselves from the Latin *crena*, slot). Type of regular folding with a wavelength of 1 cm or less. *(OU; Passchier & Trouw 1996, p. 256; Foucault & Raoult 2005, p. 89)*

CRENULATION CLEAVAGE, CRENULATION SCHISTOSITY. Type of *spaced cleavage* developed during *crenulation* of a pre-existing *foliation*, and orientated parallel to the axial plane of the *crenulation*. Proposed as a non-genetic synonym for *strain-slip cleavage* and slip cleavage. *(Knill 1960, p. 318; Rickard 1961, p. 325; Ramsay & Huber 1987, p. 442)*

Critical mineral assemblage. *Mineral assemblage* that is used to characterize a zone of a specific range of *metamorphic grade* in a *progressive metamorphic* region. The first appearance of the *mineral assemblage* (in passing from low to higher grades of metamorphism) marks the

[3] Rim structures → *corona* → *reaction rim* → *kelyphite*
Random structures →
(1) intimate intergrowth → *symplectite*
(2) crystal growth: recognizable pre-existing feature → *mimetic*
recognizable shape of previous mineral → *pseudomorph*

outer limit of the zone in question. Cf. *index mineral*. *(OU; Miyashiro, 1994, p. 348)*

CROCYDITE (Greek *krokys* gen. *krokydos*, weft, web). *Migmatite* with small (centimetre-scale) irregular or star-shaped *metatects* that are strongly interlaced with the *palaeosome*. There are gradations to larger (decimetre-scale) vein-like *metatects*. *(De Waard 1950, p. 55; Dietrich & Mehnert 1960, p. 60)*

CRUSH BRECCIA. Type of breccia formed by deformation that results in the fragmentation of the rock in place or nearly in place. Equivalent to *fault breccia*. *(Bonney 1883, p. 435; Norton 1917, p. 188; Reynolds 1928, p. 104)*

CRYPTIC METAMORPHISM (Greek *kryptos*, hidden). Metamorphism that can be detected only by special study, for example vitrinite reflectance, illite Kübler index, etc., and not by ordinary hand-specimen or microscopic study. The term is used mainly in Canada for very low-grade rocks. *(OU)*

CRYPTOEXPLOSION STRUCTURE. Term previously used for (*impact*) *craters* where an origin by *impact* or volcanic explosion (cryptovolcanic structure) was not clear. *(Bucher 1936, p. 1055; Bucher 1963, p. 597; Dietz 1959, p. 496; Dietz 1963, p. 650; French & Short 1968, p. 11)*

Crystalline limestone. Metamorphosed limestone; *marble* formed by *recrystallization* of limestone as a result of metamorphism. The term *pure marble* should be used in preference if the non-carbonate content is lower than 5% vol. and the term *impure marble* should be used in preference if the non-carbonate content is higher than 5% vol. *(Daubrée 1867; Cordier 1868, p. 285; Lemeyrie 1878, p. 260; Holmes 1928, p. 72; Pettijohn 1957, p. 407; Foucault & Raoult 2005, p. 54)*

CRYSTALLINE SCHIST. Term originally used to differentiate fissile rocks composed of metamorphic minerals from others such as *slates* which were thought to be largely composed of sedimentary grains. The distinction was important in other languages where the term schist was used to cover all fissile rocks and the distinction was made between those of metamorphic origin and those of sedimentary origin, for example 'kristalline Schiefer' and 'Thonschiefer' respectively in German, and 'schistes cristallins' and 'schistes argileux' respectively in French. *(Boué 1825, in Jung & Roques 1952; Geikie 1903, p. 785; Grubenmann 1904, p. 1; Harker 1939, p. 190; Jung 1963, p. 177; Aubouin* et al. *1968, p. 398)*

CRYSTALLOBLAST (Greek *krystallos*, crystal, and *blastos*, bud, sprout). Mineral grain or crystal that has grown under metamorphic processes as opposed to a crystallization from a melt (see *crystalloblastesis*). Hence ***crystalloblastic structure***. *(Becke 1902, p. 357; Harker 1950, p. 33; Joplin 1968, p. 28)*

CRYSTALLOBLASTESIS. Process of mineral crystallization and development in the solid state in metamorphic rocks as opposed to crystallization from an igneous melt. It includes both *neomineralization* and *recrystallization* by which terms it is largely superseded. *(Becke 1902, p. 357; Becke 1903b, p. 35)*

CRYSTALLOPHYLLIAN ROCK (Greek *krystallos*, crystal, and *phyllon*, leaf). Obsolete term for a schistose metamorphic rock resulting from *regional metamorphism*. *(Omalius d'Halloy 1806, in Jung & Roques 1936; Lapparent 1882, p. 640; Termier 1912, p. 587; Foucault & Raoult 2005, p. 93)*

C/S fabric. See *S-C fabric*.

CUCALITE (*Cucal* Nain, Plattner Pass, canton Graubünden, eastern Switzerland). Obsolete term for a local variety of *greenschist* from the so-called *chlorogrisonite* schists group (equivalent to *prasinites*), composed mainly of plagioclase and epidote, with subordinate actinolite and accessory chlorite, magnetite and haematite. *(Rolle 1879, p. 37)*

DACTYLITIC STRUCTURE (Greek *daktylos*, finger, and *lithos*, rock). Type of structure characterized by the *symplectite* intergrowth of two minerals, the one with finger-like projections into the other. *(Sederholm 1916, p. 42; Spry 1969, p. 103)*

DALMATIANITE (local miners' name after the spotted coat of a *Dalmatian* dog; Amulet mine, Noranda, western Quebec, Canada). Extensively metasomatized rock associated with mineralization, characterized by a striking *spotted* appearance and resulting from the *alteration* of Precambrian dacitic lavas. The spots may be up to 2 cm in diameter and may relate to altered cordierite. *(Cooke 1927, p. 41; Walker 1930, p. 10; Tilley 1935, p. 200)*

DAMOURITIZATION (dedicated to the French mineralogist Alexis *Damour*, 1808–1902). Process of *alteration* of feldspars and less frequently of other aluminous silicates (kyanite, topaz) into damourite, a variety of muscovite. *(Delesse 1845, p. 254; Lacroix 1897, p. 41; Deer* et al. *1962, p. 140, 147)*

DAVITE (in honour of H. *Davy*, British chemist, 1778–1829; Vulpino, Italy). Type of alabaster *marble* (oriental alabaster, onyx *marble*): fine-grained stripped semitransparent calcite or aragonite rock usually precipitated from cold water solutions and valuable as decorative or architectural material. *(Pinkerton 1811b, p. 57)*

DECORATED PLANAR DEFORMATION FEATURES. Annealed *planar deformation features* consisting of discontinuously aligned vugs and *inclusions* formed during *recrystallization* of the originally amorphous lamellae. Typically present in shocked tectosilicates in the floor of *impact craters* and in high-temperature *impact breccias* such as *suevites* and *impact melt rocks*, also known from shocked olivine in thermally annealed chondrites where the decorations consist of ultra-fine-grained troilite and metal droplets. *(Robertson* et al. *1968, p. 439; Engelhardt & Bertsch 1969, p. 206, Fig. 1; Stöffler* et al. *1991, p. 3875, Fig. 18)*

DECUSSATE STRUCTURE (Latin *decussare*, to cross in the form of the number X). Type of structure in a metamorphic rock, in which the constituent grains have a randomly oriented and interlocking arrangement. The structure is typical of contact-*metamorphosed* rocks. *(Harker 1939, p. 35; Joplin 1968, p. 29; Vernon 2004, p. 479)*

DEDOLOMITIZATION. *Metasomatic* process leading to the calcite-enrichment of dolomite or dolomitic limestone, the magnesium being used to form non-carbonate magnesium minerals. Hence ***dedolomite***, the product of dedolomitization. *(Arduino 1779, in Zirkel 1866a, p. 244; Morlot 1847, p. 313; Teall 1903, p. 514; Shand 1931, p. 204; Harker 1932, p. 84; Tatarsky 1949, p. 850; Pettijohn 1975, p. 371)*

DEFORMATION LAMELLAE, DEFORMATION BANDS. Planar defects in a crystal comprising bands of material with a slightly different refractive index (relief) from the host grain and resulting from damage to the crystal lattice or an array of submicroscopic *inclusions*. Includes kink bands. In impact petrology it was defined as planar *microstructures* oriented parallel to the basal plane (0001) in shocked quartz, but it is now regarded as an unnecessary term. *(OU; Carter 1965, p. 786–91; Passchier & Trouw 1996, p. 257; Vernon 2004, p. 479)*

DEFORMATION METAMORPHISM. Type of metamorphism in which the *recrystallization* of the rock is caused by deformational strain. Rec.syn. *dislocation metamorphism*. *(Sander 1911, p. 284)*

DEGRADATION RECRYSTALLIZATION (grain diminution, degenerative recrystallization). Type of *recrystallization* that results in a relative decrease in size of the crystals. Ant. *aggradation recrystallization*. *(OU; Bates & Jackson)*

Degranitization. Process by which a rock is depleted in chemical components that are significant in making up a granitoid, essentially silica, potash ± soda. *(Ramberg 1951, p. 27; Noë-Nygaard 1955, p. 66)*

DEHYDRATION (REACTION). In metamorphic petrology, process or reaction by which water is subtracted from minerals or mineral compounds. *(Van Hise 1904, p. 204)*

DEPOSITIONAL TEMPERATURE. Equilibration temperature of an *allochthonous impact formation* deposited by ballistic or ground surge transport within or around an *impact crater*; the temperature is achieved by heat exchange between *breccia* constituents which are at different temperatures, e.g. hot melt particles and cold lithic clasts. *(OU; Stöffler* et al. *1991, p. 3848, Fig. 3)*

Depth zones. Zones in the Earth's crust characterized by different hydrostatic pressures, stresses and temperatures, which give rise to different metamorphic conditions and produce typical *metamorphic mineral associations* and structures. The nomenclature of the zones developed gradually. Becke used the terms upper and lower zone, Van Hise zone of *katamorphism* (upper) and zone of *anamorphism* (lower), Grubenmann the uppermost, middle and deepest zones of metamorphism. Grubenmann introduced the prefixes *epi-*, *meso-*, and *kata-* for use with the names of rocks (e.g. 'Kataplagioclasegneiss') and Grubenmann and Niggli introduced the well-used terms *epizone*, *mesozone* and *katazone*. Modern usage however stresses temperature–pressure conditions rather than the depth of the zones. *(Becke 1903a, p. 562; Becke 1903b, p. 52; Grubenmann 1904, p. 57; Van Hise 1904, p. 162; Grubenmann 1907, p. 21, 172; Grubenmann & Niggli 1924, p. 374)*

DESMOSITE (Greek *desmos*, bond; Harz Mts, Germany). *Banded* soda-rich indurated schist or *slate*, with alternating lighter (albite) and darker (white mica + chlorite) layers, characterized by an indistinct *spotted* structure on the cleavage planes. Cf. *spilosite*, *adinole series*; see also *diabase contact rock*. *(Zincken 1841, p. 394; Lossen 1869, p. 286, 291; Lossen 1872, p. 728)*

DEUTERIC (Greek *deuteros*, second). Obsolete term referring to reactions between primary magmatic minerals and the volatiles that arise from the same magma at a late stage in its cooling history. *(Sederholm 1916, p. 142)*

DIABASE CONTACT ROCK (Harz Mts., Germany). *Contact-metasomatic* rock, either massive or *banded* or schistose, derived from *slate* or low-grade schist and found in metre-thick *aureoles* of diabase sheets. The particular rock association found in this type of *aureole* was contrasted with that, thought to be essentially isochemical, found around granitic plutons. Hence ***diabase contact metamorphism***, the *metasomatism* generating these rocks, and ***diabase contact aureole***. Cf. *adinole*, *spilosite*, *desmosite*, *hornschist*. *(Zincken 1845, p. 584; Lossen 1869, p. 286; Kayser 1870, p. 106, 152; Lossen 1872, p. 728; Cohen 1887, p. 251; Rosenbusch 1896, p. 1171; Milch 1917, p. 353)*

DIABLASTIC STRUCTURE (Greek *dia*, through, and *blastos*, bud, sprout). Type of structure in a metamorphic rock in which two or more minerals are intricately intergrown. The intergrowth may take a variety of forms (interfingering, *inclusions*, etc.). *(Becke 1903b, p. 46; Grubenmann 1904, p. 80; Harker 1939, p. 40)*

DIABROCHOMORPHISM (Greek *diabrechô*, to soak, and *morphê*, form). Type of *recrystallization* of a rock, with or without change in composition, under the action of solutions soaking through the rock, whatever the source of the solutions may be. Hence ***diabrochite***, the product of diabrochomorphism. *(Dunn 1942, p. 234; Dietrich & Mehnert 1960, p. 60)*

DIACHYTE (Greek *diachytos*, diffused, spread out). Rock product of marked mechanical and/or chemical contamination of *anatectic* magma by cognate basic material. Cf. *palingenite*, *lipotectite*. *(Smulikowski 1947, p. 268; Dietrich & Mehnert 1960, p. 60)*

DIADYSITE (migmatite traversée) (Greek *diadidô*, to distribute, to permeate). Type of *migmatite* with discordant or concordant granitic veins. *(Jung & Roques 1936, p. 26; Mehnert 1968, p. 354)*

DIAGENESIS ⇒ Section 2.5 (Greek *dia*, through, and *genesis*, from *gennô*, to bear, give birth to). All the chemical, mineralogical, physical and biological changes undergone by a sediment after its initial deposition, and during and after its lithification, exclusive of superficial *alteration* (weathering) and metamorphism. Diagenesis (s.l.) may be subdivided into: (i) ***shallow diagenesis*** (***diagenesis sensu stricto***) which takes place under physical conditions that do not differ significantly from those under which the sediment originated; it is characterized by the absence of *alteration* of detrital minerals; (ii) ***deep diagenesis*** which is characterized by clay mineral reactions such as the transformation of smectite to illite, kaolinite to dickite, etc., and the increase of the proportion of illite layers in interstratified clay minerals. The processes involved in diagenesis are: compaction, cementation, reworking, *authigenesis*, replacement, crystallization, leaching, hydration, *dehydration*, bacterial action, and formation of concretions. These processes occur under conditions of pressure and temperature that are normal at the Earth's surface and in the outer part of the Earth's crust. *(Gümbel 1868, p. 838; Walther 1894, p. 693; Strakhov 1958, p. 761; Chiligar* et al. *1967, p. 313; Müller 1967, p. 128; Logvinenko 1968, p. 92)*

DIAGENETIC ZONE, DIAGENETIC ILLITE KÜBLER INDEX ZONE ⇒ Section 2.5, Fig. 2.5.1. Zone of sub-metamorphic rocks characterized by illite Kübler index (KI) values greater than $0.42°\Delta 2\theta \mathrm{CuK}_{\alpha}$. *(Kübler 1967, p. 111; Kübler 1968, p. 393; Kübler 1984, p. 578; Kisch* et al. *2004, p. 323)*

DIAGENISM. Obsolete synonym of *diagenesis*. *(Grabau 1904, p. 235; Grabau 1920, p. 642)*

DIAPEPSIS (Greek *dia*, through, and *pepsis*, digestion). Complete digestion of rocks in a magma. *(Gürich 1905, p. 250)*

Diaphthoresis [diaphtoresis] (Greek *diaphtheirô*, to corrupt, destroy; southern part of the Tweng massif, Austria). Synonymous with *retrograde metamorphism*. *(OU; Harker 1939, p. 344; Foucault & Raoult 2005, p. 105)*

Diaphthorite. Metamorphic rock that is the product of *retrograde metamorphism* or *diaphthoresis*. *(Becke 1909, p. 373)*

DIAPLECTIC GLASS (Greek *dia*, through, with, and *plekô*, to weave, knit). Amorphous form of crystals (***solid state glass***) resulting from *shock wave* compression and subsequent pressure release of single crystals or polycrystalline rocks; most commonly observed in tectosilicates. *(Engelhardt* et al. *1967, p. 93; Engelhardt & Stöffler 1968, p. 163; Dence & Robertson 1989, p. 527)*

Diastathermal metamorphism (Greek *diastasis*, extension, and *thermos*, hot). Type of *burial metamorphism* found in extensional basins where enhanced heat flow leads to very low-grade metamorphism of the basin sediments. *(Robinson 1987, p. 869; Robinson & Bevins 1989, p. 81; Frey & Robinson 1999, p. 76)*

DIATECTIC STRUCTURE. Structure of rocks resulting from partial melting and characterized by the presence of corrosion borders around the minerals. *(Loewinson-Lessing 1905, p. 43)*

DIATEXIS (Greek *dia*, through, and *têkô*, to melt). Advanced stage of *anatexis* where the dark-coloured minerals are also involved in melting; the melt formed has not been removed from its place of origin. Cf. *metatexis*. *(Gürich 1905, p. 241; Fiedler 1936, p. 493; Scheumann 1936a, p. 300; Mehnert 1968, p. 253, 354)*

DIATEXITE, DIATECTITE. Type of *migmatite* where the darker and the lighter parts form *schlieren* and *nebulitic* structures which merge into one another. Cf. *diatexis*. *(Fiedler 1936, p. 493; Mehnert 1968, p. 354)*

Dictyonite (Greek *diktyon*, net). Type of *migmatite* with a *reticulated structure* formed by a network of small veins. Hence ***dictyonitic structure***. *(Sederholm 1926, p. 49; Quirke & Lacy 1941, p. 600; Mehnert 1968, p. 354; Shelley 1993, p. 109; Kornprobst 2002, p. 108)*

DIFFERENTIATION. See *metamorphic differentiation*.

DIFFUSIONAL METASOMATISM ⇒ Section 2.9. Type of *metasomatism* that takes place by the diffusion of a solute through a stagnant solution (fluid). The driving force of diffusion is the chemical potential (or chemical activity) gradients in the rock-pore solution. Diffusional metasomatic rocks form rather thinly zoned bodies (rims) along cracks, veins, and contact surfaces, and the composition of minerals may vary gradually across each *metasomatic zone*. *(OU; Lindgren 1933, p. 176; Korzhinskii 1953, p. 337; Barton* et al. *1991a, p. 321)*

DISCONTINUOUS REACTION. Metamorphic reaction that, despite compositional variability among the minerals, is at equilibrium, given a fixed pressure at a specific temperature. A *paragenetic* diagram will show a change in topology, such as a 'tie-line flip'. The reactions are facilitated by the presence of an intergranular fluid with a constant composition. *(Bowen 1928, p. 54; Thompson 1957, p. 856; Miyashiro 1994, p. 347)*

DISEQUILIBRIUM ASSEMBLAGE. Association of minerals that are not in thermodynamic equilibrium. It is common in low-grade metamorphic rocks and in the high-temperature *sanidinite facies* rocks, where adjustment to the existing conditions could not be reached owing to the low rate of chemical reactions. The existence of a great number of mineral phases, the presence of *unstable relicts*, and the mutual replacement of minerals are all indicative of disequilibrium *assemblages*. *(Turner 1948, p. 17; Turner 1981, p. 58)*

DISJUNCTIVE CLEAVAGE. Type of *spaced cleavage* that is independent of any pre-existing mineral orientation in the rock. It includes *fracture cleavage* and *pressure solution cleavage*. This definition accords with the classification proposed by Powell. However, some authors (e.g. Davis & Reynolds) regard the term as equivalent to *spaced cleavage* as defined below. *(Powell 1979, p. 29; Davis & Reynolds 1996, p. 431; Passchier & Trouw 1996, p. 65)*

DISLOCATION METAMORPHISM ⇒ Section 2.2, Fig. 2.2.1. Type of *metamorphism* of *local* extent, associated with *fault zones* or shear zones. Grain-size reduction typically occurs in the rocks, and a range of rocks largely referred to as *mylonites* and *cataclasites* is formed. Replaces terms such as *mechanical metamorphism, kinetic metamorphism, cataclastic metamorphism, dynamic metamorphism, dynamometamorphism. (Lossen 1875, p. 970; Lossen 1884, p. 501; Milch 1894, p. 121; Daly 1917, p. 400; Joplin 1968, p. 36, 84; Suk 1983, p. 19)*

DISSOLUTION CREEP. See *pressure solution.*

DISTAL IMPACTITE ⇒ Section 2.11, Tables 2.11.1, 2.11.2. *Impactite* occurring as distal *ejecta* outside the outer limit of the continuous *ejecta blanket*. It includes *tektite, microtektite* and *impactoclastic* (global) *airfall bed.* Ant. *proximal impactite. (Stöffler & Grieve Section 2.11)*

DOELLO, DUELO (Galicia, Spain). Local name for a manganese-rich *listvenite*, composed of giobertite (magnesium, iron carbonate), talc, chlorite and magnetite. *(Macpherson 1881, in Tomkeieff)*

DOLÉRINE (Greek *doleros*, deceitful, fallacious; Mt Blanc massif, French-Italian Alps). Obsolete term given to a *mylonitic*, commonly *banded*, rock composed of white mica, chlorite and microscopic feldspar. *(Jurine 1806, p. 375)*

DOMAIN. Recognizable part of a metamorphic rock at the mesoscopic or microscopic scale that has a distinctive lithological, mineralogical or chemical composition. The term is commonly used for meta-igneous rocks and in particular for *mafic* metavolcanic *rocks*. Note: the term domain is also used in sedimentary and structural petrology (see *cleavage*). *(Smith 1968, p. 194)*

DURBACHITE (*Durbach* valley, Scharzwald, Germany). Ambiguous term defined by Le Maitre as a type of melanocratic syenite with large flakes of biotite, hornblende and orthoclase, but originally considered to have developed by *metasomatism* from biotite and hornblende gneiss. *(Sauer 1890, p. 247; Jung & Chenevoy 1951, p. 868; Morche 1979, p. 1; Le Maitre, 1989, 2002)*

DYKE (IMPACT) BRECCIA. *Impact breccia* occurring in the form of a dyke. See *breccia dyke. (OU; Dence 1971, p. 5555; Lambert 1981, p. 61, Table 1; Stöffler* et al. *1988, p. 287, Table 2)*

DYKE-MYLONITE. See *gangmylonite.*

DYNAMIC METAMORPHISM, DYNAMOMETAMORPHISM (Greek *dynamis*, power). Obsolete terms synonymous with *dislocation metamorphism.*

DYNAMO-STATIC METAMORPHISM. Originally defined as a type of *burial metamorphism* in rocks lying beneath overthrust masses. Approximately equivalent to *upside-down metamorphism*. See *hot-slab metamorphism. (Daly 1917, p. 400)*

DYNAMO-THERMAL METAMORPHISM, DYNAMOTHERMAL METAMORPHISM (Greek *dynamis*, power, and *thermos*, hot). Type of *regional metamorphism* produced under dynamic conditions and with elevated temperatures. Equivalent to *orogenic metamorphism. (Daly 1917, p. 400; Tyrrell 1926, p. 255, 303)*

EBENSINITE (in honour of the Arab philosopher and physician *Ibn Sina*, Avicenna, 980–1037). Term proposed for a particular metamorphic rock, probably a *contact-metamorphosed* shale. *(Pinkerton 1811b; Tomkeieff)*

ECLOGITE ⇒ Section 2.4 (Greek *eklegô*, to choose, because composed of minerals uncommon in 'primitive' rocks, which seem to have chosen to go together; Saualpe, Carinthia, Austria). Plagioclase-free metamorphic rock composed of ≥75% vol. of omphacite and garnet, both of which are present as major constituents, the amount of neither of them being higher than 75% vol. If either omphacite or garnet is present in an amount ≥75%, the rock name is either garnet omphacite or omphacite garnetite (see the suffix *-ite*). *(Haüy 1822, p. 325, 548; Leonhard 1823, p. 137; Smulikowski 1964a, p. 27; Coleman* et al. *1965, p. 483; Church 1968, p. 757; Carswell 1990b, p. 1)*

ECLOGITE FACIES ⇒ Sections 2.2, 2.4, Fig. 2.2.4, Table 2.2.1. *Metamorphic facies* corresponding to middle to high grade and middle to high P/T values, and bounded, in rocks of suitable composition, by the omphacite + garnet-in reactions. *(Eskola 1920, p. 155, 168; Eskola 1939, p. 363; Coleman* et al. *1965, p. 506; Smulikowski 1972, p. 125; Smulikowski 1989, p. 137)*

ECTEXIS, EKTEXIS (Greek *ek*, out of, fully, and *têkô*, to melt). Formation of a *migmatite* by *in situ* generation of a melt phase. Ant. *entexis. (Scheumann 1936a, p. 302; Mehnert 1968, p. 354)*

ECTINITE (Greek *ek*, fully, and *teinô*, to stretch out; first used in Auvergne, Central Massif, France).

Obsolete term used to designate those metamorphic rocks that show the effects of *isochemical metamorphism*, i.e. without associated *metasomatism* or *migmatization*. Cf. *crystalline schist*. *(Jung & Roques 1936, p. 40; Jung* et al. *1938, p. 121; Richard 1938; Roques 1941, p. 27; Dietrich & Mehnert 1960, p. 60)*

EDOLITE (*Edolo*, Val Camonica, Adamello massif, Italian Alps). Variety of *hornfels* essentially composed of mica and feldspars. *(Salomon 1898, p. 149)*

EDUCT (Latin *ex*, out, and *ducere*, to lead). Obsolete term for *protolith*. *(Scheumann 1936b, p. 402)*

EISOMORPHISM (Greek *isos*, equal, same, and *morphê*, form). Replacement of calcite by quartz accompanied by a volume reduction which causes the development of various cavities and fissures. *(Lacroix 1936, p. 113)*

EJECTA. See (*impact*) *ejecta*.

EJECTA BLANKET, CONTINUOUS EJECTA BLANKET. Continuous *ejecta* deposit around an *impact crater*. *(OU; Melosh 1989, p. 89)*

EJECTA PLUME. Synonymous with *vapour plume*.

ELEUTHEROMORPH (Greek *eleutheros*, free, and *morphê*, form). Newly formed crystal in a metamorphic rock that has developed freely without constraints on its shape. Synonymous with *idioblast*. *(Milch 1894, p. 107)*

EMBRECHITE (Greek *en*, into, and *brechô*, to wet, to moisten). Type of *migmatite* formed by impregnation of material from outside such that the original *foliation* of the host rock is preserved as in *permeation gneiss* and *injection gneiss*; *augen* and *ribbon structure*s are common. *(Jung & Roques 1936, p. 25; Mehnert 1968, p. 354)*

EMERY ROCK. Rare type of compact metamorphic rock essentially composed of corundum and minor amounts of other Fe-Ti oxides, spinel and aluminous silicates, which may be formed by several different metamorphic processes. Originally it was assumed to form from highly aluminous *protoliths* (e.g. bauxite or lateritic deposits), although more recently emery rock bodies have been interpreted as the residual products from melting of pelitic rocks in the *aureoles* of mafic intrusions. *(Holmes 1920, p. 89; Friedman 1956)*

ENCARSIOBLASTIC (Greek *enkarsios*, transversal, and *blastos*, bud, sprout). Type of structure in metamorphic rocks in which lenticular crystals lie within the *foliation* but their basal crystallographic cleavage planes are perpendicular to the *foliation*. *(Greenly 1919, p. 43)*

ENDERBITE (*Enderby* Land, Antarctica). Plagioclase-rich member of the *charnockite* series, consisting essentially of quartz, antiperthite, hypersthene and magnetite. *(Tilley 1936, p. 312; Le Maitre 1989, 2002)*

ENDOCHORISMITE (Greek *endô*, within, and *chôrisma*, partition). Part(s) of a *chorismite* generated *in situ*. Ant. *exochorismite*. *(Huber 1943, p. 89)*

ENDOCONTACT ZONE. See *contact metasomatism*.

ENDOGENIC METAMORPHISM. Obsolete term for *endomorphism*. Ant. *exogenic metamorphism*. *(Kalkowsky 1886, p. 33)*

ENDOMETAMORPHISM (Greek *endô*, within). Synonymous with *endomorphism*. Ant. *exometamorphism*.

ENDOMIGMATIZATION (Rogart, Scotland, UK). Process of *migmatite* development in which a mobile phase is generated *in situ*. Ant. *exomigmatization*. *(Bellière 1960, p. 30)*

ENDOMORPHISM (Greek *endô*, within, and *morphê*, form). Term denoting all modifications of an igneous rock due to *contact-metamorphic* reaction with the country rock; the term comprises all related processes at the margins of the igneous rock including complex metamorphic and *metasomatic* reactions and the partial or complete assimilation of fragments of the country rock. See *endometamorphism*. Ant. *exomorphism*. *(Fournet 1847, p. 243; Naumann 1858, p. 722)*

ENDOPNEUMATOLYTIC CONTACT METAMORPHISM. *Pneumatolytic contact metamorphism* that affects the intrusive rock. *(Goldschmidt 1911, p. 211)*

ENDOSKARN (Greek *endô*, within). *Skarn* produced by the *metasomatic* transformation of the magmatic (or other silicate) rock. Ant. *exoskarn*. *(OU; Kornprobst 2002, p. 99)*

ENDOSMOTIC METAMORPHISM (Greek *endô*, within, and *ôsmos*, impulsion). Obsolete term proposed to cover all the processes adding material to an igneous body during its cooling history, including assimilation of country rocks and the transfer of components from circulating solutions or gases. Ant. *exosmotic metamorphism*. *(Johnston-Lavis 1914, p. 382)*

ENTEXIS (Greek *en*, into, and *têkô*, to melt). Formation of a *migmatite* by the introduction of a melt phase from outside. Ant. *ectexis*. *(Scheumann 1936a, p. 302; Mehnert 1968, p. 354)*

EOMETAMORPHISM (Greek *êôs*, dawn). Metamorphism of the lowest possible grade, especially as affecting hydrocarbons which are very sensitive to changes in pressure and temperature. The terms *diagenesis*, *anchimetamorphism*, very low-grade metamorphism may be used instead. *(Landes 1967, p. 832)*

Epi- (Greek *epi*, upon). (1) Prefix used to indicate an origin in the uppermost zone of metamorphism (e.g. *epizone*). Cf. *meso-*, *kata-*. (2) Prefix to names of totally altered magmatic rocks, which are classified according to their present (*secondary*) mineralogical and chemical composition, because their original nature is unrecognizable (e.g. episyenite, a rock of syenite composition whose *protolith* is unknown). (3) Prefix used as a synonym of *meta-* (e.g. epigranite, a metagranite). Because of potential ambiguity the use of (1) has 'restricted' status; the use of (2) and (3) are not recommended. *(Bascom 1893, p. 828; Grubenmann 1907, p. 21; Lacroix 1920, p. 687)*

EPIBOLITE (Greek *epi*, upon, and *bolê*, throw). Type of *migmatite* with concordant veinlets, lenses or strings of granitoid material. *(Jung & Roques 1952, p. 7; Mehnert 1968, p. 354)*

EPIDIORITE. (1) Dyke rock composed of fibrous amphibole, plagioclase, chlorite, ilmenite and magnetite (Gümbel). (2) Diabase affected by low-grade metamorphism with relics of the original minerals and structure (Rosenbusch); a metadiabase. (3) Group name for metamorphosed intrusive diabases of different *metamorphic grade* (Wiseman). *(Gümbel 1874, p. 9; Rosenbusch 1898, p. 331; Rosenbusch 1901, p. 331; Wiseman 1934, p. 355)*

EPIDOSITE. Obsolete term for a rock rich in epidote; if *monomineralic* it is equivalent to epidotite. *(Stillwell 1918, p. 115; Harker 1939, p. 268; Flawn 1951, p. 769)*

EPIDOTE-AMPHIBOLITE FACIES ⇒ Section 2.2, Fig. 2.2.4, Table 2.2.1. *Metamorphic facies* formed at low to moderate temperatures and pressures, characterized by hornblende-albite-epidote ± chlorite *assemblages* in rocks of basaltic composition. It lies between the *greenschist* and the *amphibolite facies*. Owing to the presence of epidote + plagioclase $An_{<17}$ it is considered by some workers as a high-grade *subfacies* of the *greenschist facies*. *(Eskola 1920, p. 165; Eskola 1939, p. 345; Turner 1981, p. 360)*

EPIGENESIS (Greek *epi*, upon, and *genesis* from *gennô*, to bear, give birth to). Obsolete term used especially by Russian authors to indicate changes, transformations or processes occurring at low temperatures and pressures, that affect sedimentary rocks after their compaction, excluding superficial *alteration* (weathering) and metamorphism. The realm of epigenesis is subdivided into ***early*** and ***deep*** or ***late epigenesis***. See also *catagenesis*, *katagenesis*, *diagenesis*. *(Pustovalov 1940; Kisch 1983, p. 294)*

EPIGENETIC. Term that has been used with several meanings. The two recommended definitions for metamorphic rocks are: (1) Said of processes or their products that occur at shallow levels in the Earth's crust. Cf. *epizone*. (2) Said of ore deposits that are emplaced in pre-existing rocks. Cf. *syngenetic*. *(OU; Stelzner, in Beck 1909, p. 4; Lindgren 1933, p. 155; Tomkeieff)*

EPIMETAMORPHISM (Greek *epi*, upon). Type of low-grade metamorphism occurring in the upper parts of the Earth's crust and characterized by low temperature and pressure, and abundance of water. *(Grubenmann & Niggli 1924, p. 374)*

EPIROCKS, EPI-ROCKS. Term proposed for the rocks of the uppermost zone of metamorphism. The prefix was used only in those cases where rocks of the same name were encountered in different *depth zones* (e.g. *epi-amphibolite*). Cf. *meso-*, *katarocks*. *(Grubenmann 1907, p. 21, 174)*

EPITAXIS, EPITAXY, EPITAXIAL GROWTH (Greek *epi*, upon, and *taxis*, arrangement, ordering). *Recrystallization* in which the lattice orientation of the new crystal has a systematic relationship to the crystal lattice of the parent crystal. *(OU; Spry 1969, p. 164; Yardley 1989, p. 158; Vernon 2004, p. 480)*

EPITHERMAL (Greek *epi*, upon, and *thermos*, hot). Said of a *hydrothermal* ore deposit that was formed at shallow depths and at low temperatures. Cf. *mesothermal*, *hypothermal*. *(OU; Lindgren 1933, p. 210)*

EPIZONE ⇒ Section 2.5, Fig. 2.5.1 (Greek *epi*, upon, and *zônê*, zone). Zone of low-grade metamorphic rocks characterized by illite Kübler index (KI) mean values less than 0.25° $\Delta 2\theta CuK_{\alpha}$. The term was originally proposed by Becke and by Grubenmann to indicate a shallow depth of metamorphism (including *contact metamorphism* for Grubenmann); as such it was synonymous with ***epimetamorphic zone***. This latter usage is now regarded as unnecessary and at present epizone is mainly used in the context of illite Kübler index investigations. Cf. *epimetamorphism*. *(Grubenmann 1904, p. 57; Grubenmann & Niggli 1924, p. 374, 397; Lindgren 1933, p. 95; Kübler 1967, p. 111; Kübler 1968, p. 393; Kübler 1984, p. 578)*

EQUANT (French *équant*, itself from the Latin *aequans*, being equal to). Said of mineral grains that are equidimensional. Ant. *inequant*. *(OU; Harte 1977, p. 281; Foucault & Raoult 2005, p. 125)*

EQUIGRANULAR (Latin *aequus*, equal, and *granulum*, small grain). Said of a rock in which the mineral grains are all of one size. The term was originally applied to igneous rocks and is equivalent to *homeoblastic* in metamorphic rocks. *(Cross* et al. *1906, p. 698)*

EQUILIBRATION SHOCK PRESSURE. Transient equilibration pressure achieved in a polycrystalline or porous rock via *shock wave* reverberations at grain boundaries leading to transient local pressure and temperature variations from grain to grain before a uniform *P–T*-state is achieved within all grains. *(OU; Stöffler* et al. *1991, p. 3847, Fig. 2)*

ERLAN, ERLANFELS (foundry of *Erla* near Scharzenberg, Saxony, Germany). Metamorphic rock consisting of an intimate mixture of fine-grained pyroxene with plagioclase, some vesuvianite, titanite, zoisite and mica; originally considered to be a single mineral. *(Breithaupt & Gmelin 1823, p. 77; Rosenbusch 1901, p. 545; Rosenbusch 1923, p. 679, 726; Jung & Chenevoy 1951, p. 671)*

EUHEDRAL (Greek *eu*, well, and *edra*, side, face). Said of a crystal that is wholly bounded by its own crystal faces. Term originally used to describe igneous rocks but now more widely applied. The specific term for metamorphic rocks is *idioblastic*. Ant. *anhedral*. *(Cross* et al. *1906, p. 698; Passchier & Trouw 1996, p. 258; Barker 1998, p. 233)*

EVERSE METAMORPHISM. Term now superseded by *exomorphism*. Ant. *inverse metamorphism*. *(Cotta 1846, p. 103)*

EXOCHORISMITE (Greek *exô*, outside, and *chôrisma*, partition). Part(s) of a *chorismite* introduced from outside. *(Huber 1943, p. 89)*

EXOCONTACT ZONE. See *contact metasomatism*.

EXOGENIC METAMORPHISM. Obsolete term for *exomorphism*. Ant. *endogenic metamorphism*. *(Kalkowsky 1886, p. 33)*

EXOMETAMORPHISM (Greek *exô*, outside). Synonymous with *exomorphism*. Ant. *endometamorphism*.

EXOMIGMATIZATION (Rogart, Scotland, UK). Process of *migmatite* development in which a mobile material is introduced from outside. Ant. *endomigmatization*. *(Bellière 1960, p. 30)*

EXOMORPHISM (Greek *exô*, outside, and *morphê*, form). Term denoting all modifications of the country rock due to *contact-metamorphic* reaction with an igneous body; the term comprises all the related processes including complex metamorphic and *metasomatic* reactions. Syn. *exometamorphism*. See *everse metamorphism*, *exomorphic contact metamorphism*, *exogene metamorphism*. Ant. *endomorphism*. *(Fournet 1847, p. 243)*

EXOPNEUMATOLYTIC CONTACT METAMORPHISM. *Pneumatolytic contact metamorphism* that affects the country rock. *(Goldschmidt 1911, p. 211)*

EXOSKARN (Greek *exô*, outside). *Skarn* produced by the *metasomatic* transformation of the carbonate rock. Ant. *endoskarn*. *(OU; Kornprobst 2002, p. 99)*

EXOSMOTIC METAMORPHISM (Greek *exô*, outside, and *ôsmos*, impulsion). Obsolete term proposed to cover all the processes subtracting material from an igneous body during its cooling history, including transfer of components to the country rock and to circulating fluids. Ant. *endosmotic metamorphism*. *(Johnston-Lavis 1914, p. 382)*

FABRIC ⇒ Section 2.3 (German *Gefüge* used by Sander, 1911; Latin *fabricare*, to make, through the French *fabrique*, factory, in English used especially for the product of weaving mills). Relative orientation of parts of a rock mass. Hence ***mega***, ***meso***, ***microfabric***. This is commonly used to refer to the crystallographic and/or shape orientation of mineral grains or groups of grains, but can also be used on a larger scale. Preferred linear orientation of the parts is termed ***linear fabric***, preferred planar orientation ***planar fabric***, and the lack of a preferred orientation is referred to as ***random fabric***. *(Sander 1911, p. 281; Sander 1930, p. 1; Knopf 1933, p. 438; Fairbairn 1949, p. 1; Turner & Weiss 1963, p. 19; Hobbs* et al. *1976, p. 73)*

FACIES ⇒Section 2.2. See *metamorphic facies*.

FACIES SERIES, METAMORPHIC FACIES SERIES ⇒ Section 2.2. Sequence of *metamorphic facies* developed under a particular range of P/T values. First used by Vogt, then by Eskola, and particularly developed by Miyashiro who proposed, for *regional metamorphism*, a classification in three principal facies series (see *baric types of metamorphism*) that broadly equate to radial sectors on a P–T diagram. *(Vogt 1927, p. 373; Eskola, in Barth* et al. *1939, p. 359; Miyashiro 1961, p. 277; Hietanen 1967, p. 187; Miyashiro 1973a, p. 73; Turner 1981, p. 55; Yardley 1989, p. 189; Spear 1993, p. 18; Bucher & Frey 1994, p. 105; Miyashiro 1994, p. 352; Kornprobst 2002, p. 104)*

FALLBACK. Said of an *impact ejecta* that is deposited inside the *impact crater*. Hence ***fallback formation***. *(OU; French & Short 1968, p. 11; Engelhardt 1971, p. 5566)*

FALLOUT. Said of an *impact ejecta* that is deposited outside the *impact crater*. Hence ***fallout formation***. Term originally used for radioactive airfall deposits induced by nuclear explosions. *(OU; Engelhardt 1971, p. 5566)*

FASCICULAR (Latin *fasciculus*, diminutive of *fascis*, bundle). Term used to describe bundles or aggregates of acicular crystals. Hence ***fascicular structure***, ***fascicular schist***, in which the long axes of the crystals lie in a plane. *(OU; Hatch 1888, p. 431; Spry 1969, p. 153; Barker 1990, p. 68)*

Faserkiesel (German *Faser*, fibre, and *Kiesel*, gravel, quartz). Sillimanite-quartz knots. *(Lindacker 1792; Spry 1969, p. 271)*

FAULT (French *faille*, fault, itself from *faillir*, to fail, because in coal mines faults are sites where coal is missing). Fracture surface along which rocks have moved relative to each other. *(OU; Lyell 1830, p. 418; Holmes 1978, p. 141)*

FAULT BRECCIA. Medium- to coarse-grained *cataclasite* containing more than 30% visible fragments. Supersedes *tectonic breccia*, *crush breccia*, *cataclastic breccia* and friction breccia. *(Diller 1898, p. 74; Higgins 1971, p. 73)*

FAULT GOUGE (mining term referring to the ease of 'gouging' it out). Incohesive, clay-rich, fine- to ultrafine-grained *cataclasite*, which may possess a *schistosity* and contains less than 30% visible fragments. Lithic clasts may be present. *(Lindgren 1928, p. 152; Higgins 1971, p. 73)*

FAULT ROCK ⇒ Section 2.3. Rock formed as a result of deformation in a *fault zone*. *(OU; Sibson 1977, p. 191)*

FAULT ZONE. Zone of sheared, crushed or *foliated* rock, in which numerous small dislocations have occurred, adding up to an appreciable total offset of the undeformed walls. All gradations may occur, from multiple *fault* planes to single shear zones. *(OU; Higgins 1971, p. 73; Holmes 1978, p. 142)*

FEATHER AMPHIBOLITE. Type of *amphibolite* or amphibole-rich rock in which the amphibole crystals are acicular, lie in a random orientation within the main *foliation* of the rock and intersect each other to produce a mat of crystals. The crystals commonly show stellate grouping. Cf. *garbenschiefer*. *(Adams & Barlow 1910, p. 166, Pl. 37; Harker 1950, p. 270)*

FELDSPATHIZATION (Arbresle, Rhône, France). Feldspar formation due to *metasomatism*. Hence, **to *feldspathize***. *(Fournet, in Becquerel* et al. *1837, p. 58; Michel-Lévy 1888, p. 105; Lacroix 1899, p. 247, 249; Mehnert 1968, p. 354)*

FELDSPATURALIZATION (contraction of feldspar-uralitization). Obsolete term for the formation of feldspar-amphibole *symplectite* around omphacite. *(Lacroix 1893b, p. 582; Franchi 1902, p. 113)*

FELSIC GRANULITE ⇒ Section 2.8. Type of *granulite* with less than 30% *mafic minerals* (dominantly pyroxene). *(OU)*

FELSIC MINERALS ⇒ Table 2.1.2 (mnemonic term from *fel*dspar, *fel*dspathoid, and *si*lica). Collective term for modal quartz, feldspar and feldspathoids. Hence ***felsic rock***, a rock mainly consisting of felsic minerals, generally feldspars and quartz. *(Cross* et al. *1912, p. 561)*

FELSOBLASTIC STRUCTURE. Type of structure found in a metamorphic rock, in which the rock possesses a *schistosity* and is characterized by a very fine-grained compact appearance formed by *crystalloblastic* processes. Ant. *felsoclastic structure. (Scheumann 1936c, p. 417)*

FELSOCLASTIC STRUCTURE. Type of structure found in a metamorphic rock, in which the rock possesses a *schistosity* and is characterized by a very fine-grained compact appearance formed by *cataclastic* processes. Ant. *felsoblastic structure. (Scheumann 1936c, p. 417)*

FENITE ⇒ Section 2.9, Fig. 2.9.1 (*Fen* District, S Norway). High-temperature *metasomatic* rock characterized by the presence of alkali feldspar, sodic amphibole and sodic pyroxene; nepheline, calcite and biotite/phlogopite may also be present and typical accessories are titanite and apatite. Fenites occur as zoned *aureoles* around alkaline igneous complexes, forming in a wide range of host lithologies. They occur on the metre to kilometre scale. Hence ***fenitization***, *metasomatic* process leading to the formation of *fenites. (Brögger 1921, p. 156, 171)*

FIBROBLASTIC STRUCTURE (Latin *fibra*, fibre, and Greek *blastos*, bud, sprout). Type of structure in a metamorphic rock in which the constituent grains are of equal size (*homeoblastic, equigranular*), have a fibrous form and are arranged parallel to each other. Syn. *nematoblastic. (Eskola 1939, p. 277)*

FLAKY STRUCTURE (German *schuppige Struktur* of Naumann, and *Schuppenstruktur* of Becke). Synonymous with *lepidoblastic structure. (Naumann 1858, p. 442; Becke 1903b, p. 46)*

FLAME GNEISS (German *Flammengneis*). Fine-grained gneiss with quartz-feldspar streaks resembling flames. *(Zirkel 1894b, p. 228)*

FLASER GABBRO, FLASERGABBRO, FLASER-GABBRO (German *Flaser*, streaks, lenticles). Gabbro that has suffered dominantly *mylonitic* deformation producing lensoid relicts separated by layers or streaks of more finely crushed and *recrystallized* material (see *flaser structure*). Originally defined as a stage between gabbro and gabbro-schist, the latter having a distinct fissility. *(Teall 1888, p. 177; Joplin 1968, p. 21; Honnorez* et al. *1984, p. 11381)*

FLASER STRUCTURE (German *flasrige Struktur*). Type of structure produced by dominantly *mylonitic* deformation and characterized by the presence of fissile zones or layers of highly sheared or finely crushed rocks separating lensoid relicts of relatively unaltered rock. The resultant rock may be termed a ***flaser-gneiss*** although it is generally referred to the *protolith* (e.g. *flaser gabbro*). See also *augen structure, relict structure* and *phacoidal structure. (Naumann 1849, p. 480, 564; Zirkel 1866a, p. 61; Joplin 1968, p. 29)*

FLECKSCHIEFER (German *Fleck*, spot or speck, and *Schiefer*, schist; Harz Mts., Germany). Type of *spotted schist* or *slate* in which the *concretions* are small and ill-defined. *(Zincken 1845, p. 584; Naumann 1849, p. 559; Zirkel 1866a,b, p. 474)*

FLINTY CRUSH ROCK. Dark *aphanitic fault rock* with a flint-like appearance generally occurring as veins and stringers associated with *fault zones*. The term encompasses *pseudotachylite, ultracataclasite* and *ultramylonite*. One of these more specific terms should be used as appropriate. *(Clough 1888, p. 23; Peach* et al. *1907, p. 124; Higgins 1971, p. 73)*

FLOITITE (*Floitental*, Tyrol, Austria). Metamorphic rock composed of oligoclase, biotite and zoisite, with small amounts of quartz ± amphibole. *(Becke 1903b, p. 29)*

FLOW CLEAVAGE. Type of *cleavage* characterized by the preferred orientation of all the mineral constituents of a rock and originally thought to result from rock flow. Rec.syn. *continuous cleavage. (Leith 1905, p. 23; Wilson 1961, p. 462)*

FLOW GNEISS. Layered igneous rock whose structure has been produced by viscous flow before consolidation and therefore equivalent to flow *banding* and flow layering. *(OU)*

FOLIACEOUS STRUCTURE (Latin *folium*, leaf, sheet). *Laminated structure* in which the laminae are thin and easily separated. *(Hatch 1888, p. 432)*

FOLIATED STRUCTURE (Latin *folium*, leaf, sheet). Type of structure characterized by the presence of a *foliation*. Originally used to denote a *gneissose structure* as opposed to a *schistose structure*. *(Macculloch 1821, p. 122)*

FOLIATION ⇒ Section 2.3. Any repetitively occurring or penetrative planar feature in a rock body. Examples include layering on a scale of a centimetre or less, and the *preferred* planar *orientation* of *inequant* mineral grains or grain aggregates. The surfaces to which they are parallel are called ***s-surfaces***. *(Darwin 1846, p. 141; Omalius d'Halloy 1868, p. 472; Bonney 1919, p. 198; Harker 1939, p. 203; Turner & Weiss 1963, p. 97; Passchier & Trouw 1996, p. 258)*

FORCE OF CRYSTALLIZATION (German *Kristallisationskraft*). Ability of a crystal to grow within a solid medium by forceful expansion. The property varies with the crystallographic direction, the specific mineral and the temperature. In general, the crystallizing force increases with the density of the minerals. *(Becke 1903a, p. 564; Harker 1932, p. 33)*

FORELLENGRANULITE (German *Forelle*, trout; Granulitgebirge, Saxony). *Granulite* with dark spots rich in amphibole. Syn. *spotted granulite*. *(Cotta 1862, p. 166)*

FRACTURE. General term for any break in a rock mass, whether or not it causes displacement. Fracture includes cracks, *joints* and *faults*. *(OU)*

FRACTURE CLEAVAGE. Regular set of closely spaced parallel or subparallel *fractures* along which the rock will preferentially split. The term was originally taken to include all types of *cleavage* characterized by the development of discrete parallel planes of weakness, that is, *spaced cleavage* and *crenulation cleavage*. Powell introduced the term *disjunctive cleavage* as a non-genetic alternative. *(Leith 1905, p. 119; Wilson 1961, p. 460; Powell 1979, p. 29; Passchier & Trouw 1996, p. 259)*

FRAGMENTAL (IMPACT) BRECCIA. Synonymous with *lithic (impact) breccia*. *(OU; Stöffler* et al. *1980, Table 1)*

FREE METAMORPHISM. *Regional metamorphism* independent of igneous intrusion. *(Gümbel 1888, p. 371; Tomkeieff)*

FRICTION METAMORPHISM. Equivalent to *dislocation metamorphism*. *(Gosselet 1883, 1884, in Tomkeieff)*

FRITTED ROCK ⇒ Section 2.10 (French *fritte*, fritted material, or English *to frit*, to fuse without melting, all from the Latin *frigere*, to roast). Compact, vesicular or slaggy metamorphic rock of any composition with a glass content ranging from a few per cent up to a maximum of 20% vol., and either produced by *contact metamorphism* in volcanic to subvolcanic settings or generated by *combustion metamorphism*. Hence ***fritting***, the process by which fritted rocks are formed. See *buchite*. *(Jameson 1817, p. 284; Naumann 1858, p. 737; Zirkel 1893, p. 602)*

FRUCHTSCHIEFER (German *Frucht*, fruit, and *Schiefer*, schist; Harz Mts., Germany). Type of *spotted schist* or *slate* in which the concretions are the size of beans. *(Leonhard 1824, p. 460; Naumann 1849, p. 559; Zirkel 1866a,b, p. 474; Lapparent 1882, p. 1131)*

FULGURITE ⇒ Section 2.10. Rock produced by *lightning metamorphism*, consisting of glassy crusts, tubules or drops. *(Bruckmann 1806, p. 67; Arago 1821, p. 415; Darwin 1860, p. 76; Daubrée 1879, p. 481; Essene & Fisher 1986, p. 190; Frenzel* et al. *1989, p. 265)*

GADRIOLITE (*Gadriol* river, Graubünden, eastern Switzerland). Obsolete term for a local variety of *greenschist* from the so-called *chlorogrisonite* schists group (equivalent to *prasinites*), composed mainly of plagioclase, actinolite, chlorite and magnetite, with subordinate epidote and haematite. *(Rolle 1879, p. 36)*

GANGMYLONITE (DYKE-MYLONITE) (German *Gang*, passage, dyke). Type of *ultracataclasite* that characteristically forms intrusive-looking veins and stringers around *fault zones*. It differs from *pseudotachylite* in that it lacks evidence of melting. A term such as intrusive *ultracataclasite* should be used as an alternative. *(Hammer 1914, p. 555; Waters & Campbell 1935, p. 477; Higgins 1971, p. 74)*

Garbenschiefer (German *Garbe*, sheaf, and *Schiefer*, schist). Type of *schist* characterized by *porphyroblasts* or concretions composed of radiating bunches of fibrous or *acicular* minerals resembling sheaves. The structure may also be termed *sheaf structure*. *(Naumann 1858, p. 542; Zirkel 1866a,b, p. 516; Holmes 1920, p. 215; Spry 1969, p. 269; Barker 1990, p. 135; Vernon 2004, p. 491)*

General metamorphism. Obsolete term for *regional metamorphism*, the cause of which could not be determined. Ant. *abnormal* or *special metamorphism*. *(Delesse 1857, p. 89)*

GEOBAROMETER (Greek *gê*, earth). Term used for a mineral pair whose relative chemical compositions are pressure dependent and whose compositions can therefore be used to infer the pressure conditions at the time of their formation. *(OU; Fermor, in Eskola 1920, p. 188)*

GEOTHERM. Line on a *P–T* grid showing how the temperature varies with depth below a particular point in the Earth's surface. The line may be curved, the elements of which will have different values of the *geothermal gradient*. *(OU; Yardley 1989, p. 15)*

GEOTHERMAL GRADIENT. Rate of change of temperature with depth in the Earth's crust, usually expressed as °C/km. The gradient, which is represented by a straight line on a *P–T* grid, may change from place to place and at different depths. See *geotherm*. *(OU; Yardley 1989, p. 14; Miyashiro 1994, p. 349)*

Geothermal metamorphism. Ambiguous term for a type of metamorphism broadly characterized by increased temperatures in the absence of significant deformation. As such it has been regarded as equivalent to *burial metamorphism* and also used to describe the low-grade metamorphism associated with geothermal systems and the *hydrothermal metamorphism* associated with ocean-floor settings. *(Rinne 1920, p. 101; Grubenmann & Niggli 1924, p. 193; Turner 1981, p. 214; Saggerson 1989, p. 503)*

GEOTHERMOMETER. Term used for a mineral pair whose relative chemical compositions are temperature dependent and whose compositions can therefore be used to infer the temperature conditions at the time of their formation. *(Van't Hoff 1905, p. 82)*

GHOST STRATIGRAPHY, GHOST STRUCTURE. Stratigraphy and structure of a pre-existing rock sequence which can be identified in, or traced through highly metamorphosed rocks or igneous bodies, by means of relict features. The terms were originally used to describe features in what was considered partially *granitized* rocks. *(Read 1951, p. 10)*

Giant gneiss (Malpitschgruppe, west Tyrol, Austria). Local term for an *augengneiss* composed of large dominating crystals (>2.5 cm) of feldspar and quartz and patches of biotite. *(Stache & John 1877, p. 182)*

GLAUCOPHANE-SCHIST FACIES ⇒ Section 2.4, Fig. 2.2.4, Table 2.2.1. *Metamorphic facies* corresponding to very low- to low-grade and to high *P/T* values, and bounded in rocks of suitable composition, at low *P*-values, by the lawsonite-in reactions and, at higher *P*-values, by the stability limit of iron-poor Na-amphibole. Synonymous with *blueschist facies*. *(Eskola 1929, p. 163; Eskola 1939, p. 367; Fyfe 1967, p. 43)*

Glaucophanic metamorphism. Synonymous with high-*P/T* type of metamorphism (see *facies series*). The term ***glaucophanitic metamorphism*** was originally proposed for metamorphism that resulted in the development of glaucophane. *(Miyashiro & Banno 1958, p. 98; Miyashiro 1973a, p. 77; Miyashiro 1994, p. 200)*

GLAUCOPHANITE ⇒ Section 2.4. *Monomineralic* rock consisting of glaucophane or ferroglaucophane. The term should not be used as a general name for any glaucophane-rich rock.

Glomeroblastic structure (Latin *glomus*, gen. *glomeris*, ball, e.g. of wool, and Greek *blastos*, bud, sprout). Type of structure in a metamorphic rock in which newly grown crystals are grouped together. *(Loewinson-Lessing 1911, p. 572; Fitch 1931, p. 468)*

GNEISS ⇒ Section 2.3 (*kneiss*, old word meaning nest, for the host rock of mineralized veins; Saxony, Germany). Metamorphic rock displaying a *gneissose structure*. The term gneiss may also be applied to rocks displaying a dominant linear *fabric* rather than a *gneissose structure*, but which will split on a scale of more than 1 cm; in this case the term ***lineated gneiss*** is applied. The term was originally defined by Werner as a foliated rock with quartz and feldspar; the meaning has evolved and the SCMR has adopted a structural-only definition. *(Agricola 1556; Romé de l'Isle 1783, vol. II; Werner 1786, p. 277; Brochant 1800, p. 567; Brongniart 1813, p. 34; Haüy 1822, p. 538; Beudant 1830, p. 564; Lyell 1833, p. 69; Naumann 1849–53, p. 563; Zirkel

1866a, p. 413; Cordier 1868, p. 64; Omalius d'Halloy 1868, p. 164; Van Hise 1904, p. 782; Holmes 1920, p. 107; Rosenbusch 1923, p. 651; Yardley 1989, p. 22)

GNEISSIC STRUCTURE. Synonymous with and superseded by *gneissose structure*. *(OU; Leith 1905, p. 145; Harker 1939, p. 35)*

GNEISSITE (Erzgebirge, Saxony). Obsolete local term for an orthogneiss of granitic composition (so-called 'red gneiss'). *(Cotta 1862, p. 169; Cordier 1868, p. 65)*

GNEISS-MICASCHIST (German *Gneisglimmerschiefer*). Obsolete, rarely used term for a schist whose high feldspar content makes it transitional to a *gneiss*. *(Rosenbusch 1923, p. 692)*

GNEISSOCLASTIC STRUCTURE. Type of *gneissose structure* formed by *cataclasis*. *(Erwin 1938, p. 120)*

GNEISSOSE STRUCTURE. Type of structure characterized by a *schistosity* that is either *poorly developed* throughout the rock or, if *well developed*, occurs in broadly spaced zones, such that the rock will split on a scale of more than one centimetre. Mineralogical or lithological layering is commonly present. *(OU; Holmes 1920, p. 107; Harker 1939, p. 119)*

GNEISSOSITY. Synonymous with and to be replaced by *gneissose structure*. *(OU; Vernon 2004, p. 481)*

GNÉNINE (Mt Blanc, French-Italian Alps). Obsolete term used for a gneissose metagranite (called *protogine*). *(Boubée, in Cordier 1868, p. 181)*

Gondite (after the Aboriginal tribe of *Gonds*, Central Provinces, India). High-grade metamorphic rock composed almost entirely of spessartine garnet and quartz. Rhodonite, apatite, amphibole, magnetite, rhodochrosite and microcline may also be present in small amounts. The *protolith* was a manganiferous sedimentary rock. *(Fermor 1909, p. 306)*

Gorotubite, **gorutubite** (*Gorutuba* River, State of São Paulo, Brazil). Local name for a *banded* rock composed of alternating centimetre-scale layers of fine-grained amphibole schist, metalimestone and *phyllite*. The rock was produced by the low-grade *regional metamorphism* of thin-bedded cyclic sediments. *(Barbosa 1942, p. 19; Geoffroy & Souza Santos 1942, p. 109)*

GOUGE. See *fault gouge*.

GRAIN SIZE ⇒ Table 2.1.2. No absolute grain-size values are recommended for the expressions 'coarse-grained', 'fine-grained', etc. If absolute values are required, then the most favoured values are: >16 mm: very coarse-grained; 4–16 mm: coarse-grained; 1–4 mm: medium-grained; 0.1–1 mm: fine-grained; 0.01–0.1 mm: very fine-grained; < 0.01 mm: ultra-fine-grained. However, if this scale is used the fact should be specifically stated. See also *phaneritic*, *aphanitic*.

GRANITE-GNEISS, GRANITO-GNEISS. Gneissose rock with a granitic composition normally derived from a granite *protolith*. Originally proposed as a granitic rock whose structure was transitional betwen igneous and gneissose. Rec.syn. gneissose granite, metagranite. *(Humboldt 1823, p. 34; Eichwald 1846, p. 10; Lacroix 1899; Cogné 1960, p. 59; Hameurt 1967, p. 109)*

GRANITIC FUMAROLES. See *aura granitica*. *(Élie de Beaumont 1847, p. 1295, 1314, 1316; Lacroix 1892, p. 740)*

GRANITIFICATION (Montabon near Châlons-sur-Saône, France). Obsolete term synonymous with granitization. *(Virlet d'Aoust 1847, p. 499; Delesse 1852, p. 479)*

GRANITIZATION ⇒ Section 2.6. Comprehensive term for processes by which pre-existing rocks are converted to granitoids (melting, pervasive influx of chemical components such as silica, potash, soda or other means of pervasive transformation). The term may be qualified to show the exact process, for example *metasomatic* granitization, *anatectic* granitization. *(Michel-Lévy 1888, p. 107; Sederholm 1923, p. 5; Mehnert 1968, p. 534)*

GRANOBLAST (Duluth gabbroid complex, Minnesota, USA). Term proposed for a high-grade *mafic hornfels* to replace *muscovadite*, owing to the typical *granoblastic* structure of this rock. *(Schwartz 1924, p. 117)*

GRANOBLASTIC-POLYGONAL STRUCTURE. Type of structure in a metamorphic rock in which the constituent grains are equidimensional (*equant*) and of equal size (*homeoblastic*), that is *granoblastic*, and have straight or smoothly curved boundaries (*polygonal*) meeting at triple points with *c.* 120° angles.

Synonymous with *granuloblastic*. *(OU; Spry 1969, p. 159, 186; Barker 1998, p. 73)*

GRANOBLASTIC STRUCTURE (Latin *granum*, grain, and Greek *blastos*, bud, sprout). Type of structure in a metamorphic rock in which the constituent grains are equidimensional (*equant*) and of equal size (*homeoblastic*) and have well sutured or irregular boundaries. Cf. *granoblastic-polygonal structure*. *(Becke 1903a, p. 570; Grubenmann 1904, p. 79; Joplin 1968, p. 28; Vernon 2004, p. 482)*

GRANOCLASTIC STRUCTURE (Latin *granum*, grain, and Greek *klastos*, broken). Type of *granofelsic structure* formed by *cataclasis* and in which the constituent grains are essentially equidimensional. *(Erwin 1938, p. 120)*

GRANOFELS ⇒ Sections 2.1 & 2.3 (Latin *granum*, grain, and German *Fels*, rock). Metamorphic rock displaying a *granofelsic structure*. *(Goldsmith 1959, p. 109)*

GRANOFELSIC STRUCTURE. Type of structure resulting from the absence of *schistosity* such that the mineral grains and aggregates of mineral grains are *equant* (for example quartz, feldspar, garnet and pyroxene), or if *inequant* have a random orientation. Mineralogical or lithological layering may be present. *(OU)*

GRANOLITE (Latin *granulum*, small grain). General term for rocks of high-grade *regional metamorphism*, consisting of *mineral assemblages* diagnostic of the *granulite facies*. Individual rocks may have mineral or structural qualifiers. *(Turner 1899, p. 141; Turner 1900, p. 105; Winkler & Sen 1973, p. 395)*

GRANOSCHISTOSE STRUCTURE (Latin *granum*, grain, and Greek *schistos*, split). Type of structure in a *monomineralic* metamorphic rock produced by the parallel elongation of minerals whose normal habit is *equidimensional* or nearly so. *(OU)*

GRANULAR STRUCTURE. Type of structure characterized by the aggregation of one or more varieties of mineral grains. Some authors state that the mineral grains are of approximately equal size, although this is better given as *equigranular* or more specifically as *homeoblastic*. *(Macculloch 1821, p. 142, 147; Hatch 1888, p. 433; Cross 1894, p. 232; Holmes 1920, p. 111)*

GRANULITE ⇒ Section 2.8 (Latin *granulum*, small grain; between Chemnitz and Penig, Erzgebirge in Saxony, Germany). High-grade metamorphic rock in which Fe-Mg silicates are dominantly hydroxyl-free; the presence of feldspar and the absence of primary muscovite are critical, cordierite may also be present. The mineral composition is indicated by prefixing the major constituents. The rocks with >30% *mafic minerals* (dominantly pyroxene) may be called *mafic granulites*, those with <30% *mafic minerals* (dominantly pyroxene) may be called *felsic granulites*. The term should not be applied to ultramafic rocks, *calc-silicate rocks*, *marbles*, ironstones or quartzites. Detailed names and subdivisions may be given using mineral-root names, e.g. garnet-clinopyroxene-plagioclase granulite, garnet-plagioclase *granofels*. Term originally used for fine-grained rocks of granitic mineral and chemical composition. *(Weiss 1803, p. 348; Leonhard 1823, p. 231; Mehnert 1972, p. 148; Lorenz 1998, p. 101)*

GRANULITE FACIES ⇒ Section 2.2, Fig. 2.2.4, Table 2.2.1. *Metamorphic facies* representing the highest grades of metamorphism; it forms at high temperature and at medium and high P/T values, and lies between the *eclogite* and *pyroxene-hornfels facies*. It is characterized by clinopyroxene-orthopyroxene-plagioclase (olivine not stable with plagioclase or garnet) *assemblages* in rocks of basaltic composition. *(Eskola 1929, p. 163, 167; Eskola 1939, p. 360; Turner 1948, p. 100; Turner 1981, p. 202; Miyashiro 1994, p. 188, 290)*

Granulitic. Ambiguous term; more specific terms such as *granoblastic* or *homeoblastic* should be used instead in a structural context, *granulite-facies* rock or *granulite* should be used for a high-grade rock.

GRANULITIC (IMPACT) BRECCIA. *Thermally metamorphosed* (*recrystallized*) *impact breccia* with *granoblastic* or *poikiloblastic structure*. The term is used for lunar rocks and meteorites. *(Warner et al. 1977, p. 2052; Stöffler et al. 1980, Table 1; Bischoff & Stöffler 1992, Table 1)*

GRANULITIZATION. Obsolete term for a process of grain-size reduction accompanied by *recrystallization* leading to a fine-grained, broadly *granoblastic structure. (Michel-Lévy 1888, p. 107; Harker 1932, p. 311; Spry 1969, p. 289)*

GRANULOBLASTIC (Latin *granulum*, small grain, and Greek *blastos*, bud, sprout). Synonymous with *granoblastic-polygonal*. The term is particularly used in the terminology of ultramafic rocks. *(Binns 1964, p. 297; Joplin 1968, p. 28; Harte 1977, p. 280)*

GRANULOSE STRUCTURE. Synonymous with, and superseded by *granofelsic structure*. *(Holmes 1920, p. 112)*

GRAPHITHRÈNE (Greek *threnos*, lamentation, probably because the rock is used for funeral monuments). *Crystalline limestone* containing some graphite. *(Boubée, in Cordier 1868, p. 286)*

GREENSCHIST ⇒ Section 2.5. Schist whose greenish colour is due to the presence of minerals such as actinolite, chlorite and epidote. More precise terms should be used wherever possible (e.g. epidote-bearing actinolite-chlorite schist). Note: the term should not be used as a general name for rocks of the *greenschist facies*. *(Naumann 1849–53, p. 141; Kalkowsky 1886, p. 216)*

GREENSCHIST FACIES ⇒ Sections 2.2 & 2.5, Fig. 2.2.4, Table 2.2.1. *Metamorphic facies* representing low grades of metamorphism. It forms at low to moderate temperatures and pressures and lies between the *epidote-amphibolite* and *subgreenschist facies*. It is characterized by actinolite-albite-epidote-chlorite *assemblages* in rocks of basaltic composition. *(Eskola 1920, p. 155; Turner 1981, p. 202)*

Greenstone (German *Grünstein*). Old field term used for a massive metamorphic or altered magmatic rock, i.e. a *granofels*, whose greenish colour is due to the presence of minerals such as actinolite, chlorite and epidote. More precise terms should be used wherever possible (e.g. chlorite-epidote *granofels*). Frequently used in describing metabasic rocks in Precambrian terrains: see *greenstone belt*. *(Werner 1786, p. 297; Liou* et al. *1987, p. 97)*

GREENSTONE BELT. Regional-scale elongate zone, particularly in Precambrian shield areas, that is characterized by an abundance of basic metavolcanic rocks. *(OU; Condie 1981, p. 5)*

GREISEN ⇒ Section 2.9, Fig. 2.9.1 (*Greisstein*, old Saxon mining term, meaning ash-coloured rock). Medium-temperature *metasomatic* rock characterized by the presence of quartz and white mica, commonly with topaz, fluorite, tourmaline and locally with amazonite, orthoclase, andalusite and diaspore. Typically greisens may host Be, W, Mo, Sn and Ta mineralization. They are associated with high-level late-orogenic leucogranites and form as replacements either in the granite body and/or in a wide range of country rocks. Zoning may be present. Hence ***greisenization***, *metasomatic* process leading to the formation of greisen. *(Lempe 1785, p. 105; Flett 1909, in Tomkeieff; Nakovnik 1954, p. 53; Stemprok 1987, p. 169)*

GRIOTTE (old quarrymen term, from the French *griotte*, a variety of cherry, for *aigriotte*, sour; Pyrenees, France). Type of reddish fine-grained limestone of Palaeozoic age, rich in remains of ammonoids. *(OU; Lemeyrie 1878, p. 260)*

Griquaite [griquaïte] (*Griqualand*, Kimberley, S Africa). Rock composed of pyrope-rich garnet, diopside ± orthopyroxene. It may contain diamond. Other possible minerals are biotite and magnetite. It occurs as inclusions in kimberlites. More specific terms should be used if possible, for example garnet *pyroxenite*, orthopyroxene-bearing garnet-diopside gneiss, etc. *(Beck 1907, p. 301)*

Grospydite (from *gros*sular, *py*roxene and kyanite, *dis*thene). Plagioclase-free rock mainly composed of Ca-Al clinopyroxene and calcic garnet together with some kyanite. It occurs as inclusions in kimberlites. More specific terms should be given if possible, for example, kyanite-bearing grossular-clinopyroxene gneiss. *(Bobrievich* et al. *1960, p. 23; Sobolev* et al. *1966, p. 126, 902)*

Gumbeite ⇒ Section 2.9, Fig. 2.9.1 (*Gumbeika* river, southern Ural, Russia). Medium- to low-temperature *metasomatic* rock mainly composed of quartz, orthoclase and carbonate. It forms as an *alteration* of granodiorite or syenite and is closely associated with W-Cu- or Au-Cu-bearing veins. Hence ***gumbeitization***, metasomatic process leading to the formation of gumbeite. *(OU; Korzhinskii 1953, p. 446, 448; Omel'yanenko 1978, p. 168)*

HALF-PHYLLITE. Obsolete term for a low-grade metamorphosed shale that contains newly formed biotite and quartz. *(Naumann 1858, p. 553; Loretz 1881, p. 183, 190)*

HÄLLEFLINTA (helleflint) (Swedish *hälle*, flat rock slab, and *flinte*, flint; various places in Sweden). Obsolete local name, used mainly in Sweden and Finland, for a fine-grained compact quartzofeldspathic rock of horny aspect, which may be *banded* and/or *blastoporphyritic*. It is derived from acid igneous rocks or acid tuffs and is partly synonymous with *leptynite*. Cf. *adinole*. *(Cronstedt 1758, p. 67; Geijer & Magnusson 1944, p. 19)*

HÄLLEFLINTGNEISS. Obsolete local name formerly used in Sweden for all feldspar-rich, fine-grained gneisses (cf. *hälleschist*, *leptite*). *(Sederholm 1897–9, p. 11)*

HÄLLESCHIST. Obsolete local term formerly used in Finland and Sweden for feldspar-rich, fine-grained gneiss or schist (cf. *hälleflintgneiss*, *leptite*). *(Sederholm 1897–9, p. 11)*

Hartschiefer (German *hart*, hard, and *Schiefer*, schist). Metamorphic rock with a cherty appearance and characterized by a *banded structure*, the *bands* showing a strong parallelism and equal thickness, and differing strongly in composition. The rock is the product of intense *mylonitic* deformation. *(Holmquist 1910, p. 945; Quensel 1916, p. 100)*

HELICITIC STRUCTURE (Latin *helix*, spiral). Type of structure in a metamorphic rock, in which the trails or bands of *inclusions* in *porphyroblasts* are characteristically curved or contorted and represent older folded structures. *(Becke 1903a, p. 570; Joplin 1968, p. 30; Spry 1969, p. 257; Vernon 2004, p. 482)*

HÉMITHRÈNE (Greek *hemi*, half, and *threnos*, lamentation, probably because the rock is used for funeral monuments). Metamorphic rock composed of amphibole and calcite. The terms amphibole *impure marble* should be used in preference if calcite is >50% vol., and carbonate-silicate rock if calcite is <50% vol. *(Brongniart 1813, p. 33, Cotta 1855, p. 175; Zirkel 1866a, p. 199; Omalius d'Halloy 1868, p. 167)*

HETEROBLASTIC STRUCTURE (Greek *heteros*, other, different, and *blastos*, bud, sprout). Type of structure in a metamorphic rock in which the constituent grains are of two or more different sizes. Ant. *homeoblastic*. *(Grubenmann 1904, p. 81; Spry 1969, p. 159)*

HETEROGRANULAR (Greek *heteros*, other, different, and Latin *granulum*, small grain). Synonym of *heteroblastic*. *(OU; Bates & Jackson)*

HETEROMORPHIC (Greek *heteros*, other, different, and *morphê*, form). Said of rocks of similar chemical composition but composed of different minerals. *(Lacroix 1916, p. 179)*

HIGH-RANK METAMORPHISM. Unnecessary and ill-defined term for a type of metamorphism occurring under high temperatures and pressures as opposed to *low-rank metamorphism*. It should not be confused with the use of 'rank' in very low-grade metamorphic terminology to express, for example, the thermal maturity of coals. *(OU; Bates & Jackson)*

HOLOBLAST (Greek *holos*, whole, and *blastos*, bud, sprout). Crystal in a metamorphic rock that has been developed wholly within the rock. *(Eskola 1939, p. 277)*

HOMEOBLASTIC STRUCTURE (Greek *homoeos*, similar, and *blastos*, bud, sprout). Type of structure in a metamorphic rock in which the constituent grains are essentially of equal size. Ant. *heteroblastic*. Cf. *lepidoblastic*, *granoblastic*, *nematoblastic*, etc. *(Becke 1903a, p. 570; Becke 1903b, p. 46; Grubenmann 1904, p. 79; Holmes 1920, p. 118)*

HOMOGRANULAR (Greek *homoeos*, similar, and Latin *granulum*, small grain). Synonym of *homeoblastic*. *(OU; Bates & Jackson)*

HOMOLOGOUS SERIES. Series of metamorphic rocks formed in different metamorphic terrains (provinces) under similar conditions. They may form ***homologous provinces***. *(Grubenmann & Niggli 1924, p. 483)*

HOMOPHYLLOLITE (Greek *homoeos*, similar, *phyllon*, leaf, and *lithos*, stone). Obsolete term for a schistose metamorphic rock composed of a single mineral species. *(Gümbel 1888, p. 153)*

HORNBLENDE-HORNFELS FACIES. *Facies* name proposed for *contact rocks* of medium grade lying between the *albite-epidote-hornfels facies* and the *pyroxene-hornfels facies*. *(Fyfe* et al. *1958, p. 205)*

HORNBLENDITE ⇒ Table 2.1.2. Igneous or metamorphic rock consisting of >90% modal content of hornblende. This definition accords with that of Le Maitre and as such constitutes one of the exceptions to the SCMR guidelines related to the use of the suffix *-ite*. *(Le Maitre 1989, 2002, p. 75)*.

HORNFELS ⇒ Section 2.10 (German *Horn*, horn, and *Fels*, rock). Hard, compact *contact-metamorphic* rock of any grain size, dominantly composed of silicate + oxide minerals in varying proportions, with a horny aspect and a subconchoidal to jagged fracture. It may retain some structural features inherited from its *protolith* such as bedding, sedimentary lamination, metamorphic layering. Traditionally 'hornfels' carried qualifiers referring to the nature of the *protolith* (e.g. *basic hornfels, pelitic hornfels*). However, the SCMR prefers to use qualifiers to indicate the nature of the hornfels; these may include structural (e.g. fine-grained, spotted, layered), chemical (e.g. peraluminous) or mineralogical criteria (e.g. mafic, ultramafic, cordierite-sillimanite-spinel hornfels, diopside-wollastonite-garnet hornfels). Hornfels occurs mostly, but not exclusively, in the innermost part of *contact aureoles*. *(Hausmann 1805, p. 653, in Leonhard 1823, p. 139; Macculloch 1819, p. 13; Boué 1820, p. 253; Boué 1829, p. 131; Cordier 1868, p. 101; Rosenbusch 1877; Goldschmidt 1911, p. 140; Harker 1932)*

HORNFELSED SCHIST ⇒ Section 2.10. Schistose rock converted into *hornfels* in a *contact aureole*. *(MacGregor 1931, p. 518; Harker 1932, p. 39; McIntyre & Reynolds 1947, p. 62; Spry 1969, p. 186)*

HORNSCHIST (Harz Mts. and Black Forest, Germany). (a) Schistose *diabase contact rock* resembling a siliceous schist. (b) *Adinole*-looking rock of the outermost *aureole* of a granitic intrusion. *(Lossen 1869, p. 291; Eck 1892, in Eisele 1907, p. 134, 150; Milch 1917)*

HORNSTONE (German *Hornstein*). Ambiguous term now superseded by *hornfels*, first used for hard *contact-metamorphosed* argillaceous rocks, thereafter extended to include calc-silicate *hornfels*. Also used for some hard *baked rocks*. *(Macculloch 1819, p. 13; Necker de Saussure 1821, p. 57; Barrell 1907, p. 119, 145; Tyrrell 1926, p. 302; Shand 1931, p. 182)*

HÖSBACHITE (Wenighösbach, NE of *Hösbach*, Spessart, Germany). Metamorphic rock consisting of talc, chlorite (clinochlore) and amphibole (cummingtonic hornblende). *(Matthes & Okrusch 1965, p. 100; Matthes* et al. *1995, p. 25)*

HOT-SLAB METAMORPHISM ⇒ Sections 2.2 & 2.10. Type of *metamorphism* of *local* extent, occurring beneath an emplaced hot tectonic body. The thermal gradient is inverted and usually steep. Encompasses *upside-down metamorphism*. This is a new term proposed by the SCMR.

HUGONIOT CURVE (after the French physicist Pierre *Hugoniot*, 1851–1887). Locus of all *shock states* that can be achieved by *shock wave* compression of variable intensity in any specific material. It is commonly expressed in the pressure–volume or pressure–particle velocity space. *(OU; Rice* et al. *1958, p. 9; Duvall & Fowles 1963, p. 216, Fig. 5; Ahrens & Rosenberg 1968, p. 60; Dence & Robertson 1989, p. 526; Bischoff & Stöffler 1992, p. 708)*

HUGONIOT ELASTIC LIMIT. Specific shock pressure above which the shock-compressed material behaves elastically. *(OU; Rice* et al. *1958, p. 11, Fig. 4; Duvall & Fowles 1963, p. 255, Fig. 42, Table V; Ahrens & Rosenberg 1968, p. 61; Dence & Robertson 1989, p. 526)*

HUGONIOT EQUATION-OF-STATE (RANKINE–HUGONIOT EQUATION) (after the French physicist Pierre *Hugoniot*, 1851–1887, and the British physicist William John Macquorn *Rankine*, 1820–1872). Thermodynamic equation describing a shock transition from an uncompressed state, p_0, ρ_0, e_0, to a compressed state, p_1, ρ_1, e_1, in terms of pressure (p), density (ρ), and internal energy (e). *(OU; Rice* et al. *1958, p. 9; Duvall & Fowles 1963, p. 211; Duvall 1968, p. 24; Dence & Robertson 1989, p. 526)*

HYALISTINE. Obsolete term, originally defined as a quartz-rich variety of talc-schist, although talc may have been a sack term for any whitish or greenish mica, or mixture of sheet silicates. Most probably fine-grained. *(Serres 1863, p. 146; Tomkeieff)*

HYALOMICTE (Greek *hyalos*, glass, and *mica*). Obsolete term for a rock mainly composed of hyaline quartz and bearing scattered mica. *(Brongniart 1813, p. 34; Beudant 1830, p. 565)*

HYALOMYLONITE. Glassy *fault rock* produced by melting. Equivalent to *pseudotachylite*. *(Scott & Derver 1953, p. 123)*

HYBRID GNEISS. *Banded* gneiss formed by the intrusion of a heterogeneous magma, which has partly or wholly digested fragments of the country rock. *(Harker 1903, p. 210, 212, 214)*

HYBRIDISM. Synonymous with *hybridization*. *(Harker 1909, p. 333, 356)*

HYBRIDIZATION. Intermingling of magmas or contamination of a magma by assimilation and digestion of fragments of the country or host rock. The latter meaning is applicable to some aspects of *migmatite* petrology. Hence ***hybrid***, the product of, or relating to, hybridization. (*Durocher 1857, p. 217, 676; Harker 1904, p. 169; Mehnert 1968, p. 354)*

HYDATOCAUSTIC METAMORPHISM (Greek *hydôr*, water, and *kaustikos*, burning). Type of *local metamorphism* essentially due to the combined effects of high temperature and volcanic emanations. *(Bunsen 1849, p. 16)*

HYDATOTHERMIC METAMORPHISM (Greek *hydôr*, water and *thermos*, hot). Term proposed for a type of *local metamorphism* essentially due to the activity of gases and aqueous fluids. *(Bunsen 1849, p. 16)*

HYDROCHEMICAL METAMORPHISM. Obsolete term for *hydrothermal metamorphism* or *hydrothermal metasomatism*. *(Zirkel 1893, p. 576; Tomkeieff)*

HYDROMETAMORPHISM. Term introduced to distinguish low-temperature and low-pressure processes facilitated by the presence of water (e.g. cementation), from metamorphic processes involving significant *recrystallization*; equivalent to *diagenesis*. *(Harker 1889, p. 15)*

HYDROMETASOMATISM. *Alteration* of a rock by material that is added, removed or exchanged by water solutions, without the influence of high temperature and pressure. *(OU; Bates & Jackson)*

HYDROTHERMAL ALTERATION. *Alteration* due to *hydrothermal* solutions. *(Nakovnik 1965, p. 7)*

HYDROTHERMAL CONTACT METAMORPHISM. Type of *hydrothermal metamorphism* at a magmatic contact. *(Irving 1911, p. 298; Niggli 1954, p. 523)*

HYDROTHERMAL METAMORPHISM ⇒ Section 2.2, Fig. 2.2.1 (Greek *hydôr*, water, and *thermos*, hot). Type of *metamorphism* of *local* extent caused by hot H_2O-rich fluids. It is typically of local extent in that it may be related to a specific setting or cause (e.g. where an igneous intrusion mobilizes H_2O in the surrounding rocks). However, in a setting where igneous intrusion is repetitive (e.g. in ocean-floor spreading centres) the repetitive operation of circulating hot H_2O fluids may give rise to regional effects as in some cases of *ocean-floor metamorphism*. *Metasomatism* is commonly associated with this type of metamorphism. *(OU; Naumann 1858, p. 718; Lindgren 1933, p. 91; Turner 1948, p. 5; Coombs 1961, p. 214; Yardley 1989, p. 12; Shelley 1993, p. 60; Bucher & Frey 1994, p. 10)*

Hydrothermal metasomatism. *Metasomatism* caused by hydrothermal solutions. See *hydrothermal metamorphism* and comments under *contact metasomatism*. *(Barrell 1907, p. 117; Eskola 1939, p. 373; Nakovnik 1965, p. 7)*

HYPERMETAMORPHISM (Greek *hyper*, above, higher than, with respect to 'normal' metamorphism). Term proposed for the transformation of limestones by the gaseous emanations of granitic batholiths, resulting in the formation of a wide variety of silicates and associated minerals. Equivalent to *exomorphism*, see also *exoskarn*. The term has no relationship to *ultrametamorphism*. *(Fenner 1937, p. 164)*

HYPHOLITE (Greek *hyphos*, meshwork, and *lithos*, stone; Doira, SE of Mesocco, Graubünden, eastern Switzerland). Obsolete term for a local variety of *greenschist* from the so-called *chlorogrisonite* schists group (equivalent to *prasinites*), composed mainly of actinolite forming a meshwork, and subordinate plagioclase, epidote and magnetite. *(Rolle 1879, p. 39)*

HYPIDIOBLAST (Greek *hypo*, under, *idios*, own, particular, and *blastos*, bud, sprout). Crystal formed in a metamorphic rock, that is only partly bounded by its own crystal faces. Hence ***hypidioblastic structure***, synonymous with *subidioblastic structure*. Cf. *subhedral*. *(Bowes 1989b, p. 508)*

HYPOGENE (Greek *hypo*, under, and *gennô*, to give birth to). Term proposed by Lyell to replace 'primary' for rocks that did not acquire their form and structure at the surface; it thus encompassed both plutonic and metamorphic rocks. Used by Ransome for ore deposits formed at depth from ascending hot solutions, as opposed to supergene. *(Lyell 1833, p. 374; Geikie 1879, p. 240; Ransome 1912, p. 152)*

HYPOMETAMORPHISM. Metamorphism that occurs at high pressure and temperature. Ambiguous term proposed by Fermor as an alternative to *katametamorphism*. Lindgren subsequently used ***hypometamorphic zone*** as equivalent to *katametamorphic* zone. However, the adjectival term hypometamorphic had been previously used by Callaway to describe rocks that had suffered very low-grade metamorphism. *(Callaway 1881, p. 229; Bonney 1886, p. 58; Fermor 1927, p. 335; Lindgren 1933, p. 96)*

HYPOTHERMAL (Greek *hypo*, under, and thermos, hot). Said of a *hydrothermal* ore deposit that was formed at great depths and at high temperatures. Cf. *epithermal*, *mesothermal*. *(OU; Lindgren 1933, p. 210)*

HYSTEROCRYSTALLIZATION (Greek *hysteros*, later). Obsolete term for *secondary* crystallization or *recrystallization* of igneous-*protolith* rocks during metamorphism. Hence, ***hysterocrystalline***, said of the products or process of hysterocrystallization. *(Naumann 1858, p. 695)*

HYSTEROGENIC MINERALS (Greek *hysteros*, later, and *gennô*, to give birth to). Newly formed mineral components of a metamorphic rock, which do not wholly obliterate its original metamorphic mineral composition (e.g. hornblende in an partly amphibolitized *eclogite*). Ant. *proterogenic*. *(Zirkel 1893, p. 791; Becke 1903b, p. 35)*

ICHOR (Greek *ichôr*, lymph, ethereal fluid flowing in the veins of Greek gods). Outdated term for a fluid phase considered responsible for such processes as *granitization* and *migmatization*. *(Sederholm 1926, p. 89; Mehnert 1968, p. 354)*

IDAHOAN-TYPE FACIES SERIES, IDAHOAN FACIES SERIES (Boehls Butte, *Idaho*, USA). Type of *facies series* in Hietanen's classification, which is of medium P/T style, characterized by the simultaneous occurrence of andalusite, sillimanite, kyanite, with some staurolite and cordierite, and interpreted as indicating P and T conditions close to the Al_2SiO_5 triple point. It lies between the *Pyrenean* and *Barrovian facies series*. *(Hietanen 1967, p. 195)*

IDIOBLAST (Greek *idios*, own, particular, and *blastos*, bud, sprout). Crystal formed in a metamorphic rock, which is wholly bounded by its own crystal faces. Hence ***idioblastic structure***. *(Becke 1903a, p. 564; Becke 1903b, p. 43; Joplin 1968, p. 28; Spry 1969, p. 140; Vernon 2004, p. 483)*

IDIOTOPIC STRUCTURE (Greek *idios*, own, particular, and *topos*, site). Type of structure in diagenetically altered carbonate rocks or chemically precipitated sediments, in which the constituent grains are bounded by their own crystal faces. The term has also been used in a general sense to describe all such structures in metamorphic rocks. However, the term *idioblastic structure* is recommended for general metamorphic usage. Ant. *xenotopic structure*. *(Friedmann 1965, p. 648; Spry 1969, p. 161; Barker 1998, p. 236)*

IMBIBITION (Latin *imbibere*, to soak). Obsolete term for a gneiss formed or modified by the penetration or soaking of fluid phases, now superseded by *migmatite*. *(Virlet d'Aoust 1844, p. 846; Lacroix 1899, p. 248; Mehnert 1968, p. 355)*

IMBIBITION GNEISS. Obsolete term for a gneiss formed or modified by the penetration or soaking of fluid phases. *(Lacroix 1899, p. 258)*

IMPACT. Collision of two (planetary) bodies at cosmic velocity, which causes the propagation of a *shock wave* in both the *impactor* and *target* body; also called ***hypervelocity impact***. The impact origin of craters was first proposed for the Moon by Franz von Paula Gruithuisen (1774–1852) in 1822. *(OU; Gilbert 1893, p. 256; Roddy* et al. *1977, p. 1; Melosh 1989, p. 4)*

IMPACT BRECCIA. *Monomict* or *polymict breccia* that occurs around, inside and below *impact craters*. *(OU; Dence 1964, p. 256; Stöffler* et al. *1979, p. 641, Fig. 1; Stöffler* et al. *1988, p. 282)*

IMPACT CRATER. Approximately circular crater formed either by *impact* of an interplanetary body (*projectile*) on a planetary surface or by an experimental *impact* of a *projectile* into solid matter; craters formed by very oblique *impacts* may be elliptical. See *simple impact* *crater, complex impact crater. (OU; Gilbert 1893, p. 272; Roddy* et al. *1977, p. 1; Melosh 1989, p. 6)*

(IMPACT) EJECTA. Solid, liquid and vaporized rock ejected ballistically from an *impact crater. (OU; Melosh 1989, p. 74, Figs. 5.9–5.11)*

IMPACT FORMATION. Geological formation produced by *impact*. It includes various lithological and structural units inside and beneath an *impact crater* (***inner impact formations***), the continuous *ejecta blanket* (***outer impact formations***) and distal *impactites* such as *tektites* and *impactoclastic airfall beds. (OU; Engelhardt 1971, p. 5566; Pohl* et al. *1977, p. 352)*

IMPACT GLASS. *Impact melt* quenched to glass. It includes semihyaline *impact melt rocks. (OU; Spencer 1933c, p. 394; Dence 1971, p. 5553; Stöffler 1971, Fig. 4; Stöffler 1984, p. 466, Tables 1, 2)*

IMPACTITE ⇒ Section 2.11. Rock produced by *impact metamorphism*. It includes *shocked rocks, impact breccias, impact melt rocks,* (*micro*) *tektites* and *impactoclastic airfall beds*. Term originally restricted to *impact glass. (OU; Carter 1965, p. 786; Park 1989a, p. 41)*

IMPACT MELT. Melt formed by *shock melting* of rocks in *impact craters. (OU; Dence 1968, p. 176; Dence 1971, p. 5553; Grieve* et al. *1977, p. 791; Dressler & Reimold 2001, p. 205)*

IMPACT MELT BRECCIA. *Impact melt rock* containing lithic and mineral clasts displaying variable degrees of *shock metamorphism* in a crystalline, semihyaline or hyaline matrix. See *impact melt rock. (OU; Stöffler* et al. *1979, Table 1, p. 723)*

IMPACT MELT ROCK. Crystalline, semihyaline or hyaline rock solidified from *impact melt* and containing variable amounts of clastic debris of different degree of *shock metamorphism*; should replace the previously used term *impact melt breccia. (OU; Dence 1971, p. 5553; Grieve 1987, p. 253)*

IMPACT METAMORPHISM ⇒ Sections 2.2 & 2.11, Fig. 2.2.1. Type of *metamorphism* of *local* extent caused by the passage of a *shock wave* due to the *impact* of a planetary body (*projectile* or *impactor*) on a planetary surface (*target*). It includes melting and vaporization of the *target rock*(*s*). Cf. *shock metamorphism. (Pecora 1960, p. 19; McIntyre 1962, p. 1647; Yardley 1989, p. 13; Bucher & Frey 1994, p. 10)*

IMPACTOCLAST. Rock fragment resulting from *impact*-induced comminution of rocks. It may display variable degrees of *shock metamorphism* (different *shock stages*). *(Stöffler & Grieve, Section 2.11)*

IMPACTOCLASTIC AIRFALL BED. *Pelitic* sedimentary layer containing a certain fraction of *shock-metamorphosed* material, e.g. shocked minerals and melt particles, which has been ejected from an *impact crater* and deposited by interaction with the atmosphere over large regions of a planet or globally. *(Stöffler & Grieve, Section 2.11)*

IMPACTOCLASTIC DEPOSIT. Consolidated or unconsolidated sediment resulting from ballistic excavation, transport, and deposition of rocks at *impact craters*. It may contain glassy or crystallized particles of *impact melt. (Stöffler & Grieve, Section 2.11)*

IMPACTOR. (Inter)planetary body (*projectile*) that collides with a second body (*target*) at cosmic velocity and generates a *shock wave* in both bodies. Cf. *target. (OU; Melosh 1989, p. 46)*

IMPACT PSEUDOTACHYLITE (Greek *pseudês*, false, *tachys*, quick, and *lithos*, stone). *Pseudotachylite* produced by *impact metamorphism*. *Dyke*-like *breccia* formed by frictional melting in the basement of *impact craters*, resulting often in irregular vein-like networks. Typically, it contains unshocked and shocked mineral and lithic clasts in a fine-grained *aphanitic* matrix. See also *melt vein. (Shand 1916, p. 198; Dence 1971, p. 5555; Stöffler* et al. *1988, p. 289, Fig. 8, Table 2; Reimold 1995, p. 247; Spray 1998, p. 195; Dressler & Reimold 2004, p. 2–36)*

IMPACT REGOLITH (Greek *rhegos*, blanket, and *lithos*, stone). Fine-grained *impactoclastic deposit* formed by multiple *impacts* on the surface of planetary bodies lacking an atmosphere, such as the Moon, Mercury or asteroids. The lunar *regolith* (sometimes incorrectly named *lunar soil*) contains unshocked and shocked lithic and mineral clasts, glass fragments, glass bodies of revolution (spheres, dumbbells, etc.), and agglutinate glass as well as solar-wind implanted rare gases. *(OU; Engelhardt* et al.

1970, p. 363; Shoemaker et al. *1970, p. 452; Stöffler* et al. *1980, Table 1; Warner & Simonds 1989, p. 297; Heiken* et al. *1991, p. 285)*

IMPACT STRUCTURE. Geological structure caused by *impact* irrespective of its state of preservation. *(OU; Grieve 1987, p. 246)*

INCIPIENT REGIONAL METAMORPHISM ⇒ Section 2.5. Unnecessary term broadly equivalent to *anchimetamorphism. (Kisch 1983, p. 309; Shelley 1993, p. 59)*

INCLUSION. Solid or liquid phase wholly contained within a crystal. Hence ***inclusion trails***, lines of inclusions, usually relics of a pre-existing structure (for example *poikiloblastic* and *helicitic structures*). Inclusion trails in *porphyroblasts* may be referred to as ***internal foliation*** (S_i), as opposed to the ***external foliation*** (S_e) of the matrix; the geometry and relationships of these *foliations* may indicate the growth history of the *porphyroblast*. *(OU; Hatch 1888, p. 435; Spry 1969, p. 166; Vernon 2004, p. 483)*

INDEX MINERAL (critical mineral). Mineral that is used to characterize a zone of a specific range of *metamorphic grade* in a *progressive metamorphic* region. The first appearance of the mineral (in passing from low to higher grades of metamorphism) marks the outer limit of the zone in question. See *mineral zone*. Cf. *typomorphic mineral. (Barrow 1912, p. 275; Tilley 1924a, p. 169; Turner & Verhoogen 1960, p. 191; Miyashiro 1994, p. 349)*

INEQUANT (Latin *in-* devoid of, and French *équant*, itself from the Latin *aequans*, being equal to). Said of mineral grains whose dimensions are significantly different from each other. Ant. *equant. (OU)*

INEQUIGRANULAR (Latin *in-*, devoid of, *aequus*, equal, and *granulum*, small grain). Term applied to rocks whose minerals are of two or more different sizes. The term is usually confined to igneous rocks; the specific recommended term for metamorphic rocks is *heteroblastic. (Cross* et al. *1906, p. 700)*

INFILTRATION(AL) METASOMATISM ⇒ Section 2.9. Type of *metasomatism* that takes place by the transfer of material in solution infiltrating through the host rocks. The driving force is the pressure and concentration gradients between the infiltrating and rock pore solutions. Infiltrational metasomatic rocks generally occupy much greater volumes than *diffusional metasomatic* rocks and the composition of minerals is constant across each of the *metasomatic zones*. See *diffusional metasomatism. (Korzhinskii 1953, p. 337; Korzhinskii 1970, p. 3)*

INJECTED HORNFELS (Farm Boschhoek, W of Lydenburg, Magnet Heights, South Africa). Term proposed for a shale that has been transformed into a *hornfels* by contact with ribbons of basic igneous material injected along the bedding planes. The presence of hornblende in the *hornfels* is attributed to the influence of the basic rock. *(Hall 1910, p. 66)*

INJECTION. Intrusion of a fluid phase (e.g. magma) into a rock. Ant. ***sublimation***. *(Virlet d'Aoust 1844, p. 827; Mehnert 1968, p. 355)*

INJECTION GNEISS. Gneiss whose *banding* is wholly or partly due to the *lit-par-lit*, or inter-laminar, *injection* of granitic magma into schistose, fissile or otherwise penetrable rocks. Cf. *mixto-gneiss. (Virlet d'Aoust 1884, p. 824; Gutzweiler 1912, p. 5; Holmes 1920, p. 124; Goldschmidt 1921, p. 87)*

INJECTIONITE. Obsolete term for *migmatite. (OU; Holmes 1928, p. 124)*

INJECTION METAMORPHISM. Type of metamorphism characterized by the intimate *injection* of magma as for example in *lit-par-lit* gneiss, *migmatite*, etc. It was regarded as characteristic of the deepest and/or most intense forms of metamorphism. Cf. *penetration metamorphism. (Michel-Lévy 1888, p. 104; Grubenmann & Niggli 1924, p. 325; Tyrrell 1926, p. 326; Turner 1948, p. 5)*

INTERMEDIATE ⇒ Section 2.1, Table 2.1.2. See *acid, intermediate, basic, ultrabasic*.

INVERSE METAMORPHISM. Term now superseded by *endomorphism*. Ant. *everse metamorphism*. (Not to be confused with *inverted metamorphism*.) *(Cotta 1846, p. 103)*

Inverted metamorphic zones. Superposed *metamorphic zones*, the *assemblages* of which denote an upward increasing grade. The corresponding geotherm has been named ***saw-tooth geotherm*** by Ruppel and Hodges. Hence ***inverted metamorphism***. Cf. *hot-slab metamorphism. (Medlicott 1864; Mallet 1875; Jaupart & Provost 1985, p. 385; Miyashiro 1994, p. 349; Shi & Wang 1987, p. 1048; Ruppel & Hodges 1994, p. 40)*

ISOCHEMICAL CONTACT METAMORPHISM (Greek *isos*, equal, identical). Type of *contact metamorphism* in which the original rock compositions are not significantly altered except for water and carbon dioxide. Ant. *allochemical contact metamorphism. (Reverdatto* et al. *1970, p. 311; Reverdatto* et al. *1974, p. 287)*

ISOCHEMICAL METAMORPHISM. Type of metamorphism in which the mineralogical and structural changes do not substantially modify the bulk rock composition. Ant. *allochemical metamorphism. (Scheumann 1936a, p. 402; Eskola 1939, p. 264; Yardley 1989, p. 5)*

Isochemical series. Sequence of metamorphic rocks of similar chemical composition formed under different *P–T* conditions and therefore of different mineral content. *(Grubenmann & Niggli 1924, p. 481; Tomkeieff)*

ISOFACIAL (Greek *isos*, same, identical, and Latin *facies*, face, aspect). Said of a group of rocks all belonging to the same *metamorphic facies*. Ant. *allofacial. (Tilley 1924a, p. 168)*

ISOGRAD (Greek *isos*, equal, and Latin *gradus*, degree). Surface across a rock sequence, represented by a line on a map, defined by the appearance or disappearance of a mineral, a specific mineral composition or a *mineral association*, produced as a result of a specific reaction, for example the 'staurolite-in' isograd defined by the reaction: garnet + chlorite + muscovite = staurolite + biotite + quartz + H_2O. Isograds represent mineral reactions, not rock chemical compositions. Hence, although expressions such as '*isoreaction-grad*' (Winkler) and '*reaction isograd*' (Bucher & Frey) may convey this meaning more accurately, they are unnecessary. See *isograde. (Barrow 1912, p. 275; Tilley 1924a, p. 169; Winkler 1970, p. 201; Winkler 1974, p. 65; Carmichael 1978, p. 770, note 2; Bucher & Frey 1994, p. 106)*

ISOGRADE [isogradal]. Pertaining to metamorphic rocks which, on the basis of their *mineral associations*, are considered to have reached the same *metamorphic grade* or range of grades, irrespective of their initial composition. Cf. *isofacial. (Tilley 1924a, p. 168)*

ISOPHASE METAMORPHISM (Greek *isos*, equal, and *phasis*, phase). Metamorphic *recrystallization* during which the minerals of the metamorphosed rock changed only their shape and/or orientation, but not their composition. Ant. *allophase metamorphism. (Eskola 1939, p. 264)*

ISOREACTION-GRAD, REACTION ISOGRAD. Unnecessary synonym of *isograd. (Winkler 1974, p. 65; Bucher & Frey 1994, p. 106)*

ITABIRITE (*Itabira*, and this from the Tupi *ita*, stone, and *bira*, upright, Minas Geraes, Brazil). Brazilian term originally applied to high-grade massive iron ore. Now widely used for schistose rocks composed essentially of quartz grains and scales of specular haematite with lesser magnetite or martite in variable proportions. *(Eschwege 1822, p. 28; Cordier 1868, p. 337)*

ITACOLUMITE (flexible sandstone, quartzite articulée) (Mt *Itacolumi*, Minas Geraes, Brazil, and this from the Tupi *ita*, stone, and *curumim*, child). Schistose and commonly flexible variety of quartzite containing micaceous minerals (mica, chlorite, so-called 'talc' which is probably sericite or hydromica) in addition to the chief constituent quartz. The flexibility of itacolumite seems to be due to voids between interlocking of irregular quartz grains. *(Cassendi 1655, in Tomkeieff; Eschwege 1832, p. 71; Lieber 1860, p. 359)*

-ITE ⇒ Section 2.1, Table 2.1.2. Suffix appended to a mineral to designate a rock consisting of $\geq$75% vol. of that mineral, e.g. biotitite, epidotite, glaucophanite. There are several exceptions to this guideline, namely *amphibolite*, *hornblendite*, *plagioclasite*, *olivinite* and *pyroxenite*.

JACOTINGA, JACUTINGA (probably after the *Jequetinhonha* river, Serra d'Espinhaço, Minas Geraes, Brazil; and from *Jacutinga*, Piping-Guan, a lustrous black Brazilian bird). Local term used in Brazil for a disaggregated *itabirite*, and also for variegated thin-bedded haematite iron ores containing gold. *(Heusser & Claraz 1859, p. 450)*

JASPERIZATION (Gatooma, Rhodesia now Zimbabwe). Conversion of original felsic volcanics to *banded* jaspery iron oxide rocks by *metasomatic* processes, taking place mainly at the weathering and cementation level. *(Zealley 1919, p. 45)*

JOINT (mining term). Fracture or parting surface in a rock along which there has been little or no displacement. Joints are normally planar and occur in parallel sets (***joint set***); more than one set may be present in a rock body. Hence ***jointing***, the presence of joints. Joints may be infilled during or after their formation. *(OU; Hooson 1747; Murchison 1839, p. 244; Leith 1923, p. 29; Ramsay & Huber 1987, p. 641)*

JUXTAPOSITION METAMORPHISM. Term introduced by Daubrée to designate *contact metamorphism*. His opinion was that the metamorphic effects of an igneous mass extended far beyond the contact surface, so that both the terms *contact* and *local metamorphism* were inadequate to describe the geological situation. *(Daubrée 1879, p. 153, notes 1–2)*

KAKIRITE (Lake *Kakir*, Swedish Lapland). *Fault rock* consisting of *porphyroclasts* set in a foliated and relatively fine-grained matrix. Rec.syn. *protomylonite*. *(Svenonius 1894, p. 244; Waters & Campbell 1935, p. 478; Higgins 1971, p. 74)*

KAMMSTEIN (Saxon-German *Kamm*, comb, and *Stein*, stone). Obsolete term for *serpentinite*. *(OU; Tomkeieff)*

KAMPTOMORPH (Greek *kamptos*, bent, and *morphê*, form). Crystal in a metamorphic rock that has been bent or deformed without breaking. The crystal may show *undulose extinction*. *(Milch 1894, p. 109)*

KARJALITE (*Karjala*, Finnish name for Karelia). Local name for a rock associated with *spilitized mafic rocks* and consisting essentially of albite with variable amounts of quartz, carbonate, light-coloured amphibole, chlorite and iron ore. *(Väyrynen 1938, p. 74)*

Kata- (Greek *kata*, downwards, against, completely). Prefix used to indicate an origin in the lowermost zone of metamorphism (e.g. *katazone*). Cf. *epi-*, *meso-*. The name was taken by Van Hise in the opposite sense, for a near-surface zone, but this latter meaning is not used. *(Van Hise 1904, p. 162; Grubenmann 1907, p. 21, 174)*

KATAGENESIS. See *catagenesis*.

KATAMORPHIC ZONE, ZONE OF KATAMORPHISM. Obsolete term for the zone at or near the Earth's surface where *katamorphism* takes place. The zone is divided into an upper weathering part, which extends from the surface to the groundwater level, and a lower cementation part, from groundwater level to the zone of *anamorphism*. Ant. *anamorphic zone*. *(Van Hise 1904, p. 162)*

KATAMORPHISM. Obsolete term for the process at or near the surface which leads to the *alteration* of rocks by weathering and cementation with the production of simple compounds from more complex ones. Characteristic reactions are oxidation, carbonation and hydration. Hence *katamorphic zone*, *zone of katamorphism*. Ant. *anamorphism*. *(Van Hise 1904, p. 43, 162)*

KATARANSKITE (*Kataranskii* Navolok, White Sea shore, Karelia, Russia). *Banded amphibolite* characteristically containing diallage and garnet. Terms such as diallage-amphibole-plagioclase rock/granofels/gneiss should be used in preference. *(Fyodorov 1903, p. 219)*

KATAROCKS, KATA-ROCKS. Term proposed for the rocks of the *deepest zone of metamorphism*. This prefix was used only in those cases where the rocks of a same name were encountered in different *depth zones* (e.g. kata-orthogneiss). Cf. *epi-*, *mesorocks*. *(Grubenmann 1907, p. 21, 174)*

Katazone (katametamorphic zone). Deepest zone of Grubenmann's three *depth zones* of metamorphism, where metamorphism is controlled by high to very high temperatures and pressures. According to Grubenmann and Niggli, the name also includes the effects of high-temperature *contact metamorphism*. Characteristic minerals are orthoclase, sillimanite, andalusite, cordierite, olivine, pyroxenes, spinel, garnets and basic plagioclase. Typical rocks are *granulite*, *eclogite*, etc. The concept of *depth zones* is now replaced by the *P–T* regimes of metamorphism. Cf. *epizone*, *mesozone*. *(Grubenmann & Niggli 1924, p. 376, 398)*

KATOGENE, KATOGENIC (Greek *katô*, down, below, and *gennô*, to give birth to). Obsolete and rather ambiguous term originally used for rocks derived by decomposition of rocks on or near the surface of the Earth. See *katogenic metamorphism*. Ant. *anogenic*. *(Haidinger 1845, p. 301; Kalkowsky 1886, p. 29; Tomkeieff)*

KATOGENIC DYNAMOMETAMORPHISM (Greek *katô*, down, below, and *gennô*, to give birth to). See *katogenic metamorphism*.

KATOGENIC METAMORPHISM (Greek *katô*, down, below, and *gennô*, to give birth to). Obsolete and ambiguous term: (1) for Kalkowsky, type of metamorphism operating at or near the Earth's surface and as such equivalent to *hydrothermal metamorphism* or to weathering. (2) For Becke, type of metamorphism operating in the deepest zone of the Earth's crust. In this sense it is synonymous with *katogenic dynamometamorphism*. Ant. *anogenic metamorphism*. *(Kalkowsky 1886, p. 29; Becke 1892, p. 298; Becke 1903b, p. 41)*

KAZAKHITE (*Kazakhstan*, middle Asia). Type of quartz-rich *metasomatic* rock that has formed by the replacement of acid volcanic or subvolcanic rock. *(Mashkovtsev 1937, p. 251)*

KELYPHITE (Greek *kelyphos*, rind, shell; Kremze near Ceské Budějovice, Budweis, in Bohemia, Czechoslovakia). Rim of radiating, intergrown microcrystalline minerals around garnet, variably composed of amphibole, pyroxene, plagioclase and spinel. Sometimes also used for a rim around olivine (and pyroxene). See *corona* and footnote. Hence ***kelyphitic structure***. *(Scharizer 1879, p. 244; Schrauf 1882, p. 359; Teall 1888, p. 436; Rosenbusch 1907, p. 366; Sederholm 1916, p. 47; Brière 1920, p. 163; Holmes 1920, p. 131)*

KELYPHITE-ECLOGITE (Mariànské Làzné, Marienbad, Bohemia, Czechoslovakia). Obsolete term for a *retrograded eclogite* in which garnet crystals are surrounded by *kelyphitic* rims. Should be replaced by *kelyphitic retrograded eclogite*. *(Patten 1888, p. 125)*

KERALITE (Greek *kêras*, horn, and *lithos*, stone). Obsolete term for a group of fine-grained siliceous rocks of differing origin and composition; also used for a fine-grained biotite *hornfels*. *(Pinkerton 1811a, p. 153; Pinkerton 1811b, p. 51; Cordier 1868, p. 204)*

KERATITE (Greek *kêras*, gen. *kêratos*, horn). Obsolete term for a refractory rock with a quartz aspect. Despite the etymology it was probably not a *hornfels*. *(Delamétherie 1806, p. 347)*

KEROLITE (Greek *keros*, wax, and *lithos*, stone). Obsolete term for a 'greasy' rock that is a mixture of serpentine and stevensite. *(OU; Humble 1860, p. 242)*

Khondalite (*Khonds*, population of S India). Metamorphic rock essentially composed of garnet, sillimanite, quartz and graphite; cordierite and spinel may also be present. Khondalite is the characteristic rock of the ***khondalite suite*** or ***series***, a group of high-grade predominantly metasedimentary rocks consisting of garnetiferous quartzites, graphite schists and silicate-rich *marbles*. In southern India khondalites are found in close geographical association with *charnockites*, the equivalent igneous-derived rocks. However, the term has not achieved the same international usage as *charnockite* and a systematic name should be considered in preference. *(Walker 1902, p. 11; Chacko* et al. *1992, p. 470; Cooray 1998, p. 710)*

KILLAS (Cornwall, UK). Sedimentary or schistose rock that has been metamorphosed by igneous emanations from a granite. The term is peculiar to the Cornish (SW England) mining field where the miners used it as a general term for the country rock hosting the ore field. *(OU; Woodward 1728, p. 5; Macculloch 1821, p. 344; Hawkins 1822, p. 251; Loewinson-Lessing 1893–4, p. 112; Holmes 1920, p. 132)*

KINETIC METAMORPHISM, KINEMATIC METAMORPHISM. Type of metamorphism where the pressure–temperature conditions are developed by and during regional shearing, that is with the temperature being generated by frictional heat. Also taken to mean deformation of a rock without chemical reconstitution. Rec.syn. *dislocation metamorphism*. *(Ambrose 1936, p. 257; Scheumann 1936c, p. 411; Turner & Verhoogen 1951, p. 370)*

Kinzigite (River *Kinzig*, Schwarzwald, Germany). Originally defined as a high-grade metamorphic rock, *granofelsic* or *gneissose*, mainly composed of pyrope garnet, biotite and plagioclase, smaller amounts of quartz, and graphite with, as accessories, zircon, ore minerals and sericite pseudomorphs after cordierite. Now used for a garnet-sillimanite-cordierite gneiss. *(Fischer 1860, p. 79; Fischer 1861, p. 641; Harker 1932, p. 249; Wager 1938, p. 12)*

KNOTTED SCHIST (German *Knoten*, knot and *Schiefer*, schist). Type of *schist* or *phyllite* in which mineral clots or *porphyroblasts* give the rock a knotted or nodular appearance. It is common in the outer parts of *contact-metamorphic* aureoles. *(Naumann 1849, p. 559; Naumann 1858, p. 543; Rosenbusch 1877, p. 79; Horne 1886, p. 99; Joplin 1968, p. 22)*

Kodurite (*Kodur* mines, Visakhapatnam, Andhra Pradesh, India). Metamorphic rock composed of

K-feldspar, Mn-garnet and apatite, with or without Mn-pyroxenes. Further minerals may be present, such as quartz and biotite. Kodurite is the characteristic rock of the ***kodurite series*** or ***suite***, a group of rocks probably of igneous origin, ranging from acid to basic and ultrabasic. *Alteration* of these rocks forms important residual manganese deposits. *(Holland 1907, p. 11; Fermor 1909, p. 244, 249)*

KOIKARITE (*Koikari* district, Finland). Any rock that possesses a *schistosity* and in which the *schistosity* planes lie at an angle to the bedding planes. Hence ***koikaritization***, ***koikarite structure***, etc. *(Eskola 1948, p. 159)*

KOLMITE, COLMITE (Monte *Colmo*, by Edolo, Adamello Group, Italian Alps). Mica-free fine-grained quartz-orthoclase-plagioclase gneiss forming layers in *phyllites*. *(Salomon 1908, p. 321)*

KORNITE (French *cornéenne*). Obsolete term for *hornfels*. *(Breithaupt 1830, p. 40)*

KOVDITE, KOWDITE (*Kovda*, a village on the White Sea shore, Murmansk region, Russia). Metamorphic rock composed of green clinopyroxene and orthopyroxene, with small amounts of plagioclase, mica and occasionally garnet. *(Fyodorov 1903, p. 215; Tomkeieff)*

KRITHIC STRUCTURE (Greek *krithê*, barley). Type of structure in which rounded feldspar grains are wrapped by thin sheets of mica and quartz. *(Becke 1880, p. 43)*

KUSEEVITE, KUZEEVITE (*Kuseeva* river, a tributary of Yenisey, Siberia, Russia). *Banded* variety of the *charnockitic* rock series, composed of quartz, plagioclase, K-feldspar, hypersthene, biotite, hornblende and garnet. *(Ainberg 1955, p. 111)*

KYANITE-SILLIMANITE SERIES. See *baric types of metamorphism*, *facies series*. Syn. *Barrovian-type facies series*.

KYRIOSOME (Greek *kyrios*, main, and *sôma*, body). Predominant part of a *chorismite*. Ant. *akyriosome*. *(Huber 1943, p. 90; Mehnert 1968, p. 355)*

LAMBOANITE (*Lamboany* plateau, S Madagascar). Obsolete term for a quartz-free *heteroblastic* gneissic rock composed of microcline, antiperthitic oligoclase, biotite and, locally, cordierite and garnet. The type rock derives from a sedimentary rock and is injected by cordierite pegmatite. *(Lacroix 1933, p. 62; Lacroix 1939, p. 292)*

LAMINAR STRUCTURE. Type of structure in which a rock is composed of laminae or plates into which it will preferentially break or weather. The term encompasses onion-skin exfoliation, *foliation*, *schistosity*, etc. *(Macculloch 1821, p. 119; Brongniart 1925, p. 69; Cordier 1868, p. 18)*

LAMINATED STRUCTURE. General term for a rock composed of laminae but used more specifically in the terminology of ultramafic rocks to describe a rock showing thin layers or lenticles (<2.5 mm) which are associated with modal variations. *(Harte 1977, p. 280)*

LAMPROSCHIST (Greek *lampros*, bright, shining). Metamorphosed lamprophyre with a *schistose structure*, containing brown biotite and green hornblende. *(OU; Bates & Jackson)*

LAVIALITE, LAVIAGNEISS (*Lavia*, central Finland). Local name for a metamorphosed porphyric rock of dioritic composition (*porphyroid*, metatuff) with relict xenocrysts of labradorite in a *recrystallized* groundmass composed of hornblende, biotite, oligoclase, microcline, quartz, accessory garnet, magnetite, titanite. *(Sederholm 1897–9, p. 57, 153)*

LAYER. One of a sequence of near-parallel, tabular-shaped rock bodies. Hence, ***layering***, the presence of layers; ***layered***, said of a rock that consists of a sequence of *layers*. Synonymous with *band*, *banding* and *banded*. *(OU)*

LEAF GNEISS. Type of gneiss with a leafy development of quartz. *(OU; Adams 1896, p. 93)*

LEPIDOBLASTIC STRUCTURE (Greek *lepis*, gen. *lepidos*, blade, and *blastos*, bud, sprout). Type of structure in a metamorphic rock in which the constituent grains are of equal size (*homeoblastic*), have a flaky or platy habit and are arranged parallel to each other. *(Becke 1903a, p. 570; Becke 1903b, p. 46; Grubenmann 1904, p. 79; Joplin 1968, p. 28; Passchier & Trouw 1996, p. 260)*

LEPTITE (Greek *leptos*, thin; Scandinavia). Obsolete term formerly used by Scandinavian geologists for fine-grained, *equigranular* metamorphic rocks of approximately granitic composition and probable supracrustal origin. With the increase of the mica content they pass into *phyllites* and with the increase of quartz into *quartzites*. *(Hummel 1875, p. 25; Sederholm 1897–9, p. 11, 97; Magnusson 1936, p. 333; Geijer 1944, p. 733; Geijer & Magnusson 1944, p. 20)*

LEPTITE-GNEISS (Nesodden near Oslo, Norway). Obsolete term formerly used for a metamorphic rock intermediate between a *leptite* and a *mica schist*. *(Broch 1927, p. 107; Broch 1929, p. 21)*

LEPTYNITE [leptinite] (Greek *leptynô*, to thin down, attenuate). Term used in France and some other countries and initially applied to a fine-grained high-grade rock, predominantly consisting of alkali feldspar, containing minor quartz, white mica, garnet and tourmaline, and with a planar *gneissose structure*. Later used for any white-coloured, quartzofeldspathic rock, typically forming *bands* alternating with metabasic rock, irrespective of the intensity of metamorphism. Cf. *Weissstein*. *(Haüy 1782, in Cordier 1868, p. 68; Brongniart 1813, p. 43; Coquand 1857; Lapparent 1882, p. 618; Roques 1941, p. 31; Cogné & Eller 1961; Santallier* et al. *1988, p. 3)*

LEPTYNOLITE ⇒ Section 2.10 (Greek *leptynô*, to thin down, attenuate, and *lithos*, stone). Obsolete term for a *contact-metamorphic* schistose rock, mainly composed of fine-grained mica and K-feldspar; a *contact-metasomatic* origin has been proposed for some leptynolites containing large K-feldspar *porphyroblasts*. Rosenbusch considered *cornubianite*, *proteolite* and *leptynolite* as 'schistose andalusite *hornfelses*'. Term now superseded by the special term *schistose hornfels*. *(Cordier 1842–8; Naumann 1858, p. 753; Daubrée 1867, p. 21; Cordier 1868, p. 203; Rosenbusch 1887, p. 42; Salomon 1898, p. 145; Lacroix 1899, p. 248)*

LEUCOCRATIC ⇒ Section 2.1, Table 2.1.2 (Greek *leucos*, white, and *cratô*, to predominate). According to the SCMR guidelines, adjective indicating that a metamorphic rock contains considerably less coloured minerals than would be regarded as normal for that rock type. Ant. *melanocratic*. *(Brögger 1898, p. 264; Lacroix 1916, p. 179)*

Leucophyllite (Greek *leukos*, white, and *phyllon*, leaf; Bucklinge Welt and Rosalien Mts., Austria, and Sopron Hills, Western Hungary). Local term for a type of quartz-muscovite-chlorite *phyllite*/schist, displaying a whitish colour. Also an obsolete term for a type of muscovite. *(Vendl 1929, p. 242)*

LEUCOSOME (Greek *leukos*, white, and *sôma*, body). Lightest-coloured parts of a *migmatite*. *(Dietrich & Mehnert 1960, p. 61; Mehnert 1968, p. 8)*

LIGHTNING METAMORPHISM ⇒ Sections 2.2 & 2.10, Fig. 2.2.1. Type of *metamorphism* of *local* extent that is due to a strike of lightning. The struck rock may melt (*fulgurite*) and processes such as vaporization and metal reduction may occur. *(Essene & Fisher 1986, p. 189; Frenzel* et al. *1989)*

LIME-PHYLLITE FACIES. *Metamorphic facies* proposed for very low-grade to *diagenetic* carbonate-bearing rocks containing fine-grained micaceous minerals and devoid of newly formed epidote. *(Eskola 1920, p. 165)*

LIME-SILICATE HORNSTONE. *Calc-silicate rock* with a horny aspect and produced by the *contact metamorphism* of shaly limestone. It is a calc-silicate *hornfels*. *(Barrell 1907, p. 119; Shand 1931)*

LIME-SILICATE ROCK. Carbonate-free *calc-silicate rock* of horny aspect, most commonly found in *contact aureoles* and derived from impure limestones or calcareous shales; also found in *regionally metamorphosed* complexes. *(Barrell 1907, p. 119; Harker 1932, p. 89)*

LIMURITE (in honour of Count de *Limur*, mineralogist and collector; Lesponne valley, Hautes-Pyrénées, France). Boron-rich *metasomatic calc-silicate rock* consisting of >50% axinite, associated with one or more of the following minerals: Ca-pyroxene, actinolite, epidote, vesuvianite, grossular ± quartz + calcite. The axinite results from boron-*metasomatism* overprinted on a *calc-silicate rock* generated by *contact metamorphism*. Rec.syn. axinite-rich *skarn*. *(Zirkel 1879, p. 380; Lacroix 1892, p. 740)*

LINEATION. Any repetitively occurring or penetrative visible linear feature in a rock body. It can be straight or curved and may be defined by: the alignment of the long axes of elongate mineral grains or grain aggregates (***mineral lineation***), the parallelism of small-scale folds (***crenulation lineation***), or the intersection of two foliations (***intersection lineation***).

The term is synonymous with and replaces the earlier term ***linear parallel structure***. *(OU; Naumann 1858, p. 432; Knopf & Ingerson 1938, p. 69; Cloos 1946, p. 1; Hobbs* et al. *1976, p. 267; Davis & Reynolds 1996, p. 14; Passchier & Trouw 1996, p. 260; Foucault & Raoult 2005, p. 198)*

LIPOTECTITE, LIPOTEXITE (Greek *lipô*, to miss, let out, and *têkô*, to melt). Obsolete term proposed for that part of a *migmatite* which has not passed through a melt phase and remains in a solid state after *anatexis*; the complementary melt phase was referred to as a *palingenite*. Later defined as basic material. Lipotectites were divided into *atexites/atectites*, which preserved their original structure due to their refractory nature, and *metatectites* whose original characters had been altered by partial infiltration or reaction with the melt phase. Cf. *palaeosome*, *restite*. *(Smulikowski 1947, p. 266; Smulikowski 1950, p. 131; Dietrich & Mehnert 1960, p. 61)*

Listvenite (Russian *listvenitsa*, larch; probably because of the usually green mica; Beresovsk gold deposit, Ural, Russia). Low-temperature *metasomatic* rock characterized by quartz, iron-magnesium carbonate, fuchsite and pyrite *assemblages*, resulting from the replacement of ultrabasic rocks. As such it may be regarded as a type of *ophicarbonate*. Hence ***listvenitization***, type of *metasomatism* leading to the formation of listvenites. *(Rose 1842, p. 539; Korzhinskii 1953, p. 443)*

LITHIC (IMPACT) BRECCIA. *Polymict impact breccia* with clastic matrix containing shocked and unshocked mineral and lithic clasts, but lacking cogenetic *impact melt* particles. It is synonymous with, and supersedes *fragmental breccia*. *(Stöffler & Grieve 1994, p. 1347; Stöffler & Grieve Section 2.11)*

LIT-PAR-LIT INJECTION (French *lit par lit*, bed-by-bed). Penetration of igneous material, usually granitic, along foliation planes to produce thin parallel-sided veins on the millimetre to centimetre scale and repeated throughout a rock to give a *banded structure*. Hence ***lit-par-lit gneiss***. *(Michel-Lévy 1888, p. 104)*

LOAD-CONTACT METAMORPHISM (Shuswap terrain of British Columbia, Canada). Type of *contact metamorphism* in which the superincumbent load was a significant factor. *(Daly 1917, p. 408; Grubenmann & Niggli 1924, p. 178)*

LOAD METAMORPHISM. Type of metamorphism produced under static conditions and with pressures and temperatures related to the superincumbent load. Rec.syn. *burial metamorphism*. Ant. *relief metamorphism*. *(Milch 1894, p. 121; Sander 1911, p. 284; Daly 1917, p. 400)*

LOCAL METAMORPHISM ⇒ Section 2.2, Fig. 2.2.1. Type of metamorphism of limited areal (volume) extent in which the metamorphism may be directly attributed to a localized cause, such as a magmatic intrusion, *faulting* or meteorite *impact*. *(Naumann 1849, p. 751; Teall 1888, p. 438; Daly 1917, p. 395; Joplin 1968, p. 35)*

LOHESTITE (after J. M. *Lohest*, a Belgian scientist; Ardennes, Belgium). Obsolete name for very fine-grained (submicroscopic) aggregates of andalusite occurring as almost isotropic knots in phyllites and originally taken as a stage in the formation of andalusite crystals. Cf. *knotted slate*. *(Anten 1923, p. 29)*

LOW-RANK METAMORPHISM. Ill-defined term for a type of metamorphism occurring under low to moderate temperatures and pressures as opposed to *high-rank metamorphism*. It should not be confused with the 'rank' expressing in very low-grade metamorphic terminology the thermal maturity of coals. *(OU)*

LOZÉRINE (Lozère, a mountain and a region in SW Massif Central, France). Obsolete name for a type of 'talc' gneiss, talc in the original definition probably meaning any whitish or greenish mica or mixture of sheet silicates. *(Serres 1863, p. 139; Tomkeieff)*

LUNAR SOIL. Unnecessary synonym of lunar *regolith*. It has also been defined as the submillimetre fraction of the lunar *regolith*. *(OU; Engelhardt* et al. *1970, p. 363; Warner & Simonds 1989, p. 297; Heiken* et al. *1991, p. 287)*

MACULOSE STRUCTURE (Latin *macula*, stain, spot). Type of structure in a metamorphic rock in which the development and distribution of *porphyroblasts* give the rock a *knotted* or *spotted* appearance; the structure is characteristic of *contact-metamorphic* rocks (cf. *knotted schist*, *spotted schist*, *garbenschiefer*). Holmes used the term as one of his groupings of metamorphic structures, along with schistose, gneissose, etc. The term replaced 'maculated texture' (Hatch). Rec.syn. *knotted* or *spotted structure*. *(Hatch 1888, p. 438; Holmes 1920, p. 147; Tyrrell 1926, p. 272)*

MAFIC GRANULITE ⇒ Section 2.8. Type of *granulite* with more than 30% *mafic minerals* (dominantly pyroxene). *(OU)*

MAFIC HORNFELS ⇒ Section 2.10. General term for *hornfelsed* basic igneous or meta-igneous rocks; the main components are plagioclase and one or more *mafic minerals*; pyroxene(s) and minor olivine are common under the relatively dry conditions existing in the *aureoles* of many basic intrusions, whereas hornblende ± biotite form preferentially in more hydrated conditions, as in most *aureoles* of intermediate to acid intrusions. Mafic hornfelses include, among others, rocks known in the literature as *beerbachite*, *granoblast*, *muscovadite* and *sudburite*, as well as *contact rocks* called *pyroxene-granulites* by some British petrologists. The term supersedes *basic hornfels* in accord with the SCMR guidelines. (*Callegari & Pertsev Section 2.10*)

MAFIC MINERALS ⇒ Table 2.1.2 (from *ma*gnesian and *f*erric). Collective term for modal ferromagnesian and other non-felsic minerals. Hence ***mafic rock***, a rock mainly consisting of *mafic minerals*. *(Cross* et al. *1912, p. 561; Streckeisen 1967, p. 151)*

MAGNESIAN HORNFELS. Type of *hornfels* composed predominantly of Mg-minerals, which may be derived by the *contact metamorphism* of an ultrabasic rock or by Mg-*metasomatism* associated with the metamorphic imprint. Cf. *ultramafic hornfels*. *(Williams* et al. *1982, p. 497)*

Main Donegal-type contact metamorphism (*Main Donegal* granite, Ireland). Type of *isochemical contact metamorphism* that developed contemporaneously with the country rock deformation, giving rocks ranging from biotite-sillimanite gneiss and kyanite schist down to *greenschist*. The *aureole* may be 2–3 km thick. See *contact aureole systematics*. Syn. *piezo-contact metamorphism*. *(Reverdatto* et al. *1970, p. 312)*

MANJAKITE (Volotara, east of Ambohi*manjaka*, 50 km south of Antsirabé, Madagascar). *Mafic* metamorphic *rock* found as layers in gneiss and composed of orthopyroxene (hypersthene), labradorite, biotite and large *porphyroblasts* of garnet (pyrope-almandine); accessories are magnetite and apatite. *(Lacroix 1914, p. 75)*

Mantled porphyroclast. *Porphyroclast* characterized by tapered tails or wings elongated along the *foliation* and composed of aggregates of the same material as the *porphyroclast*. The tails are derived from deformed mantles (cf. *core-and-mantle structure*) developed at the expense of the *porphyroclast*. The direction of elongation of the tail indicates the sense of shear. Syn. *porphyroclast system*. *(OU; Hanmer & Passchier 1991, p. 39; Passchier & Trouw 1996, p. 115)*

MARBLE ⇒ Section 2.7, Fig. 2.7.1 (Greek *marmaron* through the Latin *marmor*). Metamorphic rock containing more than 50% vol. of carbonate minerals (calcite and/or aragonite and/or dolomite). ***Pure marble*** contains more than 95% vol. of carbonate minerals; a marble containing less than 95% of carbonate minerals is classified as ***impure marble***. In commerce, marble is used extensively for many rock types that can take a polish. *(Herodotos* c. *440–420 BC, Lib. V-62; Theophrastos 325 BC, p. 13; Pliny the Elder 80, Lib. 36-I-2, p. 500; Geikie 1882, p. 772; Zirkel 1894b, p. 446; Joplin 1935, p. 386; Brooks 1954, p. 758)*

MARBLEIZATION. Synonymous with *marmorization*.

MARMARITE. See *marmorite*.

MARMAROSIS, MARMOROSIS. Conversion of limestone to *marble* as a result of *contact metamorphism*. *(Geikie 1882, p. 577)*

MARMOLITE (Greek *marmaron*, marble, and *lithos*, stone). Obsolete term for a variety of *serpentinite* used as an ornamental stone. *(Nutall 1822, p. 17; Cordier 1868, p. 166)*

MARMORITE, MARMARITE. Lime-silicate *hornfels*, i.e. a *contact-metamorphosed* impure limestone or a limestone to which silica has been added by magmatic solutions, containing various calcium-silicates such as wollastonite, diopside, garnet, vesuvianite, etc. Rec.syn.

calc-silicate *hornfels* or *skarn*. *(Shand 1931, p. 183; Shand 1951, p. 204)*

MARMORIZATION. Conversion of limestone into *marble* by any metamorphic process. *(Spurr 1923, p. 633)*

MASKELYNITE (in honour of the English astronomer Nevil *Maskelyne*, 1732–1811). *Diaplectic* plagioclase *glass*; term originally used for amorphous plagioclase of shocked meteorites. *(Tschermak 1872, p. 127; Milton & De Carli 1963, p. 670; Binns 1967, p. 1111)*

MAYAITE (*Maya* people; Chichen Itza, Honduras, Copan, Yucatan). Name proposed for a series of jade-like rocks grading from *tuxtlite* to nearly pure albite rock. *(Washington 1922b, p. 325)*

MECHANICAL METAMORPHISM. Synonymous with *dislocation metamorphism*. *(Baltzer 1873, in Rosenbusch 1923, p. 62; Hatch 1888, p. 438)*

MEGABRECCIA (Greek *megas*, large, old German *brecha*, fracture, and Ligurian-Italian *breccia*, crushed stone). *Polymict impact breccia* containing lithic clasts up to the size of several hundred metres as part of the *ejecta blanket* of an *impact crater*. Recognized and defined at the scale of geological field studies. Ant. *microbreccia*. *(Stöffler & Grieve Section 2.11)*

MEGAREGOLITH (Greek *megas*, large, *rhegos*, blanket, and *lithos*, stone). Layer of fractured and possibly mechanically mixed primordial planetary crust formed by multiple large *impacts* on planetary bodies during the early intense *impact* bombardment (prior to about four billion years ago). The thickness of this layer can be of the order of kilometres to tens of kilometres. *(OU; Hartmann 1973, p. 634)*

MEGASTRUCTURE. See *structure*.

MELANOCRATIC ⇒ Table 2.1.2 (Greek *melas* gen. *melanos*, black, and *cratô*, to predominate). According to the SCMR guidelines, adjective indicating that a metamorphic rock contains considerably more coloured minerals than would be regarded as normal for that rock type. Ant. *leucocratic*. *(Brögger 1898, p. 263; Lacroix 1916, p. 179)*

MELANOSOME (Greek *melas*, gen. *melanos*, black, dark, and *sôma*, body). Darkest parts of a *migmatite*, usually with prevailing dark minerals. It occurs between two *leucosomes* or, if remnants of the more or less unmodified *parent rock* (*mesosome*) are still present, it is arranged in rims around these remnants. *(Dietrich & Mehnert 1960, p. 61; Mehnert 1968, p. 355)*

MELTING METAMORPHISM (German *Einschmelzungmetamorphose*). Metamorphism produced by a partial or total melting of rocks. Equivalent to *anatexis*. *(Grubenmann 1904, p. 72)*

MELT POCKET. Region of localized quenched melt produced by shock-induced localized melting in moderately to strongly *shocked rocks*; typical of shocked *mafic rocks* including meteorites such as chondrites and achondrites. *(OU; Dodd & Jarosewich 1979, p. 335, 338, Fig. 1, Table 1; Stöffler* et al. *1991, p. 3855, Figs. 15, 16, Table 1; Bischoff & Stöffler 1992, p. 722)*

MELT VEIN. Irregular vein of quenched melt produced by shock-induced localized melting in moderately to strongly *shocked rocks*; commonly observed in the (par)autochthonous basement of *impact craters* and in allochthonous clasts of *shocked rocks* (terrestrial, lunar and meteoritic). *(OU; Dodd & Jarosewich 1979, p. 338, Table 1; Bischoff & Stöffler 1992, p. 720; Spray 1998, p. 200)*

MERISMITE (Greek *merisma*, part). *Chorismite* in which there is an irregular arrangement of the parts. *(Huber 1943, p. 90; Niggli 1948, p. 110; Dietrich & Mehnert 1960, p. 61)*

MESITIS (Greek *mesitês*, mediator). Metamorphic process tending to homogenize compositionally different parts of a rock and produce a uniform rock of intermediate composition. Mesitis is opposite to *metamorphic differentiation*. *(Hentschel 1943, p. 85)*

Meso- (Greek *mesos*, middle, median). Prefix used to indicate an origin in the middle *zone of metamorphism*. Cf. *epi-*, *kata-*. *(Grubenmann 1907, p. 21, 174)*

MESOCATACLASITE. *Cataclasite* in which the matrix forms more than 50% and less than 90% of the rock volume. *(Brodie* et al. *Section 2.3)*

MESOMETAMORPHISM. Metamorphism occurring in the middle zone of the Earth's crust under moderate pressures and temperatures. Cf. *epimetamorphism*. *(Lindgren 1928, p. 110; Tomkeieff)*

MESOMYLONITE. *Mylonite* in which more than 50% and less than 90% of the rock volume has undergone grain-size reduction. *(Brodie* et al. *Section 2.3)*

MESOROCKS, MESO-ROCKS. Term proposed for rocks of the middle *zone of metamorphism*. The prefix was used only in those cases where the rocks of the same name were encountered in different *depth zones* (e.g. meso-*amphibolite*). Cf. *epirocks*, *katarocks*. *(Grubenmann 1907, p. 21, 174)*

MESOSOME. Part of a *migmatite* that is intermediate in colour between *leucosome* and *melanosome*. If present, the mesosome is mostly a more or less unmodified remnant of the *parent rock* (*protolith*) of the *migmatite*. *(Henkes & Johannes 1981, p. 115)*

MESOSTRUCTURE. See *structure*.

MESOTHERMAL (Greek *mesos*, middle, median, and *thermos*, hot). Said of a *hydrothermal* ore deposit that was formed at moderate depths and at moderate temperatures. Cf. *epithermal*, *hypothermal*. *(OU; Lindgren 1933, p. 210)*

Mesozone (mesometamorphic zone) (Greek *mesos*, middle, median, and *zônê*, zone, belt). Middle zone of the three *depth zones* of metamorphism, where metamorphism is controlled by moderate temperatures and pressures. According to Grubenmann and Niggli the name includes also the effects of intermediate-temperature *contact metamorphism*. Characteristic minerals are staurolite, kyanite, muscovite, biotite, almandine, intermediate plagioclase, anthophyllite, etc. The concept of *depth zone* is now replaced by the *P–T* regimes of metamorphism. Cf. *epizone*, *katazone*. *(Grubenmann & Niggli 1924, p. 374, 394)*

META- ⇒ Table 2.1.2 (Greek *meta-*, indicating a change). Prefix used to indicate the presence of metamorphism. It is used in front of an igneous or sedimentary rock name to indicate that the rock has been metamorphosed. The prefix 'meta' should not be used for a former metamorphic rock (e.g. meta-*eclogite* is not an acceptable term). If the *protolith* was a metamorphic rock it should be referred to in the form 'metamorphosed *eclogite*', or more specifically, '*amphibolitized eclogite*', '*retrogressed eclogite*', '*contact-metamorphosed eclogite*', etc. *(OU; Whitman & Cross, in Bascom 1893, p. 828; Van Hise 1904, p. 776; Loewinson-Lessing 1905, p. 407; Lacroix 1920, p. 685–90)*

METABLASTESIS (Greek *meta-*, indicating a change, and *blastêsis*, germination). Preferred crystallization and growth in size of a mineral or a group of minerals by metamorphic, including *metasomatic*, processes. *(Scheumann 1936b, p. 405; Dietrich & Mehnert 1960, p. 61; Mehnert 1968, p. 355)*

METACHEMIC, METACHEMICAL. Said of a process involving chemical change in a rock. Rec.syn. *metasomatic*. Hence, ***metachemical metamorphism***, rec.syn. *metasomatism*. Cf. *methylosis*, *paramorphism*. *(Dana 1886, p. 70)*

METACONITE, METAKONITE (Greek *meta-*, indicating a change, and *konite*, a fine-grained clayey limestone). Limestone *recrystallized* under *diagenetic* conditions. *(Pinkerton 1811b, p. 429; Steinmann 1925, p. 445)*

METACRASIS (Greek *meta-*, indicating a change, and *krasis*, mixture, mixing). Obsolete term for a proposed subdivision of metamorphism dealing with the complete re-ordering of the mineral content and structure of a rock, such as the change from a sedimentary rock to a crystalline rock. Cf. *metastasis* and *methylosis*. *(Bonney 1886, p. 59)*

METACRYSTAL, METACRYST. Any large crystal grown in a metamorphic rock. Rec.syn. *porphyroblast*. *(Lane 1903, p. 388; Lindgren 1912, p. 528)*

METAGENESIS. Obsolete term used especially by Russian authors to indicate a stage of post-*diagenetic alteration* that is more advanced than in *epigenesis* or *catagenesis*. It is subdivided into ***early metagenesis***, which roughly corresponds to the *anchizone*, and ***late*** or ***deep metagenesis***, which is more or less equivalent to the *epizone* or chlorite zone of the *greenschist facies*. *(Loewinson-Lessing 1925, p. 297; Kisch 1983, p. 294–7)*

METAMORPHIC (MINERAL) ASSEMBLAGE, METAMORPHIC (MINERAL) ASSOCIATION. Totality of minerals forming a metamorphic rock, or a *domain* in a metamorphic rock. The listed minerals need not be in equilibrium with each other nor result from one *metamorphic phase*. The list is commonly complemented by the relative abundances of each mineral (but not the structure or *fabric*). See *paragenesis*. *(OU; Brongniart 1825, p. 155; Koons* et al. *1987, p. 68)*

METAMORPHIC DIFFERENTIATION ⇒ Section 2.6. Mechanical redistribution of minerals by species and/or segregation of chemical components during metamorphism to form an inhomogeneous structure of two or more species within a rock body. *(Stillwell 1918, p. 12, 62, 200; Eskola 1932, p. 68)*

METAMORPHIC EVENT ⇒ Section 2.2. Coherent sequence of metamorphic conditions (temperature, pressure, deformation) under which metamorphic reconstitution commences and continues until it eventually ceases. Typically a metamorphic event will involve a cycle of heating and cooling, which in orogenic metamorphism will be accompanied by pressure and deformation variations. It is represented by a clockwise or anticlockwise loop on a *P–T* grid. A metamorphic event may be *polyphase* or *monophase*. *(OU)*

METAMORPHIC FACIES ⇒ Section 2.2. Set of *metamorphic mineral assemblages*, repeatedly associated in time and space and showing a regular relationship between mineral composition and bulk chemical composition, such that different metamorphic facies (sets of *mineral assemblages*) appear to be related to different metamorphic conditions, in particular temperature and pressure, although other variables, such as P_{H_2O}, may also be important. The SCMR recommends the use of ten facies, namely *zeolite facies*, *subgreenschist facies*, *greenschist facies*, *epidote-amphibolite facies*, *amphibolite facies*, *pyroxene-hornfels facies*, *sanidinite facies*, *granulite facies*, *glaucophane-schist/blueschist facies* and *eclogite facies*. *(Eskola 1915, p. 17, 114; Eskola 1920, p. 146; Tilley 1925; Fyfe & Turner 1966, p. 354; Winkler 1967; Winkler 1974, p. 63; Turner 1981, p. 202; Yardley 1989, p. 50; Bucher & Frey 1994, p. 10; Miyashiro 1994, p. 188)*

METAMORPHIC FACIES SERIES ⇒ Section 2.2. See *facies series*.

METAMORPHIC GEOTHERM. Line joining the maximum temperature points (T_{max}) on the *P–T–t curves* for successive *mineral zones* in a *prograde* sequence. The line differs from a *geotherm* or *geothermal gradient* because the different T_{max} points did not exist simultaneously. *(England & Richardson 1977, p. 204; Yardley 1989, p. 199)*

METAMORPHIC GRADE ⇒ Section 2.2, Fig. 2.2.3. Term used to indicate the relative conditions of metamorphism. The SCMR recommends that metamorphic grade refer only to temperature; the whole range may be divided into five grades, that is, very low, low, medium, high and very high grade of metamorphism. *(Tilley 1924a, p. 168; Winkler 1974, p. 61; Turner 1981, p. 85; Miyashiro 1994, p. 350)*

METAMORPHIC PHASE ⇒ Fig. 2.2.2. Part of a *polyphase metamorphic event* or totality of a *monophase metamorphic event*, with only one temperature and/or pressure climax. Metamorphic phases are typically marked by changes in the *P/T* evolution of the system caused by deformation leading to increased burial or uplift. The main crystallization of a phase, for example the growth of *porphyroblasts*, may be related to the deformation by such terms as pre-kinematic, syn-kinematic, post-first deformation, etc.

METAMORPHIC ROCK. Rock that has undergone metamorphism. *(Lyell 1833, p. 375)*

METAMORPHIC SUBFACIES ⇒ Section 2.2. Subdivision of a *metamorphic facies*. *(Eskola 1939, p. 354; Turner 1948, p. 58; Turner & Verhoogen 1960, p. 505; Lambert 1965, p. 285; Fyfe & Turner 1966, p. 358; Winkler 1970, p. 193; Winkler 1974, p. 61)*

METAMORPHIC ZONE. Mappable part of a metamorphic complex, in which rocks of a same chemical composition show identical *mineral associations*. The same *metamorphic grade* can thus be inferred for them. These zones may be named according to the typical *mineral assemblage* present in a specific rock type, or by only one mineral. Zones normally occur in a series indicating *prograde* or *retrograde metamorphism*. *(Rosenbusch 1877, p. 79; Horne 1886, p. 98; Sederholm 1891, p. 140; Barrow 1893, p. 343; Becke 1903b, p. 32; Grubenmann 1904, p. 57; Barrow 1912, p. 275; D'Amico* et al. *1987, p. 374; Bucher & Frey 1994, p. 100)*

METAMORPHISM ⇒ Section 2.2 (Greek *meta-*, indicating a change, and *morphê*, form). Process involving changes in the mineral content/composition and/or *microstructure* of a rock, dominantly in the solid state. This process is mainly due to an adjustment of the rock to physical conditions that differ from those under which the rock originally formed and that also differ from the physical conditions normally occurring at the surface of the Earth and in the zone of *diagenesis*. The process may coexist with partial melting and may also involve changes in the bulk chemical composition of the rock. *(Boué 1820, p. 200; Lyell 1833, p. 375; Lyell 1854; Rosenbusch 1877, p. 178)*

METAMORPHISM BY REGENERATION. Obsolete term for a type of *isochemical contact metamorphism* in which a new *mineral assemblage* crystallized under the influence of both a high magmatic temperature and the contemporaneous presence of fluid ('crystallizing agents'), as opposed to a 'metamorphism by addition' or *additive metamorphism*. *(Spurr* et al. *1912, p. 455)*

METAMORPHITE. Synonymous with metamorphic rock, which is the recommended term.

METAPEPSIS [metapepis] (Greek *meta-*, indicating a change, and *pepsis*, digestion). Obsolete and erroneous term for *regional metamorphism*, based on the concept that the rocks had been altered by intensely heated water or steam. Ant. *paroptesis*. *(Kinahan 1878, p. 4, 175; Bonney 1886, p. 59)*

Metasomatic column ⇒ Section 2.9. Complete sequence of *metasomatic zones* characterizing an individual *metasomatic facies*. *(Korzhinskii 1953, p. 340)*

METASOMATIC CONTACT METAMORPHISM. Synonymous with *contact metasomatism* or *allochemical contact metamorphism*. *(Hatch & Rastall 1913, p. 260)*

Metasomatic facies ⇒ Section 2.9. Regular set of *metasomatic zones* (a *metasomatic column*) developed under similar physico-chemical conditions (the compositions of the original rock and the metasomatized rock, the temperature and pressure conditions and the composition of metasomatizing solutions or fluids). *(OU; Burnham 1962, p. 770)*

Metasomatic family ⇒ Section 2.9. Totality of related *metasomatic facies* typical of a given petrogenetic process. The facies differ from each other by the appearance or disappearance of mineral *parageneses* or *metasomatic zones* reflecting a difference in one (or more) of the physico-chemical parameters.

METASOMATIC ROCK (metasomatite). Metamorphic rock whose mineral and chemical bulk compositions have been substantially changed by *metasomatism*.

METASOMATIC ZONE ⇒ Section 2.9. *Metasomatic rock* defined by a specific mineral *paragenesis*. *(Zharikov* et al., *this vol)*

METASOMATISM ⇒ Section 2.9 (Greek *meta-*, indicating a change, and *sôma*, body). Metamorphic process by which the chemical composition of a rock or rock portion is altered in a pervasive manner and which involves the introduction and/or removal of chemical components as a result of the interaction of the rock with aqueous fluids (solutions); during metasomatism the rock remains in a solid state. *(Naumann 1826, p. 209; Lindgren 1912, p. 527; Lindgren 1925, p. 247; Lindgren 1933, p. 91; Turner 1948, p. 109; Barton* et al. *1991a, p. 321; Zharikov* et al. *1998, p. 17)*

METASOME. Originally defined as the new mineral or minerals in a *metasomatic* type of ore deposit, the term was later used for a metasomatized *palaeosome* in a *migmatite* and in this sense it is synonymous with *metatectite*. *(Lindgren 1912, p. 528; Scheumann 1936a, p. 302)*

METASOST (Greek *meta-*, indicating a change, and *sôzô*, to save, to retrieve). Obsolete term for the mobile or introduced part of a *migmatite*, as opposed to *metaster*. *(Scheumann 1955, p. 15)*

METASTASIS (Greek *meta-*, indicating a change, and *stasis*, position, stop). Obsolete term for a proposed subdivision of metamorphism dealing with a change of order, denoting a change of a *paramorphic* character, such as the crystallization of limestone to *marble*. Cf. *metacrasis* and *methylosis*. *(Bonney 1886, p. 59)*

METASTER (Greek *meta-*, indicating a change, and *stereos*, solid). Obsolete term for the solid (immobile or less mobile) portion of a *migmatite*, as opposed to *metasost*. Synonymous with *restite*, which is the recommended term. *(Scheumann 1955, p. 14; Dietrich & Mehnert 1960, p. 61)*

METATAXIS (Greek *meta-*, indicating a change, and *taxis*, order, arrangement). Obsolete term for a proposed subdivision of the metamorphic processes that is characterized by mechanical changes such as in the arrangement of minerals presumed to have occurred during the formation of a *slaty cleavage*. Cf. *paramorphism* and *metatropy*. Syn. *dynamic metamorphism*. *(Irving 1889, p. 5)*

METATECT (Greek *meta-*, indicating a change, and *têkô*, to melt). Discrete, mostly light-coloured body in a *migmatite* formed by *metatexis*. *(Scheumann 1936a, p. 302; Dietrich & Mehnert 1960, p. 61)*

METATECTITE. *Restite* produced during *anatexis* and whose mineral or chemical composition is further altered by a largely *metasomatic* reaction with, or impregnation by, the *anatectic* melt. See *lipotectite*. *(Smulikowski 1947, p. 267; Smulikowski 1950, p. 132; Dietrich & Mehnert 1960, p. 61)*

METATEXIS. Initial stage of *anatexis* where the *parent rock* (*palaeosome*) has been partly split into a more mobile part (*metatect*) and a non-mobilized (depleted) *restite*. See *diatexis*. *(Scheumann 1936a, p. 302; Dietrich & Mehnert 1960, p. 61; Mehnert 1968, p. 355)*

METATEXITE. Variety of *migmatite* with discrete *leucosome*, *mesosome* and *melanosome*. *(Scheumann 1936b, p. 409)*

METATROPY (Greek *meta-*, indicating a change, and *tropos*, turn). Obsolete term for a proposed subdivision of the metamorphic processes, which is characterized by changes in the physical properties of the rock without important chemical changes in either the rock or its constituents, e.g. liquefaction, *dehydration* of the minerals. Cf. *metataxis*, *paramorphism*. *(Irving 1889, p. 5)*

METEORITE CRATER (Greek *meteôros*, high in the air). Small *impact crater* with remaining fragments of the *impacted* meteoroid. Previously used as a synonym of *impact crater*. *(OU; Spencer 1932, p. 781; Spencer 1933b, p. 227; Dence 1964, p. 249, Table 1)*

METEORITE IMPACT CRATER. Synonymous with *impact crater*. *(OU; French & Short 1968, p. 636)*

METHARMOSE, METHARMOSIS [metaharmosis] (Greek *metharmosis*, change). Term proposed to designate all changes that a sediment may undergo, including *diagenesis* proper and metamorphism. Hence ***bathy-***, ***meso-*** and ***hypsi-metharmose*** depending at which depth the changes take place. *(Kessler 1921, p. 241; Tomkeieff; Bates & Jackson 1987)*

METHYLOSIS (Greek *meta-*, indicating a change, and *hylê*, substance, material). Obsolete term for a proposed subdivision of metamorphism dealing with chemical changes (change of substance) of a *pseudomorphic* character. Rec.syn. *metasomatism*. Cf. *metacrasis* and *metastasis*. *(King & Rowney, in Bonney 1886, p. 59; Dana 1886, p. 70)*

MIANTHITE (Greek *miainô*, to stain, to contaminate). Aggregates of dark-coloured refractory minerals in a *restite*. *(Smulikowski 1947, p. 269; Dietrich & Mehnert 1960, p. 62)*

MICACITE (from *mica*, to avoid using micaschist for a rock that is not a schist). Obsolete term originally used for a fissile rock composed of variously coloured mica crystals (50–70% vol.) and quartz, both constituents being separated into alternating laminae; very thin laminae are commonly folded. It may also contain calcite, graphite and other minerals. *(Cordier 1868, p. 196)*

MICA-FISH. Asymmetric or lozenge-shaped mica *porphyroclasts* found in *mylonites*. The asymmetry indicates the sense of shear. *(Lister & Snoke 1984, p. 620; Davis & Reynolds 1996, p. 522; Vernon 2004, p. 485)*

MICALCITE (contraction of *mi*ca and *calcite*). Obsolete term for a mica-bearing calcite *marble*. 'Micaceous marble' should be used in preference. *(Cordier 1868, p. 287)*

MICA SCHIST. According to SCMR rules, variety of schist with mica as the only major constituent. Mica schist ('Glimmerschiefer', 'micaschiste') was originally defined as a mixture of mica and quartz, and regarded as one of the major rock groups. Early workers differentiated mica schist from clay *slate*. The latter was regarded as retaining sedimentary affinities and mineral grains, whereas mica schist was regarded as a '*crystalline schist*', *phyllite* being introduced as a transitional rock type. *(Werner 1786, p. 279; Brochant 1800, p. 569; Omalius d'Halloy 1831; Lyell 1851, p. 465; Zirkel 1866b, p. 448; Tyrrell 1926, p. 305; Joplin 1968, p. 23)*

MICAZCITE, MICALITE, MICASYTE. Obsolete terms used for *mica schist*. *(Kinahan 1873, p. 355)*

Microbreccia (Greek *mikros*, small, old German *brecha*, fracture, and Ligurian-Italian *breccia*, crushed stone). *Polymict impact breccia* with clasts of small grain size, usually in the subcentimetre to submillimetre range; most commonly used for lunar and *meteoritic impact breccias*. Ant. *megabreccia*. *(OU; Quaide & Bunch 1970, p. 718)*

Microcrystite (Greek *mikros*, small, and *krystallos*, crystal). *Microtektite*-like spherule containing quenched crystals usually of clinopyroxene and spinel, probably derived from condensation of *impact rock* vapour. Microcrystites are found in marine sediments and are associated with iridium and other siderophile element anomalies (i.e. Cretaceous–Tertiary boundary, Late Eocene, Late Pliocene). *(Glass & Burns 1988, p. 455; Glass 1990, p. 259)*

MICROFABRIC. See *fabric*.

MICROLITHON (Greek *micros*, small, and *lithos*, stone). Laminar domain lying between the schistose planes as found with *spaced schistosity/cleavage* and *crenulation schistosity/cleavage*. *(Turner & Weiss 1963, p. 465)*

MICROSTRUCTURE. See *structure*. Recommended alternative to *texture*.

MICROTEKTITE. See *tektite*.

MICTITE (Greek *miktos*, mixed). Obsolete synonym for a *migmatite* or a composite rock formed by the partial assimilation of country rock in a magma. *(Scheumann 1936a, p. 297; Dietrich & Mehnert 1960, p. 62)*

MICTOSITE (Aberdeenshire, Scotland, UK). Obsolete synonym of *migmatite*. *(Teall 1902, p.* lxxiv*)*

MIGMA. Mobile, or potentially mobile, mixture of solid rock material and magma (injected or originated *in situ*). *(Reinhard 1935, p. 44; Dietrich & Mehnert 1960, p. 62; Mehnert 1968, p. 355)*

MIGMATITE ⇒ Section 2.6 (Greek *migma*, mixture). Composite silicate metamorphic rock, pervasively heterogeneous on a meso- to megascopic scale. It typically consists of darker and lighter parts. The darker parts usually exhibit features of metamorphic rocks whereas the lighter parts are of igneous appearance. Hence ***migmatization***, process leading to the formation of a migmatite. See also *leucosome*, *melanosome*, *mesosome*, *neosome*, *palaeosome*. Wherever minerals other than silicates and quartz are substantially involved, it should be explicitly mentioned. *(Sederholm 1907, p. 88, 110; Scheumann 1936a, p. 297; Read 1951, p. 3; Mehnert 1968, p. 355)*

MIGMATITE FONDANTE (French for melting migmatite). Obsolete term for an *anatexite*. *(Jung & Roques 1936, p. 26)*

MIGMATITE TRAVERSÉE (French for crossed migmatite). See *diadysite*. *(Jung & Roques 1936, p. 26)*

Millipede structure. Type of structure in which *inclusion trails* in a *porphyroblast* have a pattern resembling a millipede. *(Bell & Rubenach 1980, p. T11; Barker 1998, p. 238)*

MIMETIC GROWTH, MIMETIC CRYSTALLIZATION (German *Abbildungskristallisation*). Growth of a mineral in a metamorphic rock, that is influenced by, and reproduces some pre-existing structure in the rock or its component minerals. See footnote under *corona*. *(Sander 1930, p. 172; Suess 1931, p. 78; Knopf & Ingerson 1938, p. 39; Barker 1998, p. 238)*

MIMOTALCITE, PSEUDOTALCITE (Greek *mimos*, mimic, *pseudês*, false). Obsolete term for a rock composed of lamellar or scaly, coarse-grained, clasts of graphite-bearing *talcite* in a clay matrix. At the time of the original definition, talc was the name given to any fine-grained whitish or greenish mica or mixture of sheet silicates. *(Cordier 1868, p. 185)*

MINERAL ASSEMBLAGE, MINERAL ASSOCIATION. See *metamorphic assemblage, metamorphic association*.

Mineral facies. As defined by Eskola, general term for both *metamorphic* and igneous *facies*. It comprises all the rocks that have originated under temperature and pressure conditions so similar that a specific chemical composition has resulted in the same set of minerals, regardless of their mode of crystallization, whether from magma or by *recrystallization*, etc. *(Eskola 1920, p. 146)*

MINERAL ZONE. *Metamorphic zone* characterized by the presence of an *index mineral*. The *index mineral* first appears as a result of a *progressive metamorphic* reaction at the lower temperature boundary of the zone. *(Barrow 1893, p. 343; Barrow 1912, p. 275)*

MIXTO-GNEISS (Greek *miktos*, mixed). Obsolete term for a gneiss, mainly *banded*, that was formed predominantly from *slates* by *recrystallization* and *injection* by narrow veinlets of aplite. Cf. *composite gneiss*, *injection gneiss*, *lit-par-lit gneiss*. *(Luchitzky 1922, p. 323)*

MONADOBLASTIC STRUCTURE (Latin *monas*, gen. *monadis*, unit, from the Greek *monos*, single). Type of structure in a metamorphic rock in which the component minerals are evenly distributed. *(Loewinson-Lessing 1911; Tomkeieff)*

MONOCYCLIC METAMORPHISM (Greek *monos*, single). Synonymous with *monometamorphism*.

MONOMETAMORPHISM ⇒ Section 2.2 (Greek *monos*, single). Metamorphism resulting from one *metamorphic event*, as opposed to *polymetamorphism*. It can be *monophase* or *polyphase*. *(OU)*

MONOMICT IMPACT BRECCIA (Greek *monos*, single, and *miktos*, mixed). *Cataclasite* produced by *impact* and displaying weak or no *shock metamorphism*. It occurs in the (par)autochthonous floor of an *impact crater* or (up to the size of blocks and megablocks) within *allochthonous* (*polymict*) *impact breccias*. Ant. *polymict impact breccia*. *(OU; Engelhardt 1971, p. 5567; Stöffler* et al. *1979, p. 652, Fig. 6; Stöffler* et al. *1980, Table 1)*

MONOMINERALIC. Said of a rock in which ≥95% of the modal content is composed of one mineral, as opposed to *polymineralic*, *bimineralic*, *trimineralic*.

MONOPHASE METAMORPHISM ⇒ Section 2.2, Fig. 2.2.2. *Metamorphic event* with only one pressure and/or temperature climax. Ant. *polyphase metamorphism*.

MORBIHANNITE (Penboch Bay, *Morbihan* Gulf, Brittany, France). Originally defined as a fibrolitic sillimanite *mica schist*. Later described as a mica-, sillimanite- and cordierite-rich part of a *migmatite* (*embrechite*). *(Limur 1884; Barrois 1934; Cogné 1960, p. 149)*

Mortar structure. Type of structure in which a *porphyroclast* is surrounded by a fine-grained aggregate of the same mineral. This is similar to *core-and-mantle structure* but can be formed by *cataclasis*. Not recommended as it has the (often incorrect) genetic implication of a mechanically crushed rock. Rec.syn. *core-and-mantle structure*. *(Törnebohm 1880, p. 244; Joplin 1968, p. 31; Barker 1990, p. 72; Passchier & Trouw 1996, p. 261; Vernon 2004, p. 486)*

MOSAICISM. General disorientation of a crystal structure as a result of *shock metamorphism* resulting in marked, highly irregular 'mottled' extinction under the petrographic microscope. Weak mosaicism can also be caused by endogenic tectonic processes. *(OU; Stöffler 1971, p. 5542; Dence & Robertson 1989, p. 527)*

Mosaic structure. As originally defined the term is synonymous with *granoblastic-polygonal structure*, and as such it is widely used in the description of peridotites. Later the term was defined by Spry to describe *subgrains*, but this usage is not recommended. *(Teall 1888, p. 235; Spry 1969, p. 34; Harte 1977, p. 280)*

MUCRONITE (Mt *Mucrone*, Sesia zone, Italian Western Alps). Metamorphic rock mainly composed of jadeitic clinopyroxene, garnet, phengite, quartz ± K-feldspar. *(Reinsch 1977, p. 91)*

MULLION STRUCTURE. Type of structure on the mesoscopic or macroscopic scale appearing as a parallel cluster of rods or columns and produced by the presence of closely spaced fold hinges, *cleavage*-bedding intersections, *fault* grooving, etc. *(Hull* et al. *1891, p. 53; Wilson 1953, p. 119; Joplin 1968, p. 31)*

MULTI-RING (IMPACT) BASIN. *Impact crater* with relatively small depth-diameter ratio and with at least two concentric rings inside the crater; first recognized on the Moon. Syn. *multi-ring* (*impact*) *crater*. *(OU; Melosh 1989, p. 22; Grieve 1998, p. 111)*

MULTI-RING (IMPACT) CRATER. Synonymous with *multi-ring* (*impact*) *basin*.

MURKSTEIN (German, perhaps *Murk*, bit, fragment, because the rock is brittle). Metamorphic rock composed of quartz and garnet, with or without tourmaline and mica. *(Haidinger 1787, p. 40)*

MUSCOVADITE (Spanish *muscovado*, brown sugar, owing to the brown colour of the weathered rock; Duluth gabbroid complex, Minnesota, USA). Local name of a plagioclase-Ca-clinopyroxene rock, with some orthopyroxene and subordinate minerals, formed at the contact between gabbro and *greenstone*. First interpreted as a modified border facies of the gabbro, the rock has later been shown to derive from *greenstones* of the inner *aureole* of the gabbro. It is a high-grade *mafic hornfels*. Cf. *granoblast*, *sudburite*. *(Winchell 1900, p. 295; Schwartz 1924, p. 117)*

MYLONITE ⇒ Section 2.3, Fig. 2.3.1 (Greek *mylos*, mill). *Fault rock* that is cohesive and characterized by a well-developed *schistosity* resulting from tectonic reduction of grain size, and commonly containing rounded *porphyroclasts* and lithic fragments of similar composition to minerals in the matrix. Fine-scale layering and an associated mineral or stretching *lineation* are commonly present. Brittle deformation of some minerals may be present, but deformation is commonly by crystal plasticity. Mylonites may be subdivided according to the relative proportion of finer-grained matrix into *protomylonite*, *mesomylonite* and *ultramylonite*. *(Lapworth 1885, p. 559; Teall 1918, p. 2; Waters & Campbell 1935, p. 478; Higgins 1971, p. 75; Tullis* et al. *1982, p. 230; Barker 1998, p. 239)*

MYLONITE GNEISS. Type of *mylonite* with a *porphyroclastic structure*, in which the matrix has undergone significant *recrystallization*. Rec.syn. *augen mylonite*. *(Quensel 1916, p. 99; Waters & Campbell 1935, p. 478; Higgins 1971, p. 72)*

NACRITIDE (French *nacre*, mother of pearl; Funcha creek west of Pikes Peak, Kansas, USA). Muscovite-biotite granitoid rock, characterized by a pearly lustre. *(Schiel 1857, p. 119)*

NÁMIĚŠTER STEIN (*Náměšt* nad Oslavou, Bohemian massif, Moravia, Czech Republic). Obsolete local name for a *granulite*. *(Justi 1757, p. 210)*

Near-vein metasomatism. Type of *diffusional metasomatism* that forms symmetrical *metasomatic* zonation on either side of an *infiltrational metasomatic* vein (or vein in-filling). *(Korzhinskii 1953, p. 339)*

NEBULITE (Latin *nebula*, mist). *Migmatite* with diffuse relics of pre-existing rocks or rock structures. *(Gümbel 1888, p. 11; Sederholm 1926, p. 138; Mehnert 1968, p. 356)*

NEMATOBLASTIC STRUCTURE (Greek *nêma*, gen. *nêmatos*, fibre, thread, and *blastos*, bud, sprout). Type of structure in a metamorphic rock in which the constituent grains are of equal size (*homeoblastic*), have an *acicular* or rod-like form and are arranged parallel to each other. Syn. *fibroblastic structure*. *(OU; Grubenmann 1904, p. 79; Joplin 1968, p. 28; Passchier & Trouw 1996, p. 261)*

NEOBLAST (Greek *neos*, new, and *blastos*, bud, sprout). Mineral grain in a metamorphic rock that is younger than the other mineral grains in the rock. A neoblast may be of a different composition from, or the same composition as the other grains. Ant. *palaeoblast*. *(OU; Barker 1998, p. 239)*

NEOGENIC (Greek *neos*, new, and *gennô*, bear, give birth to). Said of newly formed minerals, whether they are due to *diagenetic* or metamorphic processes. Hence, ***neogenesis***, the formation of neogenic minerals. Syn. *secondary*. *(Johannsen 1931, p. 184)*

NEOMINERALIZATION. Nucleation and growth of new minerals as a result of a metamorphic reaction; type of *recrystallization* involving the formation of a new mineral species. *(Knopf 1931, p. 5; Knopf & Ingerson 1938, p. 107; Yund & Tullis 1991, p. 346)*

NEOMORPHISM. Inclusive term covering the transformation of one mineral to itself or a *polymorph* by any process. The process may result in an increase, decrease or no change in the crystal size. Cf. *aggrading neomorphism*, *degrading neomorphism*. *(Folk 1965, p. 20)*

NEOSOME (Greek *neos*, new, and *sôma*, body). Newly formed parts of a *migmatite*; it includes *metatects* and *restites*. *(Huber 1943, p. 90; Dietrich & Mehnert 1960, p. 62)*

NEWLANDITE (*Newlands* diamond pipe, west Griqualand, S Africa). Rock composed of garnet, enstatite and Cr-diopside. It may contain phlogopite. A variety of *griquaite*. *(Bonney 1899, p. 315)*

NON-CONGRESSIBLE MINERALS. Obsolete term for minerals in groups that are stable under specific conditions, but become unstable, under the same conditions, in the presence of another mineral. Cf. *kelyphite*, *corona*, *armoured relics*. *(Eskola 1920, p. 150; Eskola 1939, p. 342)*

NORMAL CONTACT METAMORPHISM. Unnecessary term for *isochemical contact metamorphism*. Not to be confused with the *normal* or *general metamorphism* of the early petrographers, terms which were used as synonyms of *regional metamorphism*. *(Goldschmidt 1911, p. 119; Grubenmann & Niggli 1924, p. 248, 283)*

NORMAL METAMORPHISM. Obsolete term for *regional metamorphism*. *(Élie de Beaumont 1841, in Tomkeieff; Naumann 1858, p. 721)*

NOVACULITE, NOVACULAR SCHIST (Latin *novaculum*, pruning-knife, razor; area of Ottré and Salm-Château, eastern Belgium). Very fine-grained rock used as a whetstone, composed of quartz, pyrophyllite and white mica; it may contain garnet and feldspar. Cf. *coticule*, *gondite*. *(Kirwan 1794, p. 238; Omalius d'Halloy 1828, p. 110; Zirkel 1866b, p. 60; Cordier 1868, p. 187; Renard 1878, p. 39)*

OCEAN-FLOOR METAMORPHISM ⇒ Section 2.2, Fig. 2.2.1. Type of *metamorphism* of *regional* or *local* extent related to the steep *geothermal gradient* occurring near spreading centres in oceanic environments. The *recrystallization*, which is mostly incomplete, encompasses a wide range of temperatures. The metamorphism is associated with circulating hot aqueous fluids (with related *metasomatism*) and typically shows an increasing temperature of metamorphism with depth. Cf. *sub-sea-floor metamorphism*. *(Miyashiro* et al. *1971, p. 602; Miyashiro 1972, p. 142; Bucher & Frey 2002, p. 8)*

OCELLAR STRUCTURE (Latin *ocellus*, small eye). Type of structure in which mineral grains are arranged in radiating growths around a central crystal. The term is predominantly used to describe igneous structures. Its use in metamorphic nomenclature is rare and alternatives such as *corona structure* are recommended. Obsolete synonyms include ***centric structure*** and ***stellated structure***. *(Judd 1886, p. 72; Hatch 1888, p. 441)*

OFISILICE, OPHISILICE (Italian *ofi*olite, ophiolite). Silicified serpentine rock. Syn. *siliciophite*, cf. *ophicarbonate*. *(Mazzuoli & Issel 1881, p. 334)*

-OID (Greek *eidos*, form, shape). Suffix that when appended to a rock term gives a meaning 'related to the original term' (e.g. eclogitoid, kelyphitoid, schistoid). The suffix is commonly used in igneous nomenclature with the same meaning (e.g. granitoid). *(OU; Le Maitre 1989, p. 22, Fig. B.9)*

OLENITE [ollenite] (*Olen* Pass between Gressoney and Olen Valleys, Italian Western Alps). Local name for a garnet-bearing epidote-green amphibole rock, the epidote being concentrated in spots and veins. *(Cossa 1881, p. 267; Artini & Melzi 1900, p. 329)*

OPHIBASE (Greek *ophis*, snake, and *basis*, base; Alps). Obsolete term for a variety of altered basalt (referred to as variolite), considered to consist of a groundmass of *ophite;* historically, *ophite* was taken to mean *serpentinite*. *(Saussure 1779, p. 133; Saussure 1796, p. 344)*

OPHICALCITE (Greek *ophis*, snake, and Latin *calx*, lime). Strictly a form of *ophicarbonate* in which calcite is the predominant carbonate. However, the term has traditionally been used in the meaning of *ophicarbonate*, that is, it is taken to include rocks containing a variety of different carbonates. *(Brongniart 1813, p. 38; Cordier 1868, p. 169; Lemeyrie 1878, p. 260; Mazzuoli & Issel 1881, p. 333; Lapparent 1882, p. 596; Zirkel 1894b, p. 452; Harker 1932, p. 78; Zanzucchi 1994, p. 375)*

OPHICARBONATE (Greek *ophis*, snake). Rock consisting of *serpentinite* and carbonate; the *serpentinite* is commonly fragmented or brecciated, and veined and impregnated by the carbonate material (calcite, dolomite or magnesite). It forms by the serpentinization of ultramafic rocks and their reaction with CO_2 solutions. Hence ***ophimagnesite*** (where the carbonate is predominantly magnesite), etc. See *ophicalcite*. *(OU; Lapparent 1882, p. 596; Bucher & Frey 1994, p. 164)*

OPHISPHERITE (Greek *ophis*, snake, and *sphaera*, sphere; first described from the Chenaillet ophiolite, Montgenèvre, Hautes-Alpes, and first named from the Col des Gets ophiolites, Haute-Savoie, France). Variety of *rodingite* that occurs in rounded, *boudinaged*, shape, with complex concentric zoning and with chlorite rims or cores. It is derived by chloritization from a predominantly basaltic *protolith*, during serpentinization of the host rock. When zoning is absent the rounded bodies are called ***para-ophispherites***. Various mineral associations may be present according to the temperature of the *metasomatic* process. *(Vuagnat 1952, p. 19; Vuagnat & Jaffé 1953, p. 413; Vuagnat & Pusztaszeri 1964, p. 12; Bertrand* et al. *1980, p. 127; Bertrand 1980, p. 143)*

OPHITE. Ambiguous term that has been used for a *serpentinite* and for a doleritic rock from the Pyrenees, commonly uralitized; hence ***ophitic structure***. These meanings are now obsolete or local. However, the name still has a restricted usage as a commercial term for an ornamental *marble* (an *ophicarbonate*) with green streaks resembling a serpent. *(Pliny 80 Lib. 36, p. 508; Wallerius 1747, in Loewinson-Lessing & Struwie 1937, p. 233; Saussure 1796, p. 344; Palassou 1798, in Tomkeieff; Brongniart 1813, p. 41;*

Palassou 1819, in Loewinson-Lessing & Struwie 1937, p. 233; Humble 1860, p. 328)

OPHTHALMITE (Greek *ophthalmos*, eye). Mixed rock (*chorismite*), usually a *migmatite*, characterized by *augen* and/or lenticular aggregates of minerals. *(Huber 1943, p. 89)*

OPTALIC METAMORPHISM (Greek *optaô*, to bake, roast). Alternative name proposed for *caustic metamorphism*. *(Tyrrell 1926, p. 301)*

OROGENIC METAMORPHISM ⇒ Section 2.2, Fig. 2.2.1 (Greek *oros*, mountain, and *gennô*, to give birth to). Type of *metamorphism* of *regional* extent related to the development of orogenic belts. The metamorphism may be associated with various phases of orogenic development and involve both compressional and extensional regimes. Dynamic and thermal effects are combined in varying proportions and timescales, and a wide range of P–T conditions may occur. *(Miyashiro 1973, p. 24; Bucher & Frey 1994, p. 6)*

ORTHO- (Greek *orthos*, straight, correct). Prefix indicating, when in front of a metamorphic rock name, that the rock was derived from an igneous rock (e.g. orthogneiss). Ant. *para-*. *(Rosenbusch 1898, p. 467)*

ORTHOMARBLE. Unnecessary and ambiguous name for a limestone whose grains are cemented and interlocking, because of *diagenetic* effects, and which is commercially valuable because it will take a polish. 'Marble' was used in the rock name to differentiate the rock from a limestone where the grains were cemented but not interlocking. See *paramarble*. This use of *ortho-* does not accord with the recommended usage. *(Tieje 1921, p. 655; Brooks 1954, p. 758)*

OVARDITE (Torre d'*Ovarda*, val d'Ala, Piemonte, Italy). Metamorphic rock from the Alps, mainly consisting of chlorite, epidote and ocellar albite. It may also contain titanite, white mica, garnet and biotite. Commonly called chlorite *prasinite*. *(Struever 1872, in Novarese 1895, p. 176)*

Paired metamorphic belts (originally defined in Japan). Parallel and coeval metamorphic belts, one of a higher P/T type and one of a lower P/T type, generally occurring along a continental margin with the higher P/T belt lying on the ocean site. *(Miyashiro 1961, p. 302; Miyashiro 1973b, p. 241; Miyashiro 1994, p. 216, 351)*

PALAEOBLAST (Greek *palaios*, ancient, and *blastos*, bud, sprout). Mineral grain in a metamorphic rock that is older than the other mineral grains in the rock. A palaeoblast may be of a composition different from or identical to the other grains. Ant. *neoblast*. *(OU)*

PALAEOSOME (Greek *palaios*, ancient, and *sôma*, body). Part of a *migmatite* representing the parent rock. See *mesosome*, *neosome*. *(Scheumann 1936a, p. 302; Huber 1943, p. 90; Dietrich & Mehnert 1960, p. 62)*

PALAIOPÊTRE (Greek *palaios*, ancient, and *petra*, stone). Obsolete French term first used for a tough, very fine-grained metamorphic rock, subsequently for *hornfels*. *(Saussure 1786, § 1194; Brochant 1800, p. 232; Naumann 1849–53, p. 566; Zirkel 1866a, p. 419; Fournet, in Loewinson-Lessing 1893)*

Palimpsest structure (Greek *palin*, again, and *ptaistos*, erased). Type of structure in a *migmatite* or *granitized* rock that can be recognized as pre-*migmatitic* (or pre-granitic); the term is also used with a more general meaning for relict features. Cf. *blasto-*. Rec.syn. *relict structure*. *(Sederholm 1893, p. 7; Sederholm 1897–9, p. 236; Becke 1903a, p. 570; Joplin 1968, p. 31; Mehnert 1968, p. 299)*

PALINGENESIS (Greek *palin*, again, and *genesis*, genesis). Formation of a new magma by complete or nearly complete melting of pre-existing rocks. Hence ***palingenite***, rock resulting from palingenesis. Syn. *anatexis*. *(Sederholm 1907, p. 37, 102; Backlund 1936, p. 295; Smulikowski 1947, p. 266; Smulikowski 1958, p. 84; Dietrich & Mehnert 1960, p. 62; Mehnert 1968, p. 356)*

PALITE (*Pfahl*, Bavaria, Germany). Local collective name for *cataclastic* rocks of dioritic to granodioritic composition, which are associated with a deep-seated regional *fault*. Related schistose rocks are termed ***Pfahlschiefer***. *(Frentzel 1911, p. 135)*

PARA- (Greek *para-*, besides or beyond). Prefix indicating, when in front of a metamorphic rock name, that the rock was derived from a sedimentary rock (e.g. paragneiss). Ant. *ortho-*. *(Rosenbusch 1898, p. 467)*

PARADIORITE (Greek *para-*, besides or beyond; *Doira*, Nufenen, canton Graubünden, eastern Switzerland).

Obsolete term for a local variety of *greenschist* from the group of so-called *chlorogrisonite* schists (equivalent to *prasinites*), composed mainly of plagioclase (oligoclase), amphibole and little subordinate epidote. The term contravenes the SCMR rules, as the prefix *para-* implies a sedimentary *protolith*. *(Rolle 1879, p. 38)*

PARAGENESIS (Greek *para*, besides or beyond, and *genesis*, genesis). Characteristic association of minerals in a metamorphic rock that are considered to have developed under the same physico-chemical conditions, and thus form an equilibrium *assemblage*. With changing metamorphic conditions new parageneses may develop and together they form a ***paragenetic sequence***. Different parageneses in the same rock can be deduced from the presence of replacement features (*pseudomorph*, *corona*, etc.). Thus, a *mineral association* in a rock may represent a disequilibrium association consisting of two or more successive parageneses, whereas a paragenesis always comprises an equilibrium *assemblage*. The term was first intended for mineral succession in ore bodies, but it is now used for all types of *mineral associations* of any origin. *(Breithaupt 1849, p. 1; Lindgren 1919, p. 562, 658; Winkler 1974, p. 27; Bucher & Frey 1994, p. 26)*

PARALAVA (Greek *para*, besides or beyond). Slag-like completely fused rock formed from a shale and sandstone sequence overlying burned coal seams. The rock may show a vesicular aspect and a flow structure similar to lavas. The glassy portions of the rock contain newly formed small mineral grains in *assemblages* typical of *pyrometamorphic* conditions. Rec.syn. coal-fire *buchite*. *(Fermor, in Tilley 1924b, p. 70; Lightfoot 1929, p. 32)*

PARAMARBLE (Greek *para*, besides or beyond). Unnecessary and ambiguous name for a limestone whose grains are cemented and interlocking because of *contact metamorphism*. See *orthomarble*. This use of *para-* does not accord with the recommended usage. *(Tieje 1921, p. 655)*

PARAMORPHISM (Greek *para*, besides or beyond, and *morphê*, form). Obsolete term for a proposed subdivision of the metamorphic processes that is characterized by chemical changes in either the rock or its constituents, with crystallization of new minerals. Rec.syn. *metasomatism*. Cf. *metataxis*, *metatropy*. *(Bonney 1886, in Irving 1889, p. 4–5; Irving 1886, p. 658)*

PARAOPHISPHERITE. See *ophispherite*.

PARENT ROCK. See *protolith*.

PAROPHITE (North Carolina, USA). Obsolete term for a variety of *serpentinite* used as a pottery stone. *(Hunt 1852, p. 95; Loewinson-Lessing & Struwie 1937, p. 238)*

PAROPTESIS (Greek *para*, besides or beyond, and *optaô*, to bake). Obsolete name for dry *contact metamorphism*. Ant. *metapepsis*. *(Kinahan 1878, p. 176)*

PARTICLE VELOCITY. Velocity of shock-compressed material moving behind the *shock front*. *(OU; Duvall & Fowles 1963, p. 211–22; Melosh 1989, p. 29)*

PATOITE (cañòn de *Pato*, departamento de Ancash, Peru). Local name for a schistose *crystalloblastic* rock variably composed of quartz, biotite, orthoclase, oligoclase, garnet, andalusite and staurolite, with accessory sillimanite, zircon and apatite, and *secondary* muscovite, limonite, ilmenite and graphite. *(Broggi 1945, p. 78, in Plaza 1945, p. 90)*

PEARL GNEISS (German *Perle*, pearl; western Tyrol, Austrian Alps). Type of gneiss in which *porphyroblasts* of albite or oligoclase occur as rounded grains. *(Hammer 1925, p. 149)*

PELIKANITE ROCKS (in honour of Dr *Pelikan*). Slightly metamorphosed rocks of plutonic origin and various compositions (granite to gabbro-norite), the feldspar of which has been altered by *pneumatolytic* and *hydrothermal* processes to a mixture of kaolinite and opal (called pelikanite). Originally used for granitic rocks, but later extended to a broader group of rocks. *(Feofilaktov 1851, p. 16; Gavrusevich 1931, p. 91)*

Pelite ⇒ Section 2.1 (Greek *pêlos*, clay). Term for a sedimentary rock composed of clay-sized particles. It has also been used locally for the metamorphosed equivalent of these sedimentary rocks, that is, a metamorphic rock with a high modal ratio of mica to quartz + feldspar. The SCMR recommends that it should be used only for sedimentary rocks and the term ***metapelite*** should be used for the metamorphosed equivalents. *(Tyrrell 1921, p. 501; Robertson 1999, p. 4)*

PELITIC HORNFELS. Type of *hornfels* derived by *contact metamorphism* of an Al-rich sedimentary or metasedimentary *protolith*. Rec.syn. hornfelsed pelite, peraluminous *hornfels* or more specific compound names such as andalusite-cordierite-biotite-quartz *hornfels*. *(Williams* et al. *1954, p. 179; Spry 1969, p. 192)*

PELITOID. Type of schist derived from a granite by *mylonitic* or *cataclastic* processes. *(Holmquist 1908, p. 292)*

PELOLITE (Greek *pêlos*, clay, and *lithos*, stone). Obsolete term for a group of homogeneous, very fine-grained rocks consisting of clastic and organic grains. The majority of the group are sedimentary rocks, but *contact-metamorphosed* rocks were also included (e.g. *spilosite*, *desmosite*, *hornfels*, *adinole*). *(Gümbel 1888, p. 91)*

Pencatite ⇒ Section 2.10 (named after Count G. Marzari-*Pencati*, 1779–1836; Predazzo, Val di Fassa, Trento, Italy). Brucite *marble* formed as a *contact rock* from dolomitic limestone and with the same CaO/MgO value as pure dolomite; its companion rock, *predazzite*, has a higher CaO/MgO value. Rec.syn. brucite *marble*. *(Roth 1851, p. 144; Holmes 1928, p. 178; Harker 1932, p. 77)*

PENETRATION METAMORPHISM. Synonym of *injection metamorphism*. *(Tomkeieff)*

PENUMBRA OF GRANITE. Rarely used expression for *contact aureole* around a granite. *(Humboldt 1831)*

PERIHEPSESIS (Greek *peri*, around, and *hepsêsis*, boiling, cooking). Obsolete term proposed for the formation of metamorphic rocks by *pneumatolytic metamorphism*. *(Gürich 1905, p. 250)*

PERIMAGMATIC (ADDITIVE) CONTACT METASOMATISM (Greek *peri*, around). Obsolete term for *contact metasomatism* that produces ore deposits in the *inner aureole*. Cf. *apomagmatic contact metasomatism*. *(Bergeat 1912, p. 11; Grubenmann & Niggli 1924, p. 312)*

PERIMETRAL METAMORPHISM. Obsolete term for *pneumatolytic* or *hydrothermal metasomatism*, especially in volcanic regions. *(Stoppani 1873, p. 42)*

PERIPHERAL METAMORPHISM. Obsolete term for *contact metamorphism*, both *isochemical* and *allochemical*, under plutonic conditions. *(Lapparent 1882, p. 1117; Gümbel 1888, p. 374; Lapparent 1893, p. 1402; Loewinson-Lessing 1893–4, p. 152)*

PERMEATION GNEISS. Variety of *migmatite* formed essentially by *feldspathization*, but with a general homogeneous character. *(Read 1931, p. 120; Read 1944, p. 75)*

PERTHISTHENE GRANULITE (from *perthi*c feldspar and hyper*sthene*). Term for *granulites* containing perthitic feldspar and hypersthene. See *quartzsthene granulite*, *plagiosthene granulite* and *plagioperthisthene granulite*. *(Nasir 1993, p. 75; Lorenz 1998, p. 112)*

PETROBLASTESIS (Greek *petra*, stone, and *blastêsis*, germination). Term for the formation of a *metasomatic rock* as a result of diffusion. *(Barth 1947, p. 181; Dietrich & Mehnert 1960, p. 62)*

PETROGENETIC GRID. Pressure–temperature grid with an intersecting set of univariant curves which allows determination of the *P–T* conditions under which a particular *mineral assemblage* evolved. *(Bowen 1940, p. 274; Bucher & Frey 1994, p. 111; Miyashiro 1994, p. 253; Kornprobst 2002, p. 70)*

Phacoid (Greek *phacos*, lentil). Used to describe a lenticular-shaped relict in a deformed body. Hence ***phacoidal structure***, a type of structure characterized by the presence of lenticular relicts, up to the metre scale, in a deformed and characteristically finer-grained matrix; equivalent to *augen structure* and *flaser structure*. *(OU; Peach* et al. *1907, p. 263; Och* et al. *2003, p. 609)*

PHAGOMORPHIC (Greek *phagô*, to eat, and *morphê*, form; Tessin, Switzerland). Obsolete adjective for corroded crystals surrounded by a *reaction rim*. *(Gutzweiler 1912, p. 52)*

PHANERITIC (Greek *phaneros*, apparent). Said of a rock in which the individual grains are visible with the unaided eye (*c.* >0.1 mm). Ant. *aphanitic*.

PHANEROBLASTIC STRUCTURE (Greek *phaneros*, apparent, and *blastos*, bud, sprout). Type of structure in a metamorphic rock in which the constituent grains are large enough to be distinguished by the naked eye. *(OU; Reinhard 1909, p. 258)*

PHENOBLAST (Greek *phainô*, to show, and *blastos*, bud, sprout). Large crystal formed in a metamorphic rock, bounded by its own faces (*idioblast*) and set in a matrix of smaller grains. The rock therefore resembles a porphyritic igneous rock. Rec.syn. *idioblastic (euhedral) porphyroblast. (Erwin 1938, p. 119; Foucault & Raoult 2005, p. 265)*

PHENOCLAST (Greek *phainô*, to show, and *klastos*, broken). Synonymous with *porphyroclast*. Also has a usage in sedimentary petrology to mean large conspicuous clasts. *(Erwin 1938, p. 119)*

PHLEBITE (Greek *phleps*, gen. *phlebos*, vein). Composite metamorphic rock or *migmatite* characterized by the presence of veins; the veins may have been formed from *injected* material or exuded *in situ*. This is a purely descriptive term, in contrast to *arterite* and *venite*, which have a genetic connotation. *(Scheumann 1936a, p. 299; Huber 1943, p. 89; Dietrich & Mehnert 1960, p. 62)*

PHYLLADE (Greek *phyllon*, leaf). Term originally used (mostly in western continental Europe) for siliceous schistose metamorphic rocks rich in fine-grained white mica (sericite). Now superseded by *phyllite. (Brongniart 1813, p. 35; Aubuisson, in Cordier 1868, p. 188; Lapparent 1882, p. 621)*

PHYLLITE (Greek *phyllon*, leaf). Fine- to medium-grained metamorphic rock characterized by a lustrous sheen and a well-developed *schistosity* resulting from the parallel arrangement of phyllosilicates. Phyllite is usually of low *metamorphic grade*. Originally introduced as an alternative to *phyllade* to describe a fine-grained schist composed predominantly of micaceous minerals, and as such forming a transition between a *(clay-)slate* and a *mica schist. (Naumann 1849, p. 553; Zirkel 1866b, p. 464; Holmes 1920, p. 183; Spry 1969, p. 270; Barker 1998, p. 240)*

PHYLLITE-MYLONITE. Obsolete term for *phyllonite. (Tomkeieff)*

PHYLLOLITE (Greek *phyllon*, leaf, and *lithos*, stone). Obsolete term proposed for *crystalline schist. (Gümbel 1888, p. 89, 152)*

PHYLLONITE (combination of *phyl*lite and my*lonite*). Phyllosilicate-rich *mylonite* that has the lustrous sheen of a *phyllite. (Sander 1911, p. 301; Knopf 1931, p. 14; Barker 1998, p. 240)*

PHYRASIS. See *tektophyrasis. (Scheumann 1955, p. 16)*

PIERRE OLLAIRE (French *pierre*, stone, Italian and Latin *olla*, cooking-pot, because the rock is used to make pots; e.g. Chiavenna in Valtelline, San Ambrosio in Piemont, Italy). Schistose rock composed of talc and unctuous to the touch. Syn. *talcite ollaire*, also called potstone, Topfstein, lavezzo. *(Delesse 1856, p. 280; Cordier 1868, p. 176)*

PIEZO-CONTACT METAMORPHISM (Greek *piezô*, to press; Central Gneiss Massifs of the Eastern Alps, Austria, and their schistose envelope, the 'Schieferhülle'). Deep-seated *contact metamorphism* that develops *contact aureoles* wider than normal, in which the *contact rocks* resemble the products of *regional metamorphism*, and contain garnet and kyanite instead of lower-pressure phases such as cordierite and andalusite. Cf. *Main Donegal-type contact metamorphism. (Weinschenk 1902a, p. 464; 1902b, p. 214; Erdmannsdörffer 1924, p. 312)*

Piezo-thermic array (Greek *piezô*, to press, and *thermos*, hot). Curve joining the least hydrated points on an array of *P–T–t* loops, each loop representing a different depth level in a metamorphic terrain. The detailed shape and slope of the curve characterize the tectonothermal evolution of the terrain. The least hydrated states do not necessarily correspond to the highest *P* or *T* attained. *(Richardson & England 1979, p. 188; Bucher & Frey 1994, p. 67)*

PILITIZATION. Obsolete term for the transformation of minerals (particularly olivine) into pilite (actinolite). *(Polenov 1897, p. 395)*

PINCH-AND-SWELL STRUCTURE. See *boudinage. (Ramberg 1955, p. 520; Ramsay & Huber 1983, p. 12; Price & Cosgrove 1990)*

PINITIZATION (in honour of Professor *Pini* and his mine; near Schneeberg, Saxony). Metamorphic process leading to the formation of pinite, a hydrous K- and Al-silicate, formed by *alteration* of vesuvianite, cordierite, etc. Also once used (unrecommended usage) with the meaning of *propylite. (Karsten 1789, p. 193; Emmerling 1799, p. 337; Karsten 1800, p. 28)*

PINOLITE (similarity of shape of the magnesite aggregate to a cross section through the *Pinus pinea* cone, and Greek *lithos*, stone; NE Alps, Austria). Obsolete and ill-defined term for a metamorphic rock composed of ferroan magnesite (breunnerite) crystals or aggregates in a schistose matrix composed of white mica. *(Rumpf 1873, p. 263–272)*

PINWHEEL STRUCTURE. See *snowball structure*.

PLAGIOCLASITE (upper valley del Sinni, Basilicate, Italy). Obsolete term originally defined as a quartz-free, plagioclase-rich, gneiss (cf. *lamboanite*), but now used only as a synonym of anorthosite. It is, thus, an igneous term. The first meaning would have been an exception to the SCMR rule on the use of the suffix *-ite*. *(Viola 1892, p. 121; Lacroix 1939, p. 298; Le Maitre 1989, 2002)*

PLAGIOPERTHISTHENE GRANULITE (from *plagio*clase, *perthi*thic feldspar, and hyper*sthene*). Term proposed for a *granulite* containing plagioclase, perthitic feldspar and hypersthene. See *quartzsthene granulite*, *plagiosthene granulite* and *perthisthene granulite*. *(Nasir 1993, p. 75; Lorenz 1998, p. 112)*

PLAGIOSTHENE GRANULITE (from *plagio*clase and hyper*sthene*). Term proposed for a *granulite* containing plagioclase and hypersthene. See *quartzsthene granulite*, *perthisthene granulite* and *plagioperthisthene granulite*. *(Nasir 1993, p. 75; Lorenz 1998, p. 112)*

PLAKITE, PLAKAITE (*Plaka*, a village close to Laurium, Attica, Greece). Locally used term for a *banded* schist that has been *contact-metamorphosed* to a feldspar-augite-scapolite *hornfelsed* schist. *(Cordella 1893, p. 126)*

PLANAR DEFORMATION FEATURES (PDF). Submicroscopic amorphous lamellae occurring in shocked minerals as multiple sets of planar lamellae (optical discontinuities under the petrographic microscope), parallel to rational crystallographic planes; indicative of *shock metamorphism*. The lamellae may be (re)crystallized by thermal annealing forming *decorated PDF*. Supersedes *planar elements*, *planar features* and *shock lamellae*. Syn. *deformation lamellae*. *(Grieve* et al. *1990, p. 1792; Stöffler & Langenhorst 1994, p. 162)*

Planar deformation structure. Synonymous with *planar microstructure*. *(Engelhardt & Bertsch 1969, p. 206)*

PLANAR ELEMENTS. Unnecessary synonym of *planar deformation features*. *(Stöffler 1971, p. 5542; Stöffler 1972, p. 82, Tables 1, 2)*

PLANAR FEATURES. Unnecessary synonym of *planar deformation features*. *(Dence 1968, p. 175, Table 1; Robertson* et al. *1968, p. 333)*

PLANAR FRACTURES. Fractures occurring in shocked minerals as multiple sets of planar fissures parallel to rational crystallographic planes, which are usually not observed as cleavage planes under normal geological (non-shock) conditions. These shock-induced fractures were originally called 'cleavage' by Englund and Roen, Bunch and Cohen, and Engelhardt and Stöffler. *(Englund & Roen 1962, p. 20; Bunch & Cohen 1964, p. 380; Carter 1965, p. 786; Engelhardt & Stöffler 1965, p. 489; Bunch 1968, p. 415; Hörz 1968, p. 243; Engelhardt & Bertsch 1969, p. 209, Fig. 6; Stöffler 1971, p. 5542; Dence & Robertson 1989, p. 527; Stöffler & Langenhorst 1994, p. 162)*

PLANAR MICROSTRUCTURE. Collective term comprising shock-induced *planar fractures* and *planar deformation features*. *(Stöffler & Langenhorst 1994, p. 162)*

PLEATED STRUCTURE. Finely folded structure in schistose rocks, equivalent to *crenulation*. *(OU; Tomkeieff)*

Plurifacial metamorphism (Latin *plures*, many, and *facies*). Said of a rock which contains evidence of two or more *metamorphic facies*. The term is intended for use when it is uncertain if the different *facies* relate to two or more phases of one *metamorphic event* (*polyphasal*) or to two or more *metamorphic events* (*polymetamorphic*). *(de Roever & Nijhuis 1963, p. 327; de Roever 1972, p. 253; Miyashiro 1994, p. 351)*

PLUTONO-METAMORPHISM (*Pluto*, Latin god of the underworld). Term introduced for high-*P* and high-*T* metamorphism which resulted in high-grade rocks with similar characteristics to plutonic rocks. Subsequently the term evolved to ***plutonic metamorphism***. *(Harker 1889, p. 17; Sederholm 1897–9, p. 242; Eskola 1914, p. 16; Suess 1937, p. 5; Read 1949, p. 148; Spry 1969, p. 447)*

PNEUMATOLYSIS (Greek *pneuma*, breath, and *lysis*, dissolution). Obsolete term for the *alteration* of rocks by the gaseous emanations from volcanic rocks. *(Bunsen 1851, p. 238, 258; Brögger 1890; Tyrrell 1926, p. 324; Tomkeieff)*

PNEUMATOLYTIC CONTACT METAMORPHISM. Type of *metamorphism* due to gaseous emanations at a magmatic contact. Hence ***pneumatolytic contact aureole***. See *pneumatolytic* *metamorphism. (Goldschmidt 1911, p. 119; Irving 1911, p. 298; Grubenmann & Niggli 1924, p. 282; Niggli 1954, p. 523)*

PNEUMATOLYTIC METAMORPHISM (pneumatolysis) (Greek *pneuma*, breath, soul, and *lysis*, dissolution; southwestern Finland). Type of metamorphism strongly influenced by emanations, mainly gases, from an igneous body, and resulting in considerable change to the chemical and mineralogical composition of the rocks. In modern usage pneumatolytic metamorphism has been largely subsumed by *hydrothermal metamorphism*, the range and associations of which have greatly increased. Synonymous with *pneumatolytic metasomatism. (Bunsen 1851, p. 238; Zirkel 1893, p. 583; Eskola 1914, p. 259; Tyrrell 1926, p. 324; Harker 1932, p. 118; Turner 1948, p. 5)*

PNEUMATOLYTIC METASOMATISM. *Contact metasomatism* caused by predominantly gaseous emanations from a magmatic body. See comments under *pneumatolytic metamorphism. (Barrell 1907, p. 117; Grubenmann & Niggli 1924, p. 368; Lindgren 1933, p. 115; Eskola 1939, p. 373)*

POIKILOBLAST (Greek *poikilos*, varied, and *blastos*, bud, sprout). Large crystal formed in a metamorphic rock (*porphyroblast*) and characterized by the presence of abundant small included grains. Hence ***poikiloblastic structure***, synonymous with *sieve structure. (Becke 1903b, p. 46; Joplin 1968, p. 28; Spry 1969, p. 169)*

POLYCYCLIC METAMORPHISM. Synonymous with *polymetamorphism. (OU; Saggerson 1989, p. 508)*

POLYGONAL STRUCTURE. Variety of structure in a metamorphic rock in which the constituent mineral grains have straight or smoothly curved crystal faces generally meeting at triple points. The structure is typical of quartz and calcite rocks. *(OU; Spry 1969, p. 186; Vernon 2004, p. 488)*

POLYMETAMORPHISM ⇒ Section 2.2, Fig. 2.2.2 (Greek *polys*, many). Metamorphism resulting from more than one *metamorphic event*; each event may be *monophase* or *polyphase*. Polymetamorphism is recognized through relics of metamorphic minerals or structures. *(Turner 1948, p. 6; Read 1949, p. 130; de Roever & Nijhuis 1963, p. 327; Kornprobst 2002, p. 40)*

POLYMICT IMPACT BRECCIA (Greek *polys*, many, and *miktos*, mixed). Breccia with clastic matrix or crystalline matrix (derived from the crystallization of *impact melt*), containing lithic and mineral clasts of different degrees of *shock metamorphism*, excavated by an *impact* from different regions of the *target rock* section, transported, mixed and deposited inside or around an *impact crater* or injected into the *target rocks* as dykes. Ant. *monomict impact breccia. (OU; Engelhardt 1971, p. 5567; Stöffler* et al. *1979, p. 652, Fig. 6; Stöffler* et al. *1980, Table 1)*

POLYMIGMATITE. *Migmatite* consisting of at least two different granitic parts and one basic part. *(Sederholm 1923, p. 144)*

POLYMINERALIC. Said of a rock in which ≥95% of the modal content is composed of two or more minerals, as opposed to *monomineralic. (Vogt 1905, p. 1)*

POLYMORPHIC TRANSFORMATION. Change in the structure of a mineral (solid phase) without change in the chemical composition. Examples are low- and high-quartz, calcite and aragonite, andalusite and kyanite. Hence ***polymorphs***, different structural states of a mineral capable of polymorphic transformation. Also ***polymorphism***, the property of a mineral to undergo polymorphic transformation. *(OU; Hatch 1888, p. 444; Buerger 1951, in Rast 1965, p. 76; Turner & Verhoogen 1951, p. 384; Shelley 1993, p. 262; Kornprobst 2002, p. 13)*

POLYPHASE METAMORPHISM ⇒ Section 2.2, Fig. 2.2.2. *Metamorphic event* with two or more temperature and/or pressure peaks. Ant. *monophase metamorphism.*

PORCELLANITE, PORCELANITE, PORCELAIN JASPER (Italian *porcellana*, fine earthenware, and a type of seashell). Dull, hard *burned rock* resembling unglazed porcelain, which breaks easily into sharply angular fragments with a subconchoidal fracture. Originally proposed for altered clays in contact with burning coal seams, the term subsequently included all *contact-metamorphic* rocks. It is also used in sedimentary petrology and because of this ambiguity it has been superseded by *buchite* or *fritted rock* according to the glass content. *(Werner 1786, p. 292; Peithner, in Kirwan 1794, p. 313; Brochant 1798, p. 336;*

Leonhard 1824, p. 557; Cordier 1868, p. 247; Tomkeieff 1940; Williams et al. *1954, p. 268)*

PORE MAGMA. Intergranular solution of granitic or, in a peridotitic environment, basaltic composition, occurring in an early stage of partial melting. *(Eskola 1936, p. 65)*

PORPHYROBLAST (*porphyry*, itself from the Greek *porphyros*, purple, and *blastos*, bud, sprout). Large crystal formed in a metamorphic rock and set in a matrix of smaller grains. Hence ***porphyroblastic structure***. *(Becke 1903a, p. 454, 570; Becke 1903b, p. 47; Joplin 1968, p. 29; Barker 1998, p. 240)*

PORPHYROCLAST (*porphyry*, itself from the Greek *porphyros*, purple, and *klastos*, broken). Large relict crystal in a metamorphic rock, set in a finer-grained matrix that was produced by deformation. The crystals in the matrix may have undergone *recrystallization*. Hence ***porphyroclastic structure***. See *mantled porphyroclast* and *porphyroclast system*. *(Becke 1903b, p. 49; Spry 1969, p. 228; Harte 1977, p. 280, 282; Barker 1990, p. 142; Vernon 2004, p. 488)*

PORPHYROCLAST SYSTEM. Term covering the geometry of a *porphyroclast* and its tails, formed by dynamically *recrystallized* material. The asymmetry of the system indicates the sense of shear. See also *mantled porphyroclast*. *(Passchier & Simpson 1986, p. 831)*

Porphyroid. Ambiguous term originally proposed for igneous rocks allied to porphyries. Later used for metamorphosed acid volcanic rocks or their tuffs. Also used for low-grade *blastoporphyritic* rocks of igneous origin with *porphyroclasts* of feldspar and quartz in a *phyllitic* groundmass. If the term is used its meaning should be made clear. *(Delamétherie 1795, p. 18; Lossen 1869, p. 330; Rosenbusch 1923, p. 371; Jung 1958, p. 189; Foucault & Raoult 2005, p. 280)*

POST-SHOCK TEMPERATURE. Temperature of gaseous, liquid or solid matter after shock pressure release. *(OU; Stöffler 1972, p. 53; Grieve 1998, p. 115)*

Prasinite (Greek *prasinos*, green; Alps). Schistose or gneissose metamorphic rock predominantly consisting of actinolite, ocellar albite and epidote, possibly with chlorite, and titanite-bearing. Prasinite (s.l.) includes *ovardite* in which chlorite occurs in a higher modal amount than amphibole. Hence ***prasinite facies***, an obsolete term, considered by Eskola as a *subfacies* of his *epidote-amphibolite facies* and in part transitional to the *glaucophane-schist facies*. *(Kalkowsky 1886, p. 217; Novarese 1895, p. 168; Angel 1929, p. 240; Eskola 1939, p. 356; Turner 1948, p. 99)*

Predazzite ⇒ Section 2.10 (*Predazzo*, Val di Fassa, Trento, Italy). Brucite *marble* formed as a *contact rock* from pure dolomitic limestone and with a higher CaO/MgO value than its companion rock, *pencatite*. *(Petzholdt 1843, p. 193; Lemberg 1872, p. 229; Zirkel 1873, p. 221; Rogers 1918, p. 582; Harker 1932, p. 77)*

PREFERRED ORIENTATION. Said of *inequant* mineral grains or grain aggregates, a statistically significant number of which have the same orientation. The term may also be used in respect of crystallographic axes. *(Knopf & Ingerson 1938, p. 17; Turner 1948, p. 158; Hobbs* et al. *1976, p. 250; Vernon 2004, p. 488)*

PREHNITE-ACTINOLITE FACIES ⇒ Sections 2.2 & 2.5, Figs. 2.2.4 & 2.5.1, Table 2.2.1. *Metamorphic facies* representing very low grades of metamorphism and characterized by the *mineral association* of prehnite-actinolite-epidote (± chlorite, albite, quartz and titanite) and by the absence of pumpellyite in rocks of appropriate bulk composition (mostly metabasic rocks and their clastic derivatives). Some authors prefer to regard this *mineral association* as a subdivision of the *subgreenschist facies*. *(Liou* et al. *1985, p. 330; Liou* et al. *1987, p. 69; Bucher & Frey 1994, p. 103)*

PREHNITE-PUMPELLYITE FACIES ⇒ Sections 2.2 & 2.5, Figs. 2.2.4 & 2.5.1, Table 2.2.1 (prehnite-pumpellyite metagreywacke facies). *Metamorphic facies* representing very low grades of metamorphism and characterized, in metasandstones and metavolcanic rocks of appropriate composition, by the presence of prehnite and/or pumpellyite in the absence of zeolites, lawsonite or jadeite. Quartz-albite-chlorite-prehnite and/or pumpellyite may coexist stably. The SCMR regards this *mineral association* as a subdivision of the *subgreenschist facies*. *(Coombs 1960, p. 341; Coombs 1961, p. 204; Seki 1961, p. 414, 421;*

Hashimoto 1966, p. 265; Liou et al. *1987, p. 69; Bucher & Frey 1994, p. 103)*

PRESSURE SHADOW (German *Streckungshof*). Area of low strain in a deformed rock that was protected from the maximum compressive stress by its proximity to a rigid body, either a *porphyroblast* or a *porphyroclast*. The area of low strain may preserve pre-existing structures or space may be created in which new minerals may grow. The growth of new minerals, typically quartz or calcite, in the low-strain area may resemble a beard, particularly if the minerals have a fibrous habit, hence *bearded structure*. Syn. *strain shadow*. *(Mügge 1930, p. 475; Spry 1969, p. 240; Ramsay & Huber 1983, p. 279; Vernon 2004, p. 477)*

PRESSURE SOLUTION. Deformation process by which material under stress goes into solution and is then transported by flow or diffusion to areas of relatively low stress. A diffusive mass-transfer deformation mechanism (in which the presence of intergranular water is inferred to accelerate the rate) whereby material is transferred from grain boundaries under high interfacial normal stress to interfaces under lower normal stress. The material may simply move around a mineral grain boundary (e.g. into a *pressure shadow*) or be transported out of the system. Hence ***pressure solution striping/cleavage***, where the development of various types of *cleavage* result in the preferential movement of material, usually quartz, from the *cleavage domains* into the intercleavage *domains* (*microlithons*), leaving the *cleavage domains* relatively enriched in phyllosilicates and giving the rock a striped appearance. Synonymous with ***solution-transfer***, ***dissolution creep***. *(OU; Sorby 1879, p. 89; Durney 1972, p. 315; Rutter 1983, p. 725; Yardley 1989, p. 168; Davis & Reynolds 1996, p. 173; Ramsay & Lisle 2000, p. 880; Vernon 2004, p. 491)*

PRIMARY GNEISS (protogneiss). Obsolete term for a gneiss derived from an igneous rock. Rec.syn. orthogneiss. *(Grout 1932, p. 358)*

PROGRADE METAMORPHISM, PROGRESSIVE METAMORPHISM. Metamorphism giving rise to the formation of minerals that are typical of a higher grade (i.e. higher temperature) than the former phase *assemblage*; a sequential increase in *metamorphic grade*; that part of the metamorphic cycle up to the thermal maximum. Note: Miyashiro used prograde metamorphism to imply a sequential increase in grade in an individual rock with time and progressive metamorphism to imply a sequential increase in grade across a region (e.g. as exhibited by *Barrow's zones*). The SCMR recommends that if the terms are used with this latter more restricted meaning then this should be made clear. Ant. *retrograde*, *retrogressive metamorphism*. *(Barrow 1893, p. 352; Knopf 1931, p. 2; Harker 1932, p. 208; Bucher & Frey 1994, p. 52; Miyashiro 1994, p. 351)*

PROGRADE SHOCK METAMORPHISM, PROGRESSIVE SHOCK METAMORPHISM. Increasing grade of *shock metamorphism* displayed either by the autochthonous rocks of the *impact crater* basement (radially increasing towards the point of *impact*) or by individual rock clasts (*impactoclasts*) of a *polymict impact breccia*. *(OU; Stöffler 1966, p. 16; Chao 1967a, p. 192; Chao 1967b, p. 205; Dence 1968, p. 170, 173, Table 1; Engelhardt & Stöffler 1968, p. 160; Stöffler 1971, p. 5541)*

PROJECTILE. Synonymous with *impactor*.

PROPYLITE ⇒ Section 2.9, Fig. 2.9.1 (Greek *propylitês*, someone near the gate). Low- to medium-temperature *metasomatic granofels* formed by the *alteration* of basic volcanic rocks. Low-temperature varieties are principally composed of albite, calcite and chlorite; higher-temperature varieties are composed of epidote, actinolite and biotite. Propylite forms at the postmagmatic stage. Hence ***propylitization***, *metasomatic* process leading to the formation of propylite. *(Richtofen 1868, p. 20; Becher 1882, p. 90; Zirkel 1894b, p. 593; Lindgren 1933, p. 457; Shelley 1993, p. 67; Foucault & Raoult 2005, p. 285)*

PROTEOLITE (*Prôteus*, a multiform Greek and Latin sea god, and *lithos*, stone; Cornwall). Obsolete term for a pelitic *hornfels* similar to *cornubianite* but easily weathered and of varied appearance; also restricted to an andalusite-quartz-mica *hornfels*. *(Boase 1832, p. 394; Naumann 1849–53; Rosenbusch 1877, p. 42; Lapparent 1882, p. 618; Bonney 1886, p. 104)*

PROTEROGENIC, PROTEROGENOUS (Greek *prôteros*, earlier, and *gennô*, to give birth to). Said of an original mineral in a metamorphic rock. The mineral may be a relict which is earlier than the main, equilibrium, *assemblage* of the rock (e.g. relict magmatic clinopyroxene in a *greenschist*). Ant. *hysterogenic*. *(Zirkel 1893, p. 791; Becke 1903b, p. 34)*

PROTOCALCITE. Variety of *pure marble*, term used in sculpture art. Syn. *statuary marble*. *(Cordier 1868, p. 286)*

PROTOCATACLASITE. *Cataclasite* in which the matrix forms less than 50% of the rock volume. *(Brögger 1890, p. 105; Spry 1969, p. 229)*

PROTOGINE [protogyne] (Greek *prôtos*, first, and *gignomai*, to be generated; Mt Blanc, French–Italian Alps). Obsolete term used in the Alps for a metagranitic rock with a gneissose structure, containing *greenschist facies* minerals such as chlorite, epidote, Mn-garnet and fine-grained white mica. The first descriptions mentioned talc, although this was actually white mica ± chlorite. *(Jurine 1806, p. 372; Beudant 1830, p. 563; Delesse 1849, p. 114)*

PROTOLITH ⇒ Section 2.1 (Greek *prôtos*, first, and *lithos*, stone). Precursor rock (igneous, sedimentary or already metamorphosed) from which a given metamorphic rock is derived. Syn. ***parent rock***. *(OU)*

PROTOMYLONITE. *Mylonite* in which less than 50% of the rock volume has undergone grain-size reduction. *(Backlund 1918, p. 195; Waters & Campbell 1935, p. 479; Spry 1969, p. 229)*

PROXIMAL IMPACTITE ⇒ Section 2.11, Fig. 2.11.2. *Impactite* occurring in the immediate vicinity of an *impact crater*, that is, inside the outer limit of the continuous *ejecta blanket*. It comprises all types of *impact breccias*, *impact melt rocks* and *shocked rocks*. Ant. *distal impactite*. *(Stöffler & Grieve, Section 2.11)*

Psammite ⇒ Section 2.1 (Greek *psammos*, sand). Term for a sedimentary rock composed of sand-sized particles. It has also been used locally for the metamorphosed equivalents of these sedimentary rocks, that is, metamorphic rocks with a low modal ratio of mica to quartz + feldspar (Tyrrell, Robertson). The SCMR recommends that it should only be used for sedimentary rocks and the term ***metapsammite*** should be used for the metamorphosed equivalents. *(Haüy, in Cordier 1868, p. 222; Tyrrell 1921, p. 501; Robertson 1999, p. 4)*

PSEUDOMORPH (Greek *pseudês*, false, and *morphê*, form). Mineral or aggregate of minerals in a metamorphic rock, whose shape is that of a pre-existing mineral which it/they has/have replaced. Hence ***pseudomorphous***, ***pseudomorphic***, pertaining to a pseudomorph. See footnote under *corona*. *(Werner 1786; Phillips 1819, p. xcii; Haüy 1822, p. 93; Blum 1843, p. 1; Teall 1888, p. 85, 445; Spry 1969, p. 90; Vernon 2004, p. 489)*

PSEUDOMORPHISM. Obsolete term for the process leading to the formation of a *pseudomorph*. *(Haüy 1822, p. 93; Omalius d'Halloy 1868, p. 474)*

PSEUDOTACHYLITE (Greek *pseudês*, false, *tachys*, quick, and *lithos*, stone; Parijs, Orange Free State). Ultrafine-grained vitreous-looking material, usually black and flinty in appearance, occurring as thin planar veins, injection veins or as a matrix to pseudo-conglomerates or breccias, and which in-fills dilation fractures in the host rock. Syn. *hyalomylonite*; cf. *trap-shotten gneiss*. See also *impact pseudotachylite*. *(Shand 1916, p. 199)*

PSEUDOTALCITE (Greek *pseudês*, false, and *talkês*, talc). See *mimotalcite*. *(Cordier 1868, p. 185)*

P–T–t **PATH**. Changing pressure–temperature conditions experienced by a mineral or a rock with time, or a line/curve on a *P–T* grid showing these changes. *(OU; Yardley 1989, p. 198; Spear 1993, p. 2; Miyashiro 1994, p. 352)*

PTYGMA (Greek *ptyssô*, to fold). Unnecessary term for (granitoid) material showing *ptygmatic folding*. *(Dietrich 1959, p. 358)*

PTYGMATIC FOLDING (Greek *ptyssô*, to fold). Originally used to describe contorted and folded granitic veins which characteristically occur in *migmatites*. The term is now used more widely to describe a form of folding where single isolated layers of relatively high competence are enclosed in a matrix of lower competence and strongly shortened. *(Sederholm 1907, p. 89, 110; Dietrich 1959, p. 358; Mehnert 1968, p. 356; Ramsay & Huber 1983, p. 12)*

PUMPELLYITE-ACTINOLITE FACIES ⇒ Sections 2.2 & 2.5, Figs. 2.2.4 & 2.5.1, Table 2.2.1. *Metamorphic facies* representing very low grades of metamorphism and characterized by the mineral association of pumpellyite-actinolite-quartz (± chlorite, albite and epidote) and by the lack of prehnite. The SCNR regards this *mineral association* as a subdivision of the *subgreenschist facies*. *(Hashimoto 1966, p. 265; Liou* et al. *1987, p. 70)*

PYRENEAN(-TYPE) FACIES SERIES (*Pyrenees*, mountain belt separating SW France and N Spain). Type of *facies series* that is of medium to low *P/T* style, characterized by the presence of staurolite, andalusite, cordierite and sillimanite, as far as chemical conditions permit. It lies between the *Buchan* and the *Idahoan facies series*. *(Hietanen 1967, p. 193)*

PYRIBOLITE ⇒ Section 2.8 (from *pyr*oxenes + amph*ibole*; Tovqussap nunâ, southern Sukkertoppen district, W Greenland). High-grade metamorphic rock composed of plagioclase, hornblende, clinopyroxene, orthopyroxene ± garnet. The presence of orthopyroxene is essential in the original definition, hornblende and pyroxenes being present in approximately equal amounts. The SCMR does not recommend the use of this term, which may be substituted by pyroxene *amphibolite* or hornblende *mafic granulite* according to the quantities of the respective minerals present. *(Berthelsen 1960, p. 41–6; Lorenz 1998, p. 110)*

PYRICLASITE ⇒ Section 2.8 (from *pyr*oxenes + plagio*clase*, 'pyri' to avoid confusion with pyroclastics; Tovqussap nunâ, southern Sukkertoppen district, West Greenland). High-grade metamorphic rock consisting mainly of feldspar (plagioclase) and pyroxene (orthopyroxene ± clinopyroxene) with or without garnet. The presence of orthopyroxene is essential according to the original definition. As redefined by the Granulite Commission, the contents of mafic constituents should be higher than 30% (in vol.). As this name was not generally accepted by the SCMR, the term *mafic granulite* should be used instead. *(Berthelsen 1960, p. 20; Mehnert 1972, p. 141; De Waard 1973, p. 384)*

PYRIGARNITE ⇒ Section 2.8 (from *pyr*oxenes + *garn*et). Initially defined by Vogel as a high-grade metamorphic rock composed of pyroxenes and garnet, in which the presence of plagioclase may be expressed by a prefix (plagio-pyrigarnite). This definition was modified by Mehnert, and plagioclase was added to the characteristic constituents of pyrigarnite. According to this revised definition pyrigarnite is composed of plagioclase, garnet and pyroxenes (clino- and/or orthopyroxene), the contents of mafic constituents being higher than 30% vol. As this name was not generally accepted by the SCMR, the term garnet-rich *mafic granulite* should be used instead. *(Vogel 1967, p. 176; Mehnert 1972, p. 142; Lorenz 1998, p. 111)*

PYROCAUSTIC METAMORPHISM [pyricaustic metamorphism] (Greek *pyr* gen. *pyros*, fire, and *kaustikos*, burning). Name given to a poorly defined type of high-temperature and essentially dry metamorphism as opposed to a wet type called *hydatothermic*. It may be interpreted as corresponding to *pyrometamorphism*. *(Bunsen 1849, p. 16)*

PYROGENIC METAMORPHISM (Greek *pyr* gen. *pyros*, fire, and *gennô*, to give birth to). Obsolete synonym for *combustion metamorphism*. *(Kalkowsky 1886, p. 34)*

PYROMETAMORPHISM ⇒ Sections 2.2 & 2.10, Fig. 2.2.1. Very high-grade type of *contact metamorphism* occurring in volcanic settings or around near-surface intrusions and characterized by *mineral assemblages* stable at or near atmospheric pressure and very high temperatures; critical minerals are spurrite, tilleyite, rankinite, larnite and merwinite in silica-deficient carbonate rocks; mullite and glass in aluminous rocks; tridymite and glass in silica-oversaturated rocks. *(Brauns 1911, p. 12; Grubenmann & Niggli 1924, p. 315; Tyrrell 1926, p. 253; Lindgren 1928, p. 104; Eskola 1939, p. 267; Reverdatto 1973, p. 36–54; Bucher & Frey 1994, p. 9)*

PYROMETASOMATISM. Type of *contact metasomatism* connected with very high temperatures and typically associated with ore deposition in predominantly carbonate rocks. *(OU; Lindgren 1928, p. 781)*

PYROMORPHISM. Name given to a *local*, very high-temperature type of *metamorphism* caused by burning coal seams or induced by contact with lava flows. It corresponds in part to *combustion metamorphism*, in part to *pyrometamorphism*. *(Lasaulx 1875, p. 444)*

PYROXENE GRANULITE. Synonymous with *mafic granulite*. *(Lehmann 1884; MacGregor 1931, p. 508)*

PYROXENE-HORNFELS FACIES ⇒ Section 2.2, Fig. 2.2.4, Table 2.2.1. *Metamorphic facies* representing very high temperatures and low or very low pressures. The temperature is lower than in the *sanidinite facies* and the pressure lower than in the *granulite facies*. It is characterized by clinopyroxene-orthopyroxene-plagioclase (olivine stable with plagioclase) *assemblages* in rocks of basaltic composition. *(Eskola 1920, p. 159)*

PYROXENITE ⇒ Table 2.1.2. Igneous or metamorphic rock consisting of >90% modal content of pyroxene. This definition accords with that of Le Maitre and as such constitutes one of the exceptions to the SCMR guideline related to the use of the suffix *-ite*. *(Le Maitre 1989, 2002)*

QUARTZ HORNFELS. Term proposed to distinguish a quartzite formed by *contact metamorphism* from other types of metamorphic quartzite. Rec.syn. *contact* quartzite. *(Eisele 1907, p. 134; Tyrrell 1926, p. 300)*

QUARTZSTHENE GRANULITE (from *quartz* and hyper*sthene*). Term proposed for quartz *charnockitic* and quartz enderbitic *granulites*. See *plagiosthene granulite*, *perthisthene granulite* and *plagioperthisthene granulite*. *(Nasir 1993, Fig. 2; Lorenz 1998, p. 111)*

QUELUZITE (*Queluz* district, Minas Geraes, Brazil). Metamorphic rock of variable composition characterized by the predominance of spessartine garnet, locally with minor contents of tephroite (manganese silicate), amphiboles, micas and manganese oxides. In some cases quartz may also be a primary constituent. Accessories include ilmenite, rutile and apatite. The rock when leached and decomposed provides important Mn-ore deposits. *(Derby 1901, p. 30; Derby 1908, p. 213)*

RANOCCHIAIA, SERPENTINA RANOCCHIAIA (Italian *ranocchia*, frog; Riviera di Levante, Italy). Local miners' name for a light green ornamental serpentinite that shows an intricate dark veinwork of serpentine and opaque minerals. *(Mazzuoli & Issel 1881, p. 332)*

REACTION RIM, REACTION BORDER. Peripheral zone around a mineral grain, composed of another mineral species and formed by reaction between the mineral and its surroundings. Cf. *corona* and footnote, *synantetic*. *(Törnebohm 1877, p. 37; Adams 1893, p. 466; Spry 1969, p. 104)*

RECRYSTALLIZATION. Nucleation and migration of high angle (grain) boundaries to produce new grains. Recrystallization does not necessarily involve any change in chemical composition. *(OU; Hobbs* et al. *1976, p. 108; Passchier & Trouw 1996, p. 263; Vernon 2004, p. 490)*

RECRYSTALLIZATION METAMORPHISM (German *Umkristallisationsmetamorphose*). Metamorphism involving the development of new crystalline mineral grains in the rock. The new grains may have the same mineralogical composition as the minerals in the *protolith*, or a different mineralogical composition. *(Eskola 1939, p. 265)*

RECRYSTALLIZATION REPLACEMENT. *Metasomatic* process by which a sedimentary rock is brecciated and *recrystallized* by *hydrothermal* fluids. *(Goodspeed & Coombs 1937, p. 23)*

REGIONAL METAMORPHISM ⇒ Section 2.2, Fig. 2.2.1. Type of metamorphism that occurs over an area of wide extent, i.e. affecting a large rock volume, and is associated with large-scale tectonic processes, such as ocean-floor spreading, crustal thickening related to plate collision, deep basin subsidence, etc. *(Naumann 1849, p. 751; Daubrée 1860, p. 155, in Tomkeieff; Teall 1888, p. 438; Geikie 1903, p. 785; Daly 1917, p. 394)*

REGIONAL METASOMATISM ⇒ Section 2.9. Type of *metasomatism* that is noticeably developed in large volumes of former rocks and does not exhibit any megascopic *metasomatic* zonation. *Metasomatic* processes such as *propylitization*, *beresitization* and serpentinization often manifest examples of regional metasomatism. *(Korzhinskii 1953, p. 431)*

REGOLITH (rhegolith) (Greek *rhegos*, blanket, and *lithos*, stone). Mantle of fragmental and unconsolidated debris. See *impact regolith. (Merrill 1897, p. 299)*

REGOLITH BRECCIA. *Regolith* lithified by shock compression due to *impact*; typically found on the Moon and asteroids. *(OU; Quaide & Bunch 1970, p. 718; Stöffler* et al. *1979, p. 660; Stöffler* et al. *1980, Table 1; Heiken* et al. *1991, p. 257, 352; Bischoff & Stöffler 1992, Table 1)*

REGRESSIVE METAMORPHISM. Synonymous with *retrogressive* and *retrograde metamorphism*. Cf. *diaphthoresis. (Tyrrell 1926, p. 263)*

RELICT STRUCTURE. Type of structure in a deformed rock characterized by the presence of remnants of the pre-existing undeformed rock. The structure usually takes the form of lenticular 'relicts' in a finer-grained matrix. See *phacoidal structure*, *flaser structure*, *augen structure*, *porphyroclastic structure* and *palimpsest structure. (OU)*

RELIEF METAMORPHISM (German *Entlastungsmetamorphose*, metamorphism by unload). Type of metamorphism that takes place when the rock is moved from a region of higher pressure to one of lower pressure; decompressional metamorphism. Ant. *load metamorphism. (OU; Grubenmann & Niggli 1924, p. 180; Tomkeieff)*

RESIDUAL STRUCTURE. See *relict structure*.

Resister ⇒ Section 2.6 (Latin *resistere*, to resist). Rock offering greater resistance to the processes of *granitization* than another by virtue of its composition or its 'impenetrable' *fabric*, for example quartzite, limestone, metadolerite or *serpentinite*. It generally forms enclaves in *granitized* (*migmatized*) rocks. *(Read 1951, p. 7; Mehnert 1968, p. 298; Shelley 1993, p. 109)*

RESTITE (Latin *restare*, to remain behind). Remnant of a metamorphic rock from which a substantial amount of the more mobile chemical components has been extracted without being replaced. *(Dietrich & Mehnert 1960, p. 62; Mehnert 1968, p. 356)*

RETICULAR STRUCTURE, RETICULATED STRUCTURE (Latin *rete*, net). Type of structure in a metamorphic rock in which the mineral grains are arranged in two sets of crossing lines like a net. *(Phillips 1819, p. xcii; Lyell 1833, p. 79)*

RETROGRADE METAMORPHISM, RETROGRESSIVE METAMORPHISM (Latin *retro*, backwards). Metamorphism giving rise to the formation of minerals that are typical of a lower grade (i.e. lower temperature) than the former phase *assemblage*. The term was introduced by Harker to replace Becke's *diaphthoresis*. Ant. *prograde*, *progressive metamorphism*. Note: Miyashiro distinguished between retrograde metamorphism which he defined as that part of a *metamorphic event* after the thermal maximum, and *retrogressive metamorphism* which he took as a second *metamorphic event* which takes place at a lower temperature than the first. The SCMR recommends that if the terms are used with this restricted meaning, then this should be made clear. *(Becke 1909, p. 373; Harker 1932, p. 342; Turner 1938, p. 172; Miyashiro 1994, p. 352)*

RETROMETAMORPHISM [retromorphism]. Synonymous with *retrograde* and *retrogressive metamorphism. (Grubenmann & Niggli 1924, p. 340)*

RHEOMORPHISM (Greek *rheô*, to flow, and *morphê*, form). Complex process or processes leading to the partial or complete melting of a rock and involving the diffusion of material. *(Backlund 1937, p. 234; Mehnert 1968, p. 356)*

RHYOLITE-GNEISS. Local name for a rhyolite metamorphosed into *gneiss*. Harker describes the rock as a *granoblastic* quartz-feldspathic groundmass enclosing partly transformed riebeckite and Na-feldspar phenocrysts. *(Harker 1932, p. 288; Weidman, in Tomkeieff)*

RIBBON STRUCTURE, RIBBON QUARTZ. Strongly elongated single crystal or very fine-grained aggregates of quartz crystals resembling ribbons in a rock. The structure is commonly the product of intense deformation at high temperature. The crystal grains may show *undulose extinction* or be *recrystallized. (OU; Spry 1969, p. 294; Vernon 2004, p. 490)*

RICOLITE (*Rico*, New Mexico, USA). *Banded* and mottled, light and dark green impure serpentine rock used as a decorative stone. *(Merrill 1891, p. 64)*

RODINGITE ⇒ Section 2.9 (*Roding* River, New Zealand). *Metasomatic* rock primarily composed of grossular-andradite garnet and calcic pyroxene; vesuvianite, epidote, scapolite and iron ores are characteristic accessories. Rodingite mostly replaces dykes or inclusions of basic rocks within serpentinized ultramafic bodies. It may also replace other basic rocks, such as volcanic rocks or *amphibolites* associated with ultramafic bodies. Hence ***rodingitization***, *metasomatic* process leading to the formation of rodingite. *(Bell* et al. *1911, p. 31; Grange 1927, p. 160; Suzuki 1953, in Tomkeieff)*

ROTATIONAL STRUCTURE. See *snowball structure.*

SACCHARITE (Greek *sakharon*, sugar; Frankenstein in Silesia). Term used in the past in two different meanings. (1) White mineral from the plagioclase group. (2) Fine-grained aggregate of plagioclase and quartz forming nests in *serpentinite. (Glocker 1845, p. 500; Lasaulx 1878, p. 629)*

SAGVANDITE (Lake *Sagvandet*, Norway). Rock composed mainly of Mg-orthopyroxene with some magnesite. Its origin is uncertain; it is generally considered as magmatic but may be, at least in part, *metasomatic. (Pettersen 1883, p. 247; Winkler 1974, p. 151; Le Maitre 1989, 2002; Bucher & Frey 1994, p. 148, 166)*

SAMMELKRYSTALLISATION (German *sammeln*, to collect). See *collective crystallization.*

SAMOSITE (*Samos*, an island in the Aegean Sea). Obsolete local term for a metamorphic rock composed of diaspore and iron oxide ores with a pisolithic structure. Diaspore bauxite should be used instead. *(Lapparent 1937, p. 30)*

SANIDINITE FACIES ⇒ Section 2.2, Fig. 2.2.4, Table 2.2.1 (Laacher See, Eifel, Germany). *Metamorphic facies* representing extreme high temperatures and low or very low pressures. It lies below the *granulite facies.* It is characterized by the occurrence of especially high-temperature varieties and *polymorphs* of minerals, for example pigeonite, K-rich labradorite, and of sanidine-rich rocks derived from pelitic rocks. See *pyrometamorphism. (Lacroix 1893a; Brauns 1911, p. 12; Eskola 1920, p. 154)*

SARRAZACITE (*Sarrazac*, Dordogne, France). Allivalite in which the anorthite and olivine have combined to form a chlorite *amphibolite. (Roques 1936, p. 334)*

SATHROLITH, SATHROLITE (Greek *sathros*, weathered, *sapros*, rotten, and *lithos*, stone; Tamerfors region, Norway). Term used for metamorphosed clastic rocks, the clastic material having been derived *in situ* by the disintegration or degradation of the local solid rock. Previously proposed as an alternative to the sedimentary term saprolite for designating this clastic material. *(Sederholm 1931, p. 77; Sederholm 1934, p. 27, 33)*

SAUSSURITIZATION (saussurite named in honour of H. B. de *Saussure*, 1740–99, Swiss naturalist and physicist, explorer of Mt Blanc). Late magmatic, metamorphic or other *alteration* process by which calcic plagioclase is altered to saussurite, a tough, compact, white, greenish or greyish mineral aggregate consisting of clinozoisite, zoisite, albite and/or epidote, with variable amounts of calcite, sericite, prehnite and calcium-aluminium silicates. *(Saussure 1806, p. 469; Beudant 1832, p. 110; Hagge 1871, p. 51; Rosenbusch 1873, p. 356; Zirkel 1873, p. 142; Becke 1878, p. 247; Cathrein 1883, p. 234; Rosenbusch 1887, p. 163; Williams 1890, p. 58; Zirkel 1894a, p. 742; Deer* et al. *1962, p. 194, 208)*

SAXON-TYPE FACIES SERIES, SAXONIAN-TYPE FACIES SERIES (*Saxony* in Germany). Type of *facies series* of high *P*/*T* style, which is characterized by the formation of kyanite *granulites.* It lies between the *Barrovian* and Alpine *facies series. (Hietanen 1967, p. 201; Carmichael 1978, Fig. 1; Turner 1981, p. 451)*

S-C FABRIC (originally **C/S fabric**). Composite *fabric* produced by the intersection of two planar fabrics (C-fabric and S-fabric) in sheared rocks. The C-fabric forms broadly parallel to the margins of the *shear zone* and the S-fabric forms oblique to the margin, the angle between the two decreasing as shearing progresses. The S-planes curve into the C-planes, the nature of the curvature reflecting the sense of shear. S-C fabrics are common in strongly *foliated* and mica-rich *mylonites* or granitoid *mylonites. (Berthé* et al. *1979, p. 33; Lister & Snoke 1984, p. 618; Passchier & Trouw 1996, p. 113; Davis & Reynolds 1996, p. 546; Barker 1998, p. 166)*

SCHALSTEIN (German *Schale*, shell; iron mining district of Nassau near Limburg, Hessen, Germany). Old term originally used for diabase tuffs and *hydrothermally* altered diabases, but now restricted to bedded Palaeozoic diabase tuffs. The name refers to the platy or flaky forms of fractures. *(Becher 1789, p. 44; Stifft 1807, p. 382; Zirkel 1894a, p. 664; Lehmann 1933, p. 80; Hentschel 1951, p. 226; Le Maitre 1989, 2002)*

SCHILLERFELS (German *schillern*, to shimmer; cf. schillerspar, diallage pyroxene; Gabbro, N Apennine, Italy). Obsolete term for slightly altered gabbro, anorthosite or peridotite containing schillerized pyroxene. *(Brochant 1800, p. 419; Raumer 1819, p. 40; Cordier 1868, p. 66; Le Maitre 1989, 2002)*

SCHINDOLITH, SKINDOLITH (Greek *skhindô*, to split, and *lithos*, stone). Obsolete term for a rock formed by the melting of older parts of the Earth's crust followed by *differentiation*. *(Loewinson-Lessing 1949, p. 451)*

SCHIST ⇒ Section 2.3 (Greek *schizô*, to tear). Metamorphic rock displaying a *schistose structure*. For phyllosilicate-rich rocks the term schist is commonly used for medium- to coarse-grained varieties, whereas finer-grained rocks may be given more specific names such as *slates* or *phyllites*. *(Plinius 80, p. 80; Boubée 1833, p. 28; Lyell 1833, p. 79; Van Hise 1904, p. 779; Holmes 1920, p. 205; Spry 1969, p. 270; Winkler 1974, p. 326; Yardley 1989, p. 22)*

SCHIST-HORNFELS ⇒ Section 2.10. Obsolete term for a schist converted to *hornfels* by *contact metamorphism* in the *inner aureole*; traces of the former *schistosity* may still be preserved. Rec.syn. *hornfelsed schist*. Cf. *cornubianite* (in part). *(Rosenbusch 1887, p. 52; Loewinson-Lessing 1893–4, p. 208; Salomon 1898, p. 145)*

SCHISTOCLASTIC STRUCTURE (Greek *schizô*, to tear, and *klastos*, broken). Type of *schistose structure* formed by *cataclasis*. *(Erwin 1938, p. 119)*

Schistoid. See *-oid*. *(OU; Cordier 1868, p. 23; Omalius d'Halloy 1868, p. 87)*

SCHISTOSE HORNFELS ⇒ Section 2.10 (German *schiefriger* Hornfels). Special term for a *contact-metamorphosed* rock of the *inner aureole* that is totally recrystallized but retains some fissility. The term replaces *leptynolite*. *(Rosenbusch 1887, p. 38)*

SCHISTOSE STRUCTURE. Type of structure characterized by a *schistosity* that is *well developed*, either uniformly throughout the rock or in narrowly spaced repetitive zones such that the rock will split on a scale of one centimetre or less. *(Phillips 1819, p.* xciii*; Macculloch 1821, p. 123; Geikie 1903, p. 244; Tyrrell 1926, p. 273)*

SCHISTOSITY ⇒ Section 2.3. Preferred orientation of *inequant* mineral grains or grain aggregates produced by metamorphic processes. A schistosity is said to be ***well developed*** if *inequant* mineral grains or grain aggregates are present in a large amount and show a high degree of *preferred orientation*, either throughout the rock or in narrowly spaced repetitive zones, such that the rock will split on a scale of less than one centimetre. A schistosity is said to be ***poorly developed*** if *inequant* mineral grains or grain aggregates are present only in small amounts or show a low degree of *preferred orientation* or, if well developed, occur in broadly spaced zones such that the rock will split on a scale of more than one centimetre. *(OU; Hatch 1888, p. 446; Holmes 1920, p. 206; Wilson 1961, p. 458; Barker 1998, p. 242)*

SCHLIEREN (German *Schliere*, flaw or streak in glass). Streaks or minor lenticular parts of a rock that differ from the main body of the rock in the mineral content or the ratio of minerals and which commonly have transitional boundaries. The term was originally used for magmatic rocks but is now also used for similar structures in *migmatites*, for example for patches of non-*leucosome* within the *leucosome*. *(OU, Hatch 1888, p. 446; Dietrich & Mehnert 1960, p. 62; Ashworth 1985, p. 3)*

SCHOLLEN (German *Scholle*, clod, flake). In a *migmatite*, blocks or rafts of *palaeosome* within the *neosome*; the structure is similar to *agmatite* but the *neosome* is more abundant so that the disrupted blocks float like rafts. *(OU; Mehnert 1968, p. 15; Ashworth 1985, p. 3)*

SECONDARY. Said of minerals formed by *alteration* of a pre-existing rock. *(Lyell 1835, p. 401; Loewinson-Lessing 1893–4, p. 212; Holmes 1920, p. 207)*

Secondary quartzite ⇒ Section 2.9, Fig. 2.9.1. Term used in Russian literature for a medium- to low-temperature *metasomatic* rock mainly composed of quartz with subsidiary high-alumina minerals such as pyrophyllite, diaspore, alunite and kaolinite. Common accessories include fluorite, dumortierite and lazulite. Secondary quartzites are associated with volcanic and subvolcanic rocks of rhyolitic to andesitic composition. Normally they form as replacements of acid igneous rocks and more rarely as replacements of sedimentary rocks. They may host mineral deposits of alunite, pyrophyllite, gold, copper, antimony and mercury. *(Nakovnik 1965, p. 9)*

SEEBENITE (*Seeben*, close to Chiusa/Klausen, Trentino-Alto Adige region, Italy). Obsolete term for a cordierite-feldspar *hornfels*. *(Salomon 1898, p. 150)*

SELLAGNEISS (Gamsbodengneiss) (*Sella* Mt, Gotthard Massif, Switzerland). Two-mica orthogneiss, commonly with *augen* structure. *(Stapff, in Heim 1891)*

Semipelite (Latin *semi*, half, and Greek *pêlos*, clay). Local term for a metamorphic rock that has a modal ratio of mica to quartz + feldspar intermediate between a *metapelite* and a *metapsammite*. *(Robertson 1999, p. 4)*

SEMISCHIST. Type of metagreywacke characterized by the presence of a weakly developed *schistosity*. *(Turner 1948, p. 30)*

SERICITIZATION (Greek *syrikon*, silk). *Hydrothermal* or other metamorphic process whereby aluminosilicate minerals are replaced by sericite. *(OU; List 1852, p. 194; Williams 1890, p. 58)*

SERPENTINITE. Metamorphic rock composed of more than 75% vol. of minerals of the serpentine group. *(OU; Boubée 1833, p. 183, 229; Cordier 1868, p. 166; Le Maitre 1989, 2002)*

SHATTER CONE (Steinheim Basin impact crater, Germany). Striated cup-and-cone structure resulting from hypervelocity *impact*; the structure occurs on the centimetre to metre scale. First described as 'Strahlenkalk'. *(Branco & Fraas 1905, p. 496; Dietz 1959, p. 496; Dietz 1960, p. 1781; Grieve 1998, p. 115, Fig. 9a)*

Sheaf structure. Synonymous with *garbenschiefer* structure. *(OU)*

SHOCK DEFORMATION. Deformation by *shock wave* compression at shock pressures above the *Hugoniot elastic limit* leading to permanent (residual) *shock effects* after pressure release. *(OU; Bunch & Cohen 1964, p. 1263; Stöffler 1972, p. 71)*

SHOCKED ROCKS. Rocks affected by *shock* (*impact*) *metamorphism*. *(OU; Chao 1967b, p. 205; Dence 1968, p. 172, Fig. 1; French 1968, p. 2; Stöffler 1971, p. 5541)*

SHOCK EFFECT. Permanent (residual) deformation and/or transformation of minerals and rocks induced by the passage of a *shock wave* after pressure release (***residual shock effect***). *(OU; Doran & Linde 1966, p. 264; Chao 1967b, p. 192; Stöffler 1972, p. 69)*

SHOCK FACIES. Unnecessary synonym of *shock stage*. *(Stöffler 1966, p. 16; Dence & Robertson 1989, p. 526)*

SHOCK FRONT. Synonymous with *shock wave*. *(OU; Duvall & Fowles 1963, p. 211; Stöffler 1972, p. 53)*

SHOCK IMPEDANCE. Thermodynamic entity defined as the product of the density of any phase (before shock compression) times the *shock wave velocity*. *(OU; Stöffler 1972, p. 60)*

SHOCK LAMELLAE. Unnecessary synonym of *planar deformation features*. *(Chao 1967a, p. 193; Chao 1967b, p. 208; Chao 1968, p. 137)*

SHOCK MELTING. Melting of solid matter by *shock wave* compression resulting from high *post-shock temperature* after pressure release; highest stage of *shock metamorphism* before *shock vaporization* is induced at still higher shock pressures. *(OU; Stöffler 1966, p. 21, Fig. 1; Chao 1967b, p. 220; Stöffler 1972, p. 100; Dence & Robertson 1989, p. 527)*

SHOCK METAMORPHISM. Type of *metamorphism* of *local* extent caused by *shock wave* compression due to the hypervelocity *impact* of a solid body or due to the detonation of high-energy chemical or nuclear explosives. Cf. *impact metamorphism*. *(OU; French 1968, p. 2; Grieve 1987, p. 250; Dence & Robertson 1989, p. 526; Bischoff & Stöffler 1992, p. 711)*

SHOCK STAGE. Degree of *shock metamorphism* of a rock achieved during *progressive* (*prograde*) *shock metamorphism*. The successive stages are characterized by specific features or sets of features in the *shocked rocks* and/or minerals. *(OU; Engelhardt & Stöffler 1968, p. 160; Stöffler 1971, p. 5545; Dence & Robertson 1989, p. 529)*

SHOCK STATE. Thermodynamic state of matter under shock compression. *(OU; Duvall & Fowles 1963, p. 214; Stöffler 1972, p. 54)*

SHOCK TEMPERATURE. Transient temperature achieved in gaseous, liquid or solid matter during *shock wave* compression. *(OU; Duvall & Fowles 1963, p. 215; Stöffler 1972, p. 53)*

SHOCK VAPORIZATION. Vaporization of solid or liquid matter by *shock wave* compression, resulting from high *post-shock temperature* after pressure release. *(OU; Chao 1968, p. 138; Stöffler 1971, p. 5546, Fig. 2; Stöffler 1972, p. 90, 100)*

SHOCK VEIN. Thin vein of quenched melt produced by shock-induced localized (frictional) melting in moderately *shocked rocks*. Called ***opaque shock veins*** in metal and troilite-rich *meteorites* (chondrites). Synonymous with *melt vein*. *(Fredriksson* et al. *1963, p. 974; Dodd & Jarosewich 1979, p. 338, Table 1; Bischoff & Stöffler 1992, p. 720; Spray 1998, p. 200)*

SHOCK WAVE. Step-like discontinuity in pressure, density, particle velocity and internal energy that propagates in gaseous, liquid or solid matter with supersonic velocity. Synonymous with *shock front*. *(Stokes 1848, p. 353; Rankine 1870, p. 278; Rice* et al. *1958, p. 1; Duvall & Fowles 1963, p. 211; Stöffler 1972, p. 54; Melosh 1989, p. 37)*

SHOCK WAVE VELOCITY. Velocity of a *shock wave* (*shock front*) propagating into material at rest. *(OU; Duvall & Fowles 1963, p. 211; Stöffler 1972, p. 54)*

SHOCK ZONE. Synonymous with *shock stage*. *(Stöffler 1966, p. 15; Dence 1968, p. 170; Dence 2004, p. 273, Fig. 8)*

SIEVE STRUCTURE. Synonymous with *poikiloblastic structure*. *(Salomon 1891, p. 483; Barker 1998, p. 85)*

SILICIOPHITE. Obsolete term for a rock consisting mainly of serpentine and opal (amorphic SiO_2), which results from the silicification of a *serpentinite*. Syn. *ofisilice*. *(Schrauf 1882, p. 352)*

SILICOFERROLITE [silicoferrolyte] (Duluth gabbroid complex, Minnesota, USA). *Exometamorphic* rock essentially composed of magnetite and fayalite, with minor pyroxene and quartz, assumed to derive from *contact metamorphism* caused by a gabbro body on a sedimentary rock composed of impure iron ore admixed with clay or quartz. *(Wadsworth 1892, in Winchell 1900, p. 353)*

SIMPLE IMPACT CRATER. Bowl-shaped *impact crater* with relatively large depth/diameter ratio. Simple impact craters generally have diameters smaller than *complex impact craters*. *(OU; Dence 1968, p. 170, 179; Grieve 1987, p. 246; Melosh 1989, p. 14; Dence 2004, p. 270)*

SKARN ⇒ Section 2.9, Fig. 2.9.1 (Swedish *skarn*, dirt, rubbish). *Metasomatic* rock formed at the contact between a silicate rock (or a magmatic melt) and a carbonate rock. It consists mainly of Ca-, Mg-, Fe-, Mn silicates, which are free from or poor in water. Hence ***magnesian skarn***, skarn formed at the contact of magmatic or other silicate rocks with calc-magnesian or magnesian carbonate rocks, and ***calc-*** (or ***lime-***) ***skarn***, skarn formed at the contact of magmatic or other silicate rocks with calcic carbonate rocks. See also *endoskarn*, *exoskarn*. *(Goldschmidt 1911, p. 213; Holmes 1920, p. 211; Semenenko 1964, p. 61; Bucher & Frey 1994, p. 24; Kornprobst 2002, p. 99)*

SKIALITH (Greek *skia*, shadow, and *lithos*, stone). Term proposed for a relict inclusion in a granitized rock, as opposed to xenolith which occurs in a magmatic rock. *(Goodspeed 1947, p. 1251; Goodspeed 1949, p. 516; Mehnert 1968, p. 40, 356)*

SKLEROPELITE (Greek *skleros*, hard, rough, and *pêlos*, clay). Obsolete term for argillaceous or allied rocks indurated by metamorphism. *(Salomon 1915, p. 404)*

SLATE (German *schleissen*, to split). Ultrafine- or very fine-grained metamorphic rock displaying *slaty cleavage*. Slate is usually of very low *metamorphic grade*, although it may also occur under low-grade conditions. Early workers also used the term *clay slate*, which was a direct parallel of the German 'Thonschiefer' and the French 'schiste argileux'. However, the qualifier is unnecessary in English. *(OU; Phillips 1818, p. 177; Lyell 1851, p. 465; Hutchings 1890, p. 264; Van Hise 1904, p. 778; Holmes 1920, p. 212; Yardley 1989, p. 22)*

SLATY CLEAVAGE. Type of *continuous cleavage* in which the individual grains are too small to be seen by the unaided eye. *(OU; Sedgewick 1835, p. 469; Geikie 1903, p. 469; Wilson 1961, p. 462; Barker 1998, p. 45)*

SNOWBALL STRUCTURE. Type of structure characterized by spiral-shaped *inclusion trails* in a *porphyroblast* and thought to be indicative of the rotation of the *porphyroblast* during growth or of the differential rotation of the *fabric* relative to the *porphyroblast* during its growth. Cf. *helicitic structure*. Synonymous with *pinwheel structure*, *rotational structure*. *(Flett 1912, p. 111; Tyrrell 1926, p. 304; Knopf & Ingerson 1938, p. 109; Spry 1969, p. 253; Barker 1998, p. 243; Vernon 2004, p. 440, 491)*

SOAPSTONE. Metamorphic rock mainly composed of talc, with subsidiary carbonates (ankerite, dolomite, magnesite); pyroxene is absent. Obsolete syn. steatite rock, steatitite (pro parte), *speckstone*, *talcite*. *(Cronstedt 1758, p. 271; Strøm 1762, p. 54; Werner 1786, p. 14)*

SODA-RICH SCHIST SERIES (German *Natronschieferreihe*; Ruhrtal, Westphalia, Germany). Name proposed for a group of Na-rich *contact-metasomatic* schists. The rocks of this series differ from *hornfelsed schists* found in normal *contact aureoles* in having a much higher sodium content than the unmetamorphosed schist; however, they also differ from the *adinole series* in that, close to the igneous contacts, they do not show any relevant chemical variations and retain their original high iron and magnesium content. *(Milch 1917, p. 357)*

SOIL. See *lunar soil*.

Solution transfer. See *pressure solution*.

SOLVUS (Latin *solvere*, to dissolve). Curved *P–T–X* line or surface that separates the field of homogeneous solid solution from the field of limited mutual solid solution. *(Edgar 1974, p. 21; Bucher & Frey 1994, p. 122)*

SONDALITE (*Sondalo*, Alto Adige, and Valtellina, N Italy). Obsolete local name for a metamorphic rock composed of garnet, cordierite, quartz, tourmaline and kyanite. Cf. *valtellinite*, *kinzigite*. *(Stache 1876, p. 358; Stache & John 1877, p. 194)*

SPACED CLEAVAGE. Type of *cleavage* in which the *cleavage* planes are spaced at regular intervals and separated by zones known as *microlithons*. The structure is visible to the unaided eye. Spaced cleavage encompasses *crenulation cleavage* and *disjunctive cleavage*. This definition accords with the classification proposed by Powell. However, some authors (e.g. Davis & Reynolds) regard spaced cleavage as equivalent only to *disjunctive cleavage*. *(Powell 1979, p. 33; Davis & Reynolds 1996, p. 432)*

SPACED SCHISTOSITY. Type of *spaced cleavage* characterized by regularly spaced zones with *schistose structure* that are structurally distinct and separate from rock layers (called *microlithons*). The structure is visible to the unaided eye. *(Chidester 1962, p. 22)*

SPECIAL METAMORPHISM. Synonymous with *abnormal* or *local metamorphism*. *(Delesse 1857, p. 90)*

SPECKSTONE (German *Speck*, ham, bacon, and *Stein*, rock, alluding to its greasy appearance). Obsolete term for a metamorphic rock composed dominantly of talc. Cf. *soapstone*. Syn. bacon stone, lard-stone, lardite. *(OU)*

SPILITE ⇒ Section 2.5 (Greek *spilos*, stain, spot). Altered basic to intermediate, volcanic or subvolcanic rock in which the feldspar is partially or completely composed of albite and is typically accompanied by chlorite, calcite, quartz, epidote, prehnite and low-temperature hydrous crystallization products. Preservation of eruptive (volcanic and subvolcanic) features is an important characteristic of spilites. The term may be replaced by metabasalt or meta-andesite, as appropriate, regardless of the origin of the rock. *(Bonnard 1819, in Brongniart 1827, p. 93; Flett 1907, p. 95; Amstutz 1968, p. 737; Amstutz 1974, p. 1; Le Maitre 1989, 2002)*

SPILOSITE (Greek *spilos*, stain, spot; Harz Mts., Germany). Soda-rich *spotted schist* or *slate*, with alternating lighter (albite) and darker (white mica + chlorite) layers, generally associated with *contact adinole*. The millimetre-sized spots are albite-chlorite-white mica *pseudomorphs* after cordierite. Cf. *desmosite*, *adinole series*; see also *diabase contact rock*. *(Zincken 1841, p. 394; Lossen 1872, p. 735; Rosenbusch 1896, p. 1174; Milch 1917, p. 354; Rosenbusch 1923, p. 615)*

SPILYTE. Schistose metacarbonate rock. *(Kinahan 1873, p. 355)*

SPOTTED GRANULITE (German *Fleckgranulit*; Granulitgebirge, Saxony, Germany). See *Forellengranulite*. *(Cotta 1862, p. 166)*

SPOTTED SCHIST, SPOTTED SLATE. Type of fine-grained schist or *slate* characterized by the presence of concretions or clots representing incipient *porphyroblasts* that may range in size from minute spots to the size of beans. The rocks are characteristic of the outer part of *contact metamorphic aureoles*. The term encompasses the German terms *Fleckschiefer* and *Fruchtschiefer*. Cf. *knotted schist*. *(OU; Harker 1950, p. 48)*

STABLE RELICT (stable relic). Mineral phase that was stable in an early *assemblage* and is also stable under a later set of imposed conditions and thus persists in the new *assemblage*. If the mineral phase is metastable under the later set of conditions, it is said to be an ***unstable relict***. *(Eskola 1915, p. 24, 118; Eskola 1920, p. 149; Eskola 1939, p. 341)*

STATIC, STATICAL METAMORPHISM. Type of metamorphism produced under static conditions. *(Judd 1889, p. 343; Daly 1917, p. 375)*

STATOHYDRAL METAMORPHISM. Type of metamorphism produced under static conditions and relatively low temperature, equivalent to *burial metamorphism*. *(Daly 1917, p. 375)*

STATOTHERMAL METAMORPHISM. Type of metamorphism produced under static conditions and relatively high temperature. *(Daly 1917, p. 375)*

STATUARY MARBLE (Italian *marmo statuario*). Term used mainly in architecture and sculpture for a white or cream-coloured *pure marble*, which is homogeneous and semi-translucent. Cf. *protocalcite*. *(Ferber 1776, p. 328; Tomkeieff)*

STEANITE (Greek *stear*, grease, fat). Obsolete term for *steatite*, in the sense of a talc-rich rock. *(Rosière 1826, in Tomkeieff)*

STEASCHIST (from *steatite*, the talc mineral). Talc schist, originally defined as a rock composed of a talc groundmass including copper and other disseminated minerals, with a *foliated* structure. Cf. *talcite*, *pierre ollaire*. *(Brongniart 1813, p. 37; Omalius d'Halloy 1868, p. 171; Teall 1888, p. 169)*

STEREOGENETIC, STEREOGENIC (Greek *stereos*, solid, and *gennô*, to give birth to). Said of the solid or immobile phase of a *migmatite*. Ant. *chymogenetic*. *(Huber 1943, p. 90; Mehnert 1968, p. 357)*

STEREOSOME (Greek *stereos*, solid, and *sôma*, body). Solid or immobile part of a mixed rock (*chorismite*). *(Dietrich & Mehnert 1960, p. 62)*

STICTOLITE (Greek *stiktos*, spotted, and *lithos*, stone). Obsolete term for a type of *migmatite* in which spots of relict minerals are the only traces of assimilated material. *(Sederholm 1926, p. 136; Mehnert 1968, p. 357)*

STILPNOLITE (Greek *stilpnos*, shining, and *lithos*, stone). Obsolete term for a group of schistose rocks composed of mica, talc, chlorite or haematite, and with a characteristic shiny appearance. *(Senft 1857, p. 55, 228)*

STRAIN EXTINCTION. See *undulose extinction*.

STRAIN SHADOW. Ambiguous term, now usually taken as meaning *pressure shadow*, but used by earlier workers to mean *undulose extinction*. *(OU; Holmes 1920, p. 218; Harker 1939, p. 163; Vernon 2004, p. 491)*

STRAIN-SLIP CLEAVAGE. Synonymous with, and largely superseded by *crenulation cleavage*. *(Bonney 1886, p. 95; Turner & Weiss 1963, p. 98; Ramsay 1967, p. 389)*

STRATIC METAMORPHISM. Type of metamorphism peculiar to a specific stratum or horizon in a sedimentary series. *(Gosselet 1888; Tomkeieff)*

STRESS MINERALS. Minerals such as chlorite, kyanite and amphiboles which were formed in metamorphic rocks and which were thought to have developed where shearing stress was present. As such these minerals were regarded as typical products of *regional metamorphism*. Ant. *anti-stress minerals*. *(Harker 1918, p. 77)*

STROMATITE (Greek *strôma*, layer). Type of *migmatite* with regular layers, the layers having two or more different compositions or appearances, for example the alternation of *mesosome* and *leucosome*. Hence ***stromatitic structure***. *(Huber 1943, p. 89; Niggli 1948, p. 109; Dietrich & Mehnert 1960, p. 62; Mehnert 1968, p. 357; Shelley 1993, p. 110; Kornprobst 2002, p. 108)*

Stronalite (*Strona* valley, near Novara, Italy). Local term for a high-grade metamorphic rock mainly composed of garnet, feldspar and quartz. Biotite and cordierite may be present, as well as kyanite and sillimanite. The name may be replaced by the term *granulite* or, according to the SCMR general rules, by garnet-quartz-plagioclase gneiss/granofels. *(Artini & Melzi 1900, p. 284; Bertolani 1964, p. 32)*

STRUCTURE ⇒ Section 2.3. Arrangement of the parts of a rock mass irrespective of scale, including spatial relationships between the parts, their relative size and shape and the internal features of the parts. Hence *microstructure*, structure at the microscopic scale, *mesostructure*, structure at the hand-specimen or outcrop scale, *megastructure*, structure at the regional or mappable scale. *(OU; Brongniart 1813, p. 23; Brongniart 1825, p. 69; Holmes 1920, p. 218; Turner 1948, p. 149; Hobbs* et al. *1976, p. 141; Passchier & Trouw 1996, p. 265)*

SUBFABRIC. One of a number of elements constituting the total *fabric* of a rock. *(Paterson & Weiss 1961, p. 863; Turner & Weiss 1963, p. 34)*

SUBFACIES. See *metamorphic subfacies*.

SUBGRAIN. See *subgrain boundary*.

SUBGRAIN BOUNDARY. Relatively planar array of dislocations separating two volumes of crystalline material (*subgrains*) with the same composition but with small (usually $<5°$) angular misorientations of their crystal lattices. Usually formed by recovery during or after deformation. *(OU; Hobbs* et al. *1976, p. 93; Passchier & Trouw 1996, p. 265; Vernon 2004, p. 323)*

SUBGREENSCHIST FACIES ⇒ Sections 2.2 & 2.5, Figs. 2.2.4 & 2.5.1, Table 2.2.1 (sub-blueschist-greenschist facies). Part of the field of very low-grade metamorphism that is characterized by pressures lower than those of the *glaucophane-schist facies*; the term refers to various *metamorphic facies* lower in grade than the *greenschist facies*, including the *prehnite-pumpellyite*, *prehnite-actinolite* and *pumpellyite-actinolite facies*. Originally, the *zeolite facies* was also included. *(Liou* et al. *1985, p. 321; Liou* et al. *1987, p. 68; Bucher & Frey 2002, p. 111)*

SUBHEDRAL (Latin *sub*, under, almost, and Greek *edra*, side, face). Said of a crystal that is only partly bounded by its own crystal faces. Term originally used to describe igneous rocks but now more widely applicable. The recommended specific terms for metamorphic rocks are *hypidioblastic* and *subidioblastic*. *(Cross* et al. *1906, p. 698; Passchier & Trouw 1996, p. 265)*

SUBIDIOBLAST, SUBIDIOBLASTIC STRUCTURE (Latin *sub-*, under, almost, and Greek *idios*, own, particular, and *blastos*, bud, sprout). Synonymous with *hypidioblast* and *hypidioblastic structure*. Cf. *subhedral*. *(OU, Joplin 1968, p. 28; Vernon 2004, p. 492)*

SUB-SEA-FLOOR METAMORPHISM. Synonymous with *ocean-floor metamorphism*. *(Spooner & Fyfe 1973, p. 293)*

SUBTRACTION ROCKS. Term proposed for residual basic masses produced during the process of *granitization*. It was considered that the more acid components had been removed from the mass leaving it relatively enriched in basic material. Equivalent to *restite*. *(Read 1951, p. 11)*

SUBTRACTIVE METASOMATISM. Type of *contact metasomatism* characterized by the loss of material originally present in the rock body. Ant. *additive metasomatism*. *(Barrell 1907, p. 117)*

SUDBURITE (*Sudbury* Ni-Cu mining district, Ontario, Canada). Fine-grained pyroxene-plagioclase *mafic hornfels* found along the contacts of the plutonic complex at Sudbury, the so-called nickel intrusion, from high-temperature *contact metamorphism* of *greenstones* and other rocks. Syn. *muscovadite* as used in Minnesota. *(Thomson 1935, p. 427)*

SUEVITE, suevite breccia, suevitic impact breccia, SUEVITIC MIXED BRECCIA (*Suevia*, Latin name for the province of Schwaben, southern Germany; Nördlinger Ries, Bavaria). *Polymict impact breccia* with particulate matrix containing lithic and mineral clasts in all stages of *shock metamorphism* including cogenetic *impact melt* particles which are in a glassy or crystallized state. The first detailed description was of a 'rhyolitic tuff'. *(Gümbel 1870, p. 156; Sauer 1901, p. 88; Dence 1971, p. 5553; Engelhardt 1971, p. 5568; Pohl* et al. *1977, p. 359)*

SURREITE STRUCTURE (Greek *surrheô*, to flow together). Type of structure in a *migmatite* where the mobilized material (*leucosome*) intruded into dilatation spaces of the more competent portions of the rock. Cf. *metatectite*. *(Holmquist 1920, p. 202, 211)*

SUTURED BOUNDARY. Mutual boundary between two mineral grains that is interlocking as opposed to straight or smoothly curved. The feature is characteristic of strained crystals. *(OU; Vernon 2004, p. 492)*

SWELLING METAMORPHISM. Name given to a change in a rock considered to have been produced by some local pressure, such as the increase in volume of the olivine in an ultramafic rock on its transformation into serpentine. *(Steinmann 1908, p. 12)*

SWIRLED GNEISS. Type of gneiss in which the *foliation* is intensely folded and contorted in all three dimensions, giving the rock a swirled or convoluted appearance. *(Waters & Krauskopf 1941, p. 1384)*

SYMPHRATTISM (Greek *symphrazomai*, to meditate, to press together). Proposed synonym for both *regional* and *dynamometamorphism*. Hence ***symphrattic rocks***. Cf. *aethobalism. (Grabau 1904, p. 236)*

SYMPLECTITE, **SYMPLECTITIC INTERGROWTH** (Greek *symplekô*, to interlace, interweave). Type of *(micro)structure* characterized by the intimate intergrowth of two or more different minerals, one of them commonly having a vermicular habit. Symplectites form as a result of the instability and decomposition of previous minerals. The term is used both for a rock exhibiting such a *(micro)structure* and for the intergrowth itself. Hence, ***symplectic***. See footnote under *corona. (Haüy 1822, p. 574; Naumann 1849–53, p. 667; Sederholm 1916, p. 46)*

SYNANTETIC (Greek *synantô*, to meet). Said of a mineral formed at the boundary between two primary minerals and resulting from a reaction between these two minerals. Cf. *corona, reaction rim, kelyphitic rim. (Sederholm 1916, p. 1; Shand 1945, p. 247; Spry 1969, p. 104)*

SYNGENETIC (Greek *syn*, with, and *genesis*, genesis). Said of ore deposits that were formed by similar processes and simultaneously to their enclosing rocks. Cf. *epigenetic. (OU; Lindgren 1933, p. 154)*

SYNTAXIAL (Greek *syn*, with, and *taxis*, arrangement, ordering). Said of a vein-infilling in which fibres grow from the walls towards the centre. Hence ***syntaxial vein***, ***syntaxial growth***. Ant. *antitaxial. (Ramsay & Huber 1983, p. 241; Shelley 1993, p. 288)*

SYNTECTITE, SYNTEXITE (Greek *syn*, with, and *têkô*, to melt). Rock resulting from *syntexis. (Backlund 1936, p. 295)*

SYNTEXIS. Obsolete term covering the sum of processes leading to the modification of a magma by assimilation of crustal fragments or the mixing of *anatectic* melts generated from two or more different rock types. *(Loewinson-Lessing 1899, p. 375; Smulikowski 1958, p. 78; Dietrich & Mehnert 1960, p. 63; Mehnert 1968, p. 357)*

SYSTIL (perhaps Greek *systellô*, to contract, referring to the columnar jointing). Dull, hard rock, with a conchoidal to splintery fracture, derived by *contact metamorphism* of claystone or marly sandstone by basalt. Also used as a synonym of *basalt jasper. (Zimmermann, in Nöggerath 1822, p. 109; Zirkel 1866a, p. 620)*

TABULAR. Descriptive term for an object with two of its dimensions significantly greater than the third. Hence ***tabular structure***, a term applicable to a single body or a parallel set of tabular bodies; the term has been applied to minerals and to rock bodies, for example the structure produced by sets of horizontal *joints. (OU; Phillips 1819, p.* xciv; *Macculloch 1821, p. 129; Cordier 1868, p. 23; Hatch 1888, p. 449)*

TACTITE (Latin *tactus*, touched). *Calc-silicate rock* formed by *contact metasomatism* of a carbonate or marly rock. Also used as a synonym of calc-silicate *hornfels*. It may host ore minerals containing gold, silver, lead, copper, bismuth, zinc, tungsten and molybdenum. Rec.syn. *skarn. (Hess 1919, p. 377; Aubouin* et al. *1968, p. 469; Foucault & Raoult 2005, p. 85)*

Tagamite. Term used in Russia for *impact melt rock. (Masaitis* et al. *1975, p. 86; Masaitis 1994, p. 155, Table 1, Figs. 1, 7)*

TALCITE, TALCITE OLLAIRE (French, from the Italian and Latin *olla*, cooking-pot). Originally defined as a schistose rock mainly composed of talc. According to SCMR rules, a rock essentially composed of talc. Syn. *steaschist, pierre ollaire. (Werner 1786; Cordier 1868, p. 175; Omalius d'Halloy 1868, p. 171)*

TANGIWAI, TANGAWAITE, TANGIWAITE (*Tangiwai* in New Zealand). Name given by the Maoris in New Zealand to *bowenite*, a variety of *serpentinite. (Bates & Jackson)*

Tanohata-type contact metamorphism (*Tanohata* granitic pluton, Japan). Type of *isochemical contact metamorphism* mostly observed around granitoid plutons and in which the *mineral assemblages* reflect very low-pressure conditions. The thickness of the aureole may reach 2–3 km. See *contact aureole systematics*. *(Reverdatto* et al. *1970, p. 312)*

TAPANHOACANGA, CANGA (*tupi tàpui una a kâga*, blackman's head; Brazil). Tough well-consolidated ferruginous breccia or conglomerate composed of fragments of the metamorphic rock *itabirite*, and of haematite and magnetite principally cemented together by limonite and haematite. *(Eschwege 1822, p. 30; Cordier 1868, p. 338)*

TARGET. Volume on a planetary body that is *impacted* by a second body (*projectile*) travelling at cosmic velocity and generating a *shock wave* in both bodies. *(OU; Melosh 1989, p. 46)*

TARGET LITHOLOGY. Synonym of *target rock*.

TARGET ROCK. Rock (lithology) exposed at the site of an *impact* before crater formation. *(OU; Melosh 1989, p. 46)*

TASPINITE (Alp *Taspin*, Schamsertal, eastern Switzerland). Obsolete term for a metamorphic rock composed of sericite, quartz, muscovite and feldspar that is commonly characterized by large *relict phenoblasts* (up to 6 cm) and which has formed from granite porphyry or porphyric granite. *(Heim 1891, p. 387)*

TECTONIC BRECCIA. Term originally used to include all breccias formed by deformation, including *fault* processes, folding, etc.; now equivalent to *fault breccia*. *(Norton 1917, p. 186; Reynolds 1928, p. 104; Higgins 1971, p. 76)*

TECTONITE. Metamorphic rock that possesses a penetrative *fabric* recording the deformational history of the rock, the *fabric* having been produced by flow in the solid state. Hence ***L-tectonite***, tectonite with a dominantly *linear fabric*; ***S-tectonite***, tectonite with a dominantly *planar fabric*; ***L-S tectonite***, tectonite with mixed *linear* and *planar fabrics*. *(Knopf 1933, p. 433; Turner & Weiss 1963, p. 38; Davis & Reynolds 1996, p. 414; Barker 1998, p. 45)*

TECTOSPHERE, TEKOSPHERE (Greek *têkô*, to melt, and *sphaera*, sphere). Obsolete term for what was believed to be a zone of melting in the Earth's crust. *(Gürich 1905, p. 248; Sederholm 1907, p. 36, 101)*

TEKTITE (Greek *têktos*, molten). *Impact glass* formed at terrestrial *impact craters* from melt ejected ballistically and deposited as aerodynamically shaped bodies in a strewn field outside the continuous *ejecta blanket*. The size of tektites ranges from the submillimetre range (*microtektites*, generally found in deep-sea sediments) to the subdecimetre range, rarely to decimetres. *(Suess 1900, p. 191; Spencer 1933a, p. 117; Park 1989b, p. 554; Glass 1990, p. 393)*

TEKTOPHYRASIS, PHYRASIS (Greek *tektôn*, carpenter, builder, hence tectonic, and *phyrasis*, mixing). Obsolete term for tectonic intermingling or interleaving of different rocks resulting in a *banded* or lenticular structure. *(Scheumann 1936c, p. 463)*

Texture ⇒ Section 2.3 (Latin *texere*, to weave, or *textura*, woven material). (a) Term defining the relative size, shape and spatial interrelationship between grains and internal features of grains. Synonymous with *microstructure*. (b) Term denoting the presence of *preferred orientation* on the microscopic scale. Synonymous with *microfabric*. The SCMR encourages the use of (micro)structure and (micro)fabric to avoid ambiguity. If texture is used, then its meaning must be clear. *(OU; Brongniart 1825, p. 70; Omalius d'Halloy 1868, p. 85; Turner 1948, p. 149; Spry 1969, p. 5; Barker 1990, p. 146; Passchier & Trouw 1996, p. 265; Vernon 2004, p. 493)*

THERMAL CONTACT METAMORPHISM, THERMO-CONTACT METAMORPHISM (Greek *thermos*, hot). *Contact metamorphism* due to heating alone, contrasting with both *hydrothermal* and *pneumatolytic contact metamorphism*. Syn. *normal contact metamorphism*. *(Irving 1911, p. 298; Niggli 1954, p. 521)*

Thermal metamorphism ⇒ Section 2.2 (Greek *thermos*, hot). Type of *local metamorphism* resulting from a rise in temperature without a significant rise in pressure, mostly used as synonymous with *contact metamorphism*, and opposed to *regional metamorphism*. Sometimes also used as a collective name for temperature-dominated metamorphic processes including *contact metamorphism*, *combustion metamorphism* and *pyrometamorphism. (Harker 1889, p. 16; Daly 1917, p. 408; Tyrrell 1926, p. 289; Harker 1932, p. 2, 21; Turner 1948, p. 6)*

THERMANTIDE, THERMANTIDE PORCELLANITE (Greek *thermos*, hot, or *thermantir*, cauldron). Obsolete terms proposed as alternatives to the German 'Porzellanjaspis' (see *porcellanite*, *porcelain jasper*); now included among the *burned rocks* as a *buchite* or *fritted rock* according to the glass content. *(Haüy, in Brochant 1798, p. 336; Haüy 1822, p. 510, 582; Leonhard 1824, p. 587; Cordier 1868, p. 247)*

THERMOCALCITE (Greek *thermos*, hot). *Contact-metamorphosed* limestone. *(Cordier 1868, p. 290)*

THERMOCHRON (Greek *thermos*, hot, and *chronos*, time; Scottish Caledonides, UK). Conceptual line or surface in a rock body joining points that had the same temperature at the same time. *(Harper 1967, p. 129; Dewey & Pankhurst 1970, p. 377)*

THERMODYNAMIC METAMORPHISM. Metamorphism considered to have been produced by high temperature and directed pressure. *(Teall 1902, p. lxxvii)*

THERMOMETAMORPHISM. Term mostly used for *thermal metamorphism*, also for *pyrometamorphism*, and for a process of mineral grain coarsening in a dry solid system. Rec.syn. *contact metamorphism. (Harker 1889, p. 16; Barrow 1893, p. 336; Loewinson-Lessing 1893–4, p. 231; Brauns 1911, p. 2; Milch 1922, p. 288)*

THETOMORPHIC GLASS (Greek *thêtos*, placed, and *morphê*, form). Unnecessary synonym of *diaplectic* glass. *(Chao 1967b, p. 228, Table 2; Chao 1968, p. 148; Dence & Robertson 1989, p. 529)*

TOPAZ-BROKENFELS, TOPAZ HORNFELS (Schneckenstein in Saxony). Obsolete term for a brecciated tourmaline *contact-metamorphic* rock with a cement of quartz and topaz. The topaz commonly penetrates the tourmaline rock and seems to have locally absorbed tourmaline. *(Rosenbusch 1896, p. 105–106)*

TOPOCHEMICAL CHANGE (Greek *topos*, site). Mineral change not involving any change in the composition of the system, as opposed to *metasomatic* change. It includes *polymorphic transformation* and many mineral reactions. Equivalent to *isochemical metamorphism. (OU; Kornprobst 2002, p. 13)*

TOPOTAXIS, TOPOTAXY, TOPOTAXIAL GROWTH (Greek *topos*, site, and *taxis*, arrangement, ordering). *Recrystallization* in which the new crystal grows with minimal change to the lattice of the parent crystal. *(OU; Spry 1969, p. 89; Yardley 1989, p. 158)*

TOURMALITE. Rock composed of tourmaline and quartz in undetermined proportions, with biotite, feldspar and topaz possibly present. To be distinguished from tourmalinite which, according to SCMR rules, is a rock composed of $\geq$75% tourmaline. *(Daubrée 1841, in Tomkeieff; Zirkel 1894a, p. 125; Johannsen 1939, p. 22; Cordier 1968, p. 205)*

TRANSFUSION. Obsolete and ambiguous term originally proposed for the transport of matter from one solid to another under the influence of heat, without fusion or the action of water or other mineralizers. Later taken to include the transfer of gaseous or liquid material from or into a solid body to produce a rock of igneous appearance. *(Adams 1930, p. 153, 160; Reynolds 1936, p. 368, 403)*

TRANSIENT CRATER, TRANSIENT CAVITY. Transient, bowl-shaped cavity of an *impact crater* that is produced as a result of the compression and excavation stages of the crater-forming process. It is unstable in the case of all large craters and collapses immediately in a highly dynamic process leading to a flat *complex impact crater* with either a central peak (uplift) or a peak-ring or multiple rings for increasing diameters of the transient cavity. First called 'primary crater'. *(Dence 1968, p. 179; Grieve* et al. *1977, p. 803; Melosh 1989, p. 77)*

TRAPPGRANULITE (Granulitgebirge, Saxony, Germany). Obsolete term for *mafic granulites* ('*pyroxene granulites*') mostly of basaltic composition and consisting of plagioclase, quartz, pyroxene (originally described as a micaceous mineral), pyrrhotine and garnet. *(Stelzner 1871, p. 245)*

TRAP-SHOTTEN GNEISS. Type of gneiss characterized by a network of *pseudotachylite* veins (originally identified as trap rock, that is, an igneous rock). *(King & Foote 1865, p. 49)*

TREPTOMORPHISM (Greek *treptos*, turned, converted, and *morphê*, form). Obsolete term for a type of *isochemical metamorphism* resulting in the formation of an igneous-looking rock. *(Dietrich 1963, p. 114)*

TRIMINERALIC. Said of a rock in which ≥95% of the modal content is composed by three minerals, as opposed to *polymineralic*, *monomineralic*, *bimineralic*.

TRIPOLI (Tarabalus, former *Tripoli*, northern Lebanon). Powdery deposit of refractory siliceous material remaining after the spontaneous combustion of coal seams or bituminous rocks; rec.syn. *coal-fire ash*. Also a term used in sedimentary petrology. *(Wallerius 1747, in Holmes 1920; Brochant 1798, p. 381; Haüy 1822, p. 582; Cordier 1868, p. 249; Bates & Jackson 1987, p. 702)*

TUFFOID (Westphalia, Germany). Metamorphic rock derived from pyroclastic and detrital material; a metamorphosed tuffite. The term was used earlier by Loewinson-Lessing for rocks of tuff appearance that originated by weathering ('Verwitterungstuff') or by *dislocation metamorphism* ('Klastotuff'). The SCMR name for these composite rocks is metatuffite. *(Mügge 1893, p. 708; Loewinson-Lessing 1888, p. 532)*

TUXTLITE (from a Mayan statuette found near *Tuxtla*, SE Mexico, and made of that rock). *Monomineralic* jade-like rock composed of omphacitic pyroxene. Cf. *mayaite*. *(Washington 1922a, p. 4; Washington 1922b, p. 321)*

TYPES OF METAMORPHISM ⇒ Sections 2.2 & 3.4, Fig. 2.2.1.

Typomorphic minerals (Greek *typos*, type, and *morphê*, form). Major mineral constituents of a metamorphic rock that are in chemical equilibrium. *(Becke 1903b, p. 34)*

ULTRABASIC ⇒ Section 2.1, Table 2.1.2. See *acid*, *intermediate*, *basic*, *ultrabasic*.

ULTRABASIC HORNFELS. Type of *hornfels* derived by the *contact metamorphism* of an ultrabasic rock. Rec.syn. *ultramafic hornfels*, *magnesian hornfels*. *(Williams* et al. *1954, p. 197; Spry 1969, p. 201)*

ULTRACATACLASITE (Latin *ultra*, beyond). *Cataclasite* in which the matrix forms more than 90% of the rock volume. *(Spry 1969, p. 229)*

ULTRA-CONTACT METAMORPHISM. Type of *contact metamorphism* accompanied by partial melting of the rocks immediately surrounding the igneous body with the formation of *migmatites* and contamination of the magma. *(Niggli 1954, p. 523)*

ULTRAHIGH-PRESSURE METAMORPHISM. That part of the metamorphic *P–T* field where pressures exceed the minimum necessary for the formation of coesite. Pressure may be sufficient for the formation of diamond. *(Coleman & Wang 1995, p. 2; Carswell & Compagnoni 2003, p. 3)*

ULTRAMAFIC HORNFELS. General term for *hornfelsed* ultrabasic igneous or meta-igneous rocks; the main components are one or more *mafic minerals*. The term supersedes *ultrabasic hornfels* in accord with the SCMR guidelines. (*Callegari & Pertsev Section 2.10*)

ULTRAMETAMORPHISM. Metamorphism under the extreme upper range of temperatures and pressures, as a result of which rocks suffer complete or almost complete *in situ* melting. Cf. *palingenesis*. *(OU; Holmquist 1909, p. 108; Holmquist 1920, p. 203; Holmquist 1921, p. 269; Mehnert 1968, p. 357)*

ULTRAMYLONITE. *Mylonite* in which more than 90% of the rock volume has undergone grain-size reduction. *(Staub 1915, p. 71; Quensel 1916, p. 99, 103; Waters & Campbell 1935, p. 481; Spry 1969, p. 229; Higgins 1971, p. 77)*

UNDULOSE EXTINCTION. Poorly defined and variable extinction across a strained crystal caused by local distortion of the crystal lattice. The phenomenon may occur in a variety of minerals but is most commonly seen in quartz. Synonyms include ***undulatory*** and *strain extinction*. *(OU; Harker 1954, p. 269; Spry 1969, p. 62; Barker 1998, p. 122; Vernon 2004, p. 493)*

UNSTABLE RELICT. See *stable relict*.

UPSIDE-DOWN METAMORPHISM. Synonymous with *hot-slab metamorphism*. Cf. *dynamo-static metamorphism*. *(Blake* et al. *1967, p. C1)*

URALITIZATION (*Ural*). Late magmatic or metamorphic process by which a primary pyroxene is altered to uralite, a green amphibole (hornblende or actinolite or tremolite), which is generally, but not necessarily, fibrous. *(Rose 1831, p. 342; Williams 1890, p. 58; Duparc & Hornung 1904, p. 223)*

VALENTINITE (after the mediaeval alchemist Basilius *Valentine*). (1) Obsolete term for a variety of talc schist or *mica schist* with dolomite, serpentine or amphibole. (2) Orthorhombic Sb_2O_3 mineral. *(Pinkerton 1811b, p. 48)*

VALRHEINITE (Rhaeto-Romanic *Valrhein*, Hinterrhein, Graubünden, eastern Switzerland). Obsolete term for a local variety of *greenschist* from the so-called *chlorogrisonite* schists group (equivalent to *prasinites*), composed mainly of oligoclase, epidote, chlorite and magnetite, with subordinate haematite and actinolite. *(Rolle 1879, p. 32)*

VALTELLINITE (*Valtellina*, Alto Adige, Italy). See *veltlinite*. Cf. *sondalite*, *kinzigite*.

VAPOUR PLUME. Vaporized, melted and crushed *target rock* and *impactor* material being ejected from a hypervelocity *impact crater* after the *impactor* has reached its stagnation point. The plume forms in the central region of the *impact crater* and the material involved is ejected very early in the cratering process and at very high speed. Upon collapse of the plume, *polymict breccias* of the *suevite* type are formed and deposited inside and around the crater. At very large craters (> about 100 km in diameter) the plume material is distributed globally and forms *impactoclastic airfall beds*. Syn. *ejecta plume*. *(OU; Melosh 1989, p. 68)*

VELTLINITE (*Veltlin* valley, now Valtellina, Alto Adige, Italy). Obsolete local term for a garnet-rich rock also containing plagioclase, quartz, hornblende and diallage pyroxene; biotite may also be present locally. *(Stache 1876, p. 358; Stache & John 1877, p. 194)*

VENITE (Latin *vena*, vein). Originally defined as a term for a veined *migmatite* irrespective of the genesis of the *leucosome* (veins), but later taken to mean a type of *migmatite* in which the material of the lighter veins (*leucosome*) has been extracted from the *parent rock*. Cf. *arterite*, *phlebite*. *(Holmquist 1921, p. 629; Dietrich & Mehnert 1960, p. 63; Mehnert 1968, p. 357)*

VERDE DI PRATO (Italian *verde*, green; *Prato*, Italy). Variety of the *ophicalcite* rock group (strictly an *ophicarbonate*), containing olivine, ortho-clinopyroxene and hornblende, set in a serpentine groundmass and cut by carbonate veinlets. It can be polished and it is described by the commercial stone industry as a *marble*. Cf. *vert antique*. *(Ferber 1776, p. 400; Zirkel 1894a, p. 764; Zirkel 1894b, p. 396)*

Verdite (Italian *verde*, green; southern Africa). Green ornamental stone consisting chiefly of impure fuchsite and other phyllosilicates. *(OU)*

Verdolite (Italian *verde*, green, and the Greek *lithos*, stone; New Jersey, USA). Variety of *ophicalcite* rock group (strictly an *ophicarbonate*) which is used as a decorative stone. *(Peck 1901, p. 425)*

Vert antique (Italian *verde*, green, and *antico*, ancient). Variety of *ophicalcite* rock group (strictly an *ophicarbonate*) that can be polished; it is described by the commercial stone industry as a *marble*. *(Ferber 1776, p. 332; Zirkel 1894b, p. 452)*

Vogtland-type contact metamorphism (*Vogtland* area, Saxony, Germany). Type of *isochemical contact metamorphism* usually associated with near-surface igneous bodies and characterized by *mineral assemblages* including albite and epidote. The *aureole* thickness does not usually exceed one metre. See *contact aureole systematics*. *(Reverdatto* et al. *1970, p. 312)*

VULLINITE (Allt a' Mhuillin, pronounced Aultivullin; Assynt, Scotland, UK). Obsolete local name for a massive to slightly schistose fine-grained *spotted contact rock* composed of orthoclase, plagioclase, diopside, hornblende and biotite, interpreted as a metamorphosed calcareous sedimentary rock. *(OU; Shand 1910, p. 406)*

WALLERITE (after the Swedish mineralogist J.G. *Wallerius*). Obsolete name for a rock composed of feldspar and amphibole (*greenstone*). *(Pinkerton 1811a, p. 10, 16)*

WEISSSTEIN (German *weiss*, white, and *Stein*, rock; Granulitgebirge, Saxony, Germany). Obsolete term for a *granulite* from the earliest descriptions of these rocks in Saxony. *(Engelbrecht 1802, p. 26)*

Whiteschist. Light-coloured schist containing kyanite and talc. More precise rock terms should be used wherever possible (e.g. kyanite-talc-phengite schist). Whiteschists represent Al-Mg-rich rocks metamorphosed under the conditions of the *eclogite facies*. *(Schreyer 1973, p. 736)*

WHITE TRAP (Scotland, UK). Local name for a soft, carbonate-kaolinite, clay-like material formed in the margin of a doleritic body by the action of organic gases emanating from surrounding *contact-metamorphosed* coals. *(Flett 1910, p. 311)*

XENOBLAST (Greek *xenos*, foreign, and *blastos*, bud, sprout). Crystal formed in a metamorphic rock without developing any of its own crystal faces. Hence ***xenoblastic structure***. Cf. *xenotopic structure*. Ant. *idioblast*. *(Becke 1903a, p. 264; Becke 1903b, p. 43; Joplin 1968, p. 28; Spry 1969, p. 140)*

XENOTOPIC STRUCTURE (Greek *xenos*, foreign, and *topos*, place). Type of structure in *diagenetically* altered carbonate rocks or chemically precipitated sediments, in which the majority of the mineral grains lack their own crystal faces. In general, in metamorphic rocks such a structure should be described as *xenoblastic structure*. Ant. *idiotopic structure*. *(Friedman 1965, p. 648)*

ZEOLITE FACIES ⇒ Sections 2.2 & 2.5, Figs. 2.2.4 & 2.5.1, Table 2.2.1 (Taringatura Hills, South Island, New Zealand). *Metamorphic facies* that embraces all *mineral assemblages* that include various zeolites plus quartz, irrespective of the mode of origin, whether metamorphic (including *hydrothermal*) or *diagenetic*. *(Eskola 1929, p. 163; Eskola 1939, p. 345; Fyfe* et al. *1958, p. 216; Coombs* et al. *1959, p. 91; Coombs 1971, p. 319; Boles & Coombs 1977, p. 982)*

ZEOSPHERE (Greek *zeô*, to boil, and *sphaira*, sphere). Zone of *pneumatolytic* mineral formation. *(Gürich 1905, p. 250)*

ZOBTENITE (*Zobtenberg*, now Mt Sleza, Silesia, Poland). Metagabbroid rock with an *augen structure*, composed of augite rimmed by uralite, enclosed in a matrix of epidote and *saussuritized* plagioclase. *(Roth 1887, p. 611)*

ZOSIMITE, ZOZIMITE (*Zôsimos*, a Greek alchemist from Panopolis in Egypt, third century BC; St-Gervais, Haute-Savoie, France). Bright red micaceous and ferruginous rock with irregular nodules of quartz, the two colours of the rock being reminiscent of Egyptian granite. *(Saussure 1786, p. 595; Pinkerton 1811b, p. 43)*

4 References

ABDULLAEV, K. M., 1947. Geology of scheelite-bearing skarns of Central Asia (in Russian). *Monographs Uzbek SSR Academy of Science Publications*, 1–399.

ADAMS, F. D., 1893. Ueber das Norian oder Ober-Laurentian von Canada. *Neues Jahrbuch für Mineralogie, Geologie und Paläontologie*, Beilageband, **8**, 419–98.

ADAMS, F. D., 1896. Laurentian area in north-west corner of the sheet. *Geological Survey of Canada, Annual Report* (for 1894), **7**, report J, 93–157.

ADAMS, F. D., 1930. The transfusion of matter from one solid to another, a new factor in the process of metamorphism. *Canadian Journal of Research*, **2**, 153–61.

ADAMS, F. D. & BARLOW, E. B., 1910. Geology of the Haliburton and Bancroft areas, Province of Ontario. *Geological Survey of Canada, Memoirs*, **6**, 1–420.

AGRICOLA (BAUER, G.), 1556. De re metallica, Libri XII, posthumous edition illustrated by H. R. Manuel. *Basel: Frobenius*. (1st edn in 1530)

AHRENS, T. J. & ROSENBERG, J. T., 1968. Shock metamorphism: experiments on quartz and plagioclase. In B. M. French & N. M. Short (eds.), Shock Metamorphism of Natural Materials. *Baltimore: Mono Book Corporation*, pp. 59–81.

AINBERG, L. F., 1955. On the problem of genesis of the charnockites and rocks of the charnockite series (in Russian). *Isvestiya Akademii Nauk SSSR, Seriya Geologicheskaya*, **4**, 102–20.

ALLEN, T. & CHAMBERLAIN, C. P., 1989. Thermal consequences of mantled dome emplacement. *Earth and Planetary Science Letters*, **93**, 392–404.

ALT, J. D., 1999. Very low-grade hydrothermal metamorphism of basic igneous rocks. In M. Frey & D. Robinson (eds.), Low-grade Metamorphism. *Oxford: Blackwell Scientific Publications*, pp. 169–201.

AMBROSE, J. W., 1936. Progressive kinetic metamorphism in the Missi series near Flinflon, Manitoba. *American Journal of Science*, **232**, 257–86.

AMSTUTZ, C., 1968. Spilites and spilitic rocks. In H. H. Hess (ed.), Basalts: The Poldervaart Treatise on Rocks of Basaltic Composition. *New York: Wiley Interscience*, pp. 737–53.

AMSTUTZ, C., 1974. Spilites and Spilitic Rocks. *Berlin-Heidelberg-New York: Springer*.

ANGEL, F., 1929. Gesteine von südlichen Grossvenediger. *Neues Jahrbuch für Mineralogie, Geologie und Paläontologie*, Beilageband **59-A**, 223–72.

ANTEN, J., 1923. Le Salmien métamorphique du sud du massif de Stavelot. *Mémoires de l'Académie royale de Belgique, classe des Sciences*, 2e sér., **5**, fasc. 3, 1–34. Also *Mineralogical Magazine*, **21**, 569.

ARAGO, F., 1821. Sur des tubes vitreux qui paraissent produits par des coups de foudre. *Annales de Chimie et de Physique*, **19**, 290–303. Also in: J. A. Barral, 1854–1862. Oeuvres complètes de François Arago, vol. 4. 13 vols. Notices Scientifiques: Le tonnerre, ch. XXI, Tubes de foudre et fulgurites. *Paris: Gide*, and *Leipzig: Weigel*, pp. 415–21.

ARNOLD, R. & ANDERSON, R., 1907. Metamorphism by combustion of hydrocarbons in the oil bearing shale of California. *Journal of Geology*, **15**, 750–8.

ARTINI, E., 1952. Le rocce. Concetti e nozioni di petrografia, 3rd edn. *Milano: Hoepli*.

ARTINI, E. & MELZI, G., 1900. Ricerche petrografiche e geologiche sulla Valsesia. *Memorie del Reale Istituto Lombardo di Scienze e Lettere*, **18**, 219–390.

ASAY, J. R. & SHAHINPOOR, M., 1993. High-pressured shock compression of solids. *New York, Berlin, Heidelberg: Springer*.

ASHWORTH, J. R., 1985. Introduction. In J. R. Ashworth (ed.), Migmatites. *Glasgow & London: Blackie*, pp. 1–35.

AUBOUIN, J., BROUSSE, R. & LEHMAN, J.-P., 1968. Précis de géologie, vol. 1, Pétrologie. *Paris: Dunod*.

AUBUISSON DE VOISINS, J. F. D', 1819. Traité de géognosie, vol. 2. *Strasbourg: Levrault*.

BACKLUND, H., 1918. Petrogenetische Studien an Taimgrgesteinen. *Geologiska Föreningens i Stockholm Förhandlingar*, **40**, 101–203.

Metamorphic Rocks: A Classification and Glossary of Terms. Recommendations of the International Union of Geological Sciences, eds. Douglas Fettes and Jacqueline Desmons. Published by Cambridge University Press. © Cambridge University Press 2007.

Backlund, H. G., 1936. Der Magmaaufstieg in Faltengebirgen. *Comptes Rendus de la Société géologique de Finlande*, **9**, 293–347.

Backlund, H. G., 1937. Die Umgrenzung der Svecofenniden. *Bulletin of the Geological Institute of the University of Uppsala*, **27**, 219–69.

Bailey, E. H., 1962. Metamorphic facies of the Franciscan Formation of California and their geologic significance (abstract). *Geological Society of America, Special Paper*, **68**, 4–5.

Bailey, E. H., Irwin, W. P. & Jones, D. L., 1964. Franciscan and related rocks, and their significance in the geology of western California. *Californian Division of Mines and Geology, Bulletin*, **183**, 89–112.

Banno, S., 1970. Classification of eclogites in terms of physical conditions of their origin. *Physics of the Earth and Planetary Interiors*, **3**, 405–21.

Barbosa, O., 1942. Geologia e petrologia na Regiâo de Apiaì, Estado de Sâo Paulo. Unpublished thesis.

Bardina, N. Yu. & Popov, V. S., 1994. Fenites: Systematics, conditions of formation and significance for crustal magmatism (in Russian). *Zapiski Vsesoyuznogo Mineralogicheskogo Obshchestva*, **123** (6), 1–19.

Barker, A. J., 1990. Introduction to Metamorphic Textures and Microstructures. *Glasgow and London: Blackie*.

Barker, A. J., 1998. Introduction to Metamorphic Textures and Microstructures, 2nd edn. *Cheltenham: Thorne*.

Barrell, J., 1907. Geology of the Marysville mining district, Montana. A study of igneous intrusion and contact metamorphism. *US Geological Survey Professional Paper*, **57**, 1–178.

Barrois, Ch., 1934. Note sur les gisements de staurotide de Bretagne. *Annales de la Société géologique du Nord*, **59**, 29–65.

Barrow, G., 1893. On an intrusion of muscovite-biotite gneiss in the southeastern Highlands of Scotland, and its accompanying metamorphism. *Quarterly Journal of the Geological Society of London*, **49**, 330–58.

Barrow, G., 1912. On the geology of lower Dee-side and the southern Highland Border. *Proceedings of Geologists' Association*, **23**, 268–84.

Barrow, G. & Thomas, H. H., 1908. On the occurrence of metamorphic minerals in calcareous rocks in the Bodmin and Camelford areas, Cornwall. *Mineralogical Magazine*, **15**, 113–23.

Barth, T. F., Correns, C. W. & Eskola, P. (eds.), 1939. Die Entstehung der Gesteine. *Heidelberg: Springer*.

Barth, T. F. W., 1947. The Birkland granite, a case of petroblastesis. *Bulletin de la Commission géologique de Finlande*, **140**, 173–82.

Barth, T. F. W., 1952. Theoretical Petrology. *New York: Wiley*.

Barth, T. F. W., 1962. Theoretical Petrology, 2nd edn. *New York, London: Wiley*.

Barth, T. J. W., 1978. In D. N. Lapedes (ed.), Encyclopedia of the Geological Sciences. *New York: McGraw-Hill*.

Barton, M. D., Ilchik, R. P. & Marikos, M. A., 1991a. Metasomatism. In D. M. Kerrick (ed.), Contact Metamorphism, *Mineralogical Society of America, Reviews in Mineralogy*, **26**, 321–50.

Barton, M. D., Staude, J.-M., Snow, E. A. & Johnson, D. A., 1991b. Aureole systematics. In D. M. Kerrick (ed.), Contact metamorphism. *Mineralogical Society of America, Reviews in Mineralogy*, **26**, 723–847.

Bascom, F., 1893. The structures, origin, and nomenclature of the acid volcanic rocks of South Mountain. *Journal of Geology*, **1**, 813–32.

Bates, R. L. & Jackson, J. A., 1979. Glossary of Geology, 2nd edn. *Virginia: American Geological Institute*.

Bates, R. L. & Jackson, J. A., 1987. Glossary of Geology, 3rd edn. *Virginia: American Geological Institute*.

Bates, R. L. & Jackson, J. A. (eds.), 1997. Glossary of Geology, 4th edn. *Virginia: American Geological Institute*.

Bauluz, B., Mayayo, M. J., Yuste, A., Fernandez-Nieto, C. & Gonzalez Lopez, J. M., 2004. TEM study of mineral transformations in fired carbonated clays: relevance to brick making. *Clay Minerals*, **39**, 333–44.

Bayly, B., 1968. Introduction to Petrology. *New Jersey: Prentice-Hall*.

Becher, G. F., 1882. Geology of the Comstock Lode and the Washoe District. *US Geological Survey Monograph*, **3**, 1–422.

Becher, J. P., 1789. Mineralogische Beschreibung der Oranien-Nassauischen Lande nebst einer Geschichte des Siegen'schen Hütten- und Hammerwesens. *Marburg*.

Beck, R., 1892. Die Contacthöfe der Granite und Syenite im Schiefergebiet des Elbtalgebirges.

Tschermak's mineralogische und petrographische Mitteilungen, **13**.

BECK, R., 1907. Untersuchungen über einige südafrikanische Diamantenlagerstätten. *Zeitschrift der deutschen geologischen Gesellschaft*, **59**, 275–307.

BECK, R., 1909. The Nature of Ore Deposits. Translated from the German by W. H. Weeden. *New York: Hill Publishing Co.*

BECKE, F., 1878. Gesteine der Halbinsel Chalcidice. *Mineralogische und petrographische Mitteilungen*, Neue Folge, 1. Also *Sitzungsberichte der Kaiserlichen Akademie der Wissenschaften, Mathematisch-Naturwissenschaftliche Klasse, Wien*, 1878, **77**, 609–15.

BECKE, F., 1880. Gesteine von Griechenland, part 2. *Mineralogische und Petrographische Mitteilungen*, 17–77.

BECKE, F., 1892. Vorläufiger Bericht über den geologischen Bau und die krystallinischen Schiefer des Hohen Gesenkes (Altvatergebirge). *Sitzungsberichte der Kaiserlichen Akademie der Wissenschaften, Mathematisch-Naturwissenschaftliche Klasse, Wien*, **101**, 286–300.

BECKE, F., 1902. Notizen: Mittheilungen der Wiener mineralogischen Gesellschaft. *Mineralogische und petrographische Mitteilungen*, **21**, 356–7.

BECKE, F., 1903a (1904). Über Mineralbestand und Struktur der krystallinischen Schiefer. *Report of the 9th International Geological Congress, Vienna*, **2**, 553–70.

BECKE, F., 1903b (1913). Ueber Mineralbestand und Struktur der krystallinischen Schiefern. *Denkschrifte der Kaiserlichen Akademie der Wissenschaften, Mathematisch-Naturwissenschaftliche Klasse*, **75**, 1–53.

BECKE, F., 1909. Ueber Diaphtorite. *Mineralogische und petrographische Mitteilungen*, **28**, 369–75.

BECQUEREL, A. C., ÉLIE DE BEAUMONT, J. & BRONGNIART, A., 1837. Rapport sur un travail de M. Fournet intitulé: Mémoire sur les filons métallifères et le terrain des environs de l'Arbresle (département du Rhône). *Comptes rendus de l'Académie des Sciences, Paris*, **5**, 51–60.

BEHR, H. J., DEN TEX, E., DE WAARD, D. *et al.*, 1971. Granulites. Results of a discussion. *Neues Jahrbuch für Mineralogie*, Monatsheft, 97–123.

BELL, J. M., CLARK, E. DE C. & MARSHALL, P., 1911. The geology of the Dun Mountain Subdivision, Nelson. *Bulletin of the New Zealand Geological Survey*, **12**, 31–5.

BELL, T. H. & RUBENACH, M. J., 1980. Crenulation cleavage development, evidence for progressive bulk inhomogeneous shortening from 'millipede' microstructures in the Robertson River metamorphics. *Tectonophysics*, **68**, T9–T15.

BELLIÈRE, J., 1960. Signification des structures de corrosion dans les roches migmatitiques. *Report of the 21st International Geological Congress, Copenhagen*, **14**, 30–6.

BELYAEV, G. M. & RUDNIK, V. A., 1978. Zonation and relationship of metasomatic products as the basis for an analysis of their formation (in Russian). In Metasomatism and ore formation. *Moscow: Nauka Press*, pp. 34–47.

BENTOR, Y. K. & KOSTNER, M., 1976. Combustion metamorphism in Southern California. *Science*, **193**, No. 4252, 486–8.

BENTOR, Y. K., GROSS, S. & HELLER, L., 1963. High-temperature minerals in non-metamorphosed sediments in Israel. *Nature*, **199**, No. 4892, 478–9.

BENTZ, A., 1925. Die Entstehung der 'Bunten Breccie', das Zentralproblem im Nördlinger Ries und Steinheimer Becken. *Zentralblatt für Mineralogie, Geologie und Paläontologie*, Abt. B, 97–104 and 141–5.

BERGEAT, A., 1912. Epigenetische Erzlagerstätten und Eruptivgesteine. *Fortschritte der Mineralogie, Kristallographie und Petrographie*, **2**, 9–23.

BERGERON, J., 1888. Note sur les roches éruptives de la Montagne Noire. *Bulletin de la Société géologique de France*, (3) **17**, 54–63.

BERTHÉ, D., CHOUKROUNE, P. & JEGOUZO, P., 1979. Orthogneiss, mylonite and non-coaxial deformation of granites: the example of the South Armorican Shear Zone. *Journal of Structural Geology*, **1**, 31–42.

BERTHELSEN, A., 1960. Structural studies in the Precambrian of western Greenland. 2. Geology of Tovqussap nunâ. *Grönlands geologiske Undersogelse, Bulletin*, **25**, 1–223. Also *Meddelelser om Grønland*, **123**, 1–223.

BERTOLANI, M., 1964. Le stronaliti. *Rendiconti della Società Mineralogica Italiana*, **20**, 31–70.

BERTRAND, J., 1980. Caractéristiques des diverses inclusions associées aux serpentinites de la nappe des Gets (Haute-Savoie, France). *Archives des Sciences, Genève*, **33**, 139–60.

BERTRAND, J., STEEN, D., TINKLER, C. & VUAGNAT, M., 1980. The melange zone of the Col du Chenaillet (Montgenèvre ophiolite,

Hautes-Alpes, France). *Archives des Sciences, Genève*, **33**, 117–38.

Bessmertnaya, M.S. & Gorzhevsky, D.I., 1958. Alterations surrounding ores in polymetallic ore deposits in Rudny Altay. *Izvestiya (Proceedings) of the Academy of Science of the USSR, Geological Series*, **10**, 21–36.

Beudant, F.S., 1824. Traité de minéralogie. *Paris: Verdière*.

Beudant, F.S., 1830. Traité élémentaire de minéralogie, 2nd edn. *Paris: Verdière*, vol. 1.

Beudant, F.S., 1832. Traité élémentaire de minéralogie, 2nd edn. *Paris: Verdière*, vol 2.

Beus, A.A. & Scherbakova, T.F., 1993. On the geochemistry of the rare-metal-bearing albitised granites (apogranites). In A.A. Beus (ed.), Geochemical Evolution of Granitoids in the Lithosphere History (in Russian). *Moscow: Nauka Press*, pp. 218–43.

Beyenberg, E., 1930. Kalkaugenphyllit, ein neues Gestein aus dem oberen Gedinne des Hunsrücks. *Zeitschrift der deutschen geologischen Gesellschaft*, **82**, 318–20.

Bideaux, A.J.W., Bladh, K.W. & Nichols, M.C., 1995. Handbook of Mineralogy, vol. 2, parts 1–2. *Tucson: Mineral Data Publishing*.

Binns, R.A., 1964. Zones of progressive regional metamorphism in the Willyama Complex, Broken Hill District, New South Wales. *Journal of the Geological Society of Australia*, **11**, 283–330.

Binns, R.W., 1967. Stony meteorites bearing maskelynite. *Nature*, **214**, 1111–12.

Bischoff, A. & Lange, M.A., 1984. Experimental shock-lithification of chondritic powder: implications for ordinary chondrite regolith breccias. *Proceedings of the 15th Lunar Planetary Science Conference, Abstracts*, 60–1.

Bischoff, A. & Stöffler, D., 1992. Shock metamorphism as a fundamental process in the evolution of planetary bodies: information from meteorites. *European Journal of Mineralogy*, **4**, 707–55.

Blake, M.C., Irwin, W.P. Jr & Coleman, R.G., 1967. Upside-down metamorphic zonation, blueschist facies, along a regional thrust in California and Oregon. *US Geological Survey Professional Paper*, **575-C**, 1–9.

Blum, J.H., 1843. Die Pseudomorphosen des Mineralreichs. *Stuttgart: Schweizerbart*.

Boase, H.S., 1832. Contribution towards a knowledge of the geology of Cornwall. *Transactions of the Royal Geological Society of Cornwall*, **4**, 166–474.

Bobrievich, A.P., Smirnov, G.I. & Sobolev, V.S., 1960. Mineralogy of the xenoliths of a grossular-pyroxene-kyanite rock (grospydite) from kimberlites in Yakutia (in Russian). *Sibirskoye otdelenie Akademii Nauk SSSR, Geologiya i Geofisika*, **3**, 18–24.

Boles, J.R. & Coombs, D.S., 1977. Zeolite facies alteration of sandstones in the Southland Syncline, New Zealand. *American Journal of Science*, **277**, 982–1012.

Bonney, T.G., 1883. On some breccias and crushed rocks. *Geological Magazine*, **20**, 435–8.

Bonney, T.G., 1886. Anniversary address of the President: Metamorphic rocks. *Quarterly Journal of the Geological Society of London, Proceedings*, **42**, 38–115.

Bonney, T.G., 1899. The parent-rock of the diamond in South Africa. *Geological Magazine*, **6**, 309–21.

Bonney, T.G., 1919. Foliation and metamorphism. *Geological Magazine*, **56**, 196–203.

Borodaevskii, N.I. & Borodaevskaya, M.B., 1947. Berezovskoye Ore Field (in Russian). *Moscow: Metallurgist Publications*.

Boubée, N., 1833. Géologie populaire à la portée de tout le monde appliquée à l'agriculture et à l'industrie. *Paris: Bureau du Nouveau Bulletin d'Histoire naturelle*.

Boué, A., 1820. Essai géologique sur l'Écosse. *Paris: Courcier*.

Boué, A., 1829. Geognostiches Gemälde von Deutschland, mit Rücksicht auf die Gebirgs-Beschaffenheit nachbarlicher Staten. *Frankfurt-am-Main: Hermann*.

Bouillier, A.-M. & Nicolas, A., 1975. Classification of textures and fabrics from peridotite xenoliths from South African kimberlites. *Physics and Chemistry of the Earth*, **9**, 467–75.

Bowen, G.T., 1822. Analysis of a variety of nephrite, from Smithfield, Rhode Island. *American Journal of Science*, Ser.1, **5**, 346–8.

Bowen, N.L., 1928. The Evolution of the Igneous Rocks. *Princeton University Press*.

Bowen, N.L., 1940. Progressive metamorphism of siliceous limestone and dolomite. *Journal of Geology*, **48**, 225–74.

Bowes, D.R. (ed.), 1989a. The Encyclopedia of Igneous and Metamorphic Petrology. *New York: Van Nostrand Reinhold*.

Bowes, D. R., 1989b. Textures of metamorphic rocks. In D. R. Bowes (ed.), The Encyclopedia of Igneous and Metamorphic Petrology. *New York: Van Nostrand Reinhold*, pp. 503–8.

Branco, W. & Fraas, E., 1905. Das kryptovulkanische Becken von Steinheim. *Akademie der Wissenschaften Berlin, physikalisch-mathematische Klasse, Abhandlungen*, **1**, 1–64.

Brauns, R., 1911. Die Kristallinenschiefer des Laacher See Gebietes und ihre Umbildung zu Sanidinit. *Stuttgart: Schweizerbart*.

Breislak, S., 1822. Traité sur la structure extérieure du globe, ou institutions géologiques, vol. 3. *Milano: Giegler*.

Breithaupt, A., 1847. Handbuch der Mineralogie, vol. 3. *Dresden & Leipzig: Arnoldische Buchhandlung*.

Breithaupt, A., 1849. Die Paragenesis der Mineralien: mineralogisch, geognostisch und chemisch beleuchtet mit besonderer Rücksicht auf Bergbau. *Freiberg: Engelhardt*.

Breithaupt, A. & Gmelin, C. G., 1823. Vollständige Beschreibung des Erlans, eines lange verkannten und neu bestimmten Minerals. *Annalen für Physik und Chemie*, **37**, 76–82.

Breithaupt, J. F. A., 1830. Uebersicht des Mineral-Systems. *Freiberg: Engelhardt*.

Bresson, A., 1903. Étude sur les formations anciennes des Hautes et Basses-Pyrénées (haute chaîne). *Bulletin du Service de la Carte géologique de la France*, no. 93, **14**, 45–322.

Brière, Y., 1920. Les éclogites françaises, leur composition minéralogique et chimique, leur origine. Unpublished thesis, Faculté des Sciences de Paris. Cf. *Mineralogical Abstracts*, **1**, 163.

Broch, O. A., 1927. Ein suprakrustaler Gneiskomplex auf der Halbinsel Nesodden bei Oslo. *Norsk Geologisk Tidsskrift*, **9**, 81–219.

Broch, O. A., 1929. Ein suprakrustaler Gneiskomplex auf der Halbinsel Nesodden bei Oslo. *Neues Jahrbuch für Mineralogie, Geologie und Paläontologie*, **29–2**, 20–5.

Brochant, A. J. M., 1798. Traité élémentaire de minéralogie, vol. 1. *Paris: Villier*.

Brochant, A. J. M., 1800. Traité élémentaire de minéralogie suivant les principes du professeur Werner, conseiller des Mines de Saxe, vol. 2. *Paris: Villier*.

Brögger, W. C., 1890. Mineralien der Syenitpegmatitgänge der südnorwegischen Augit- und Nephelinsyenite; allgemeiner Teil. *Zeitschrift für Krystallographie und Mineralogie*, **16**, 104–20.

Brögger, W. C., 1898. Die Eruptivgesteine des Kristianiagebietes, 3. *Skrifter udgit av Videnskabsselskabet i Kristiania. I. Matematisk-naturvidenskabelig Klasse*, **6**, 1–377.

Brögger, W. C., 1921. Die Eruptivgesteine des Kristianiagebietes, 4. Das Fengebiet in Telemark, Norwegen. *Skrifter udgit av Videnskabsselskabet i Kristiania. I. Matematisk-naturvidenskabelig Klasse* (1920), **9**, 150–67.

Brongniart, A., 1813. Essai d'une classification minéralogique des roches mélangées. *Journal des Mines, Paris*, **34**, (199), 5–48.

Brongniart, A., 1825. Introduction à la minéralogie. Dictionnaire des Sciences naturelles. *Paris: Levrault*, p. 31.

Brongniart, A., 1827. Classification et caractères minéralogiques des roches homogènes et hétérogènes. *Paris: Levrault*.

Brooks, H. K., 1954. The rock and stone terms limestone and marble. *American Journal of Science*, **252**, 755–60.

Brothers, R. N. & Blake, M. C., 1987. Comment on blueschists and eclogites, and reply by B. W. Evans & E. H. Brown. *Geology*, **15**, 773–5.

Bruckmann, U. F. B., 1806. Von den sogenannten Blitzstein oder pierre foudroyée am Montblanc. *Voigt's Magazin für die neue Zustande der Naturkunde, Weimar*, **2**, 11, 67–8.

Bucher, K. & Frey, M., 1994. Petrogenesis of Metamorphic Rocks: A complete Revision of Winkler's Textbook, 6th edn. *Berlin-Heidelberg-New York: Springer*.

Bucher, K. & Frey, M., 2002. Petrogenesis of Metamorphic Rocks, 7th edn. *Berlin-Heidelberg-New York: Springer*.

Bucher, W. H., 1936. Cryptovolcanic structures in the United States. *Report of the 16th International Geological Congress, Washington DC, 1933*, **2**, 1055–84.

Bucher, W. H., 1963. Cryptoexplosion structures caused from without or from within the earth ('astroblemes' or 'geoblemes')? *American Journal of Science*, **261**, 597–649.

Buddington, A. F., 1939. Adirondack igneous rocks and their metamorphism. *Geological Society of America Memoir* **7**, 1–354.

Bunch, T. E., 1968. Some characteristics of selected minerals from craters. In B. M. French & N. M. Short (eds.), Shock Metamorphism of Natural

Materials. *Baltimore: Mono Book Corporation*, pp. 413–32.

BUNCH, T. E. & COHEN, A. J., 1964. Shock deformation of quartz from two meteorite craters. *Geological Society of America Bulletin*, **75**, 1263–6.

BUNSEN, R., 1849. Ueber den innern Zusammenhang der pseudovulkanischen Erscheinungen Islands. *Annalen der Chemie und Pharmacie*, **62**, 1–59.

BUNSEN, R., 1851. Ueber die Prozesse der vulkanischen Gesteinsbildungen Islands. *Annalen der Physik und Chemie*, **83**, 197–272.

BURNHAM, C. W., 1962. Facies and types of hydrothermal alteration. *Economic Geologist*, **57**, 768–84.

CALLAWAY, C., 1881. The Archaean geology of Anglesey. *Quarterly Journal of the Geological Society of London*, **37**, 210–38.

CANN, J. R., 1979. Metamorphism in the ocean crust. In M. Talwani, C. G. Harrison & D. E. Hayes (eds.), Deep Drilling Results in the Atlantic Ocean, Oceanic Crust. *American Geophysical Union*, pp. 230–8.

CANNON, R. T., 1963. Classification of amphibolites. *Geological Society of America, Bulletin*, **74**, 1087–8.

CAPACCI, C., 1881. La formazione ofiolitica del Monteferrato presso Prato (Toscana). *Bolletino del Regio Comitato Geologico d'Italia*, **12**, 275–312.

CARMICHAEL, D. M., 1974. Metamorphic bathograds: a measure of the depth of post-metamorphic erosion on the regional scale. *Geological Society of America, Abstracts with Programs*, **6**, 680–1.

CARMICHAEL, D. M., 1978. Metamorphic bathozones and bathograds: a measure of the depth of post-metamorphic uplift and erosion on a regional scale. *American Journal of Science*, **278**, 769–97.

CARSWELL, D. A. (ed.), 1990a. Eclogite Facies Rocks. *Glasgow and London: Blackie*.

CARSWELL, D. A., 1990b. Eclogites and the eclogite facies: definitions and classification. In D. A. Carswell (ed.), Eclogite Facies Rocks. *Glasgow and London: Blackie*, pp. 1–13.

CARSWELL, D. A. & COMPAGNONI, R. (eds.), 2003. Ultrahigh Pressure Metamorphism. *European Mineralogical Union, Notes in Mineralogy 5. Eötvös University Press.*

CARTER, N. L., 1965. Basal quartz deformation lamellae, a criterion for recognition of impactites. *American Journal of Science*, **263**, 786–806.

CASSIDY, W. A., 1968. Meteorite impact structures at Campo del Cielo, Argentina. In B. M. French & N. M. Short (eds.), Shock Metamorphism of Natural Materials. *Baltimore: Mono Book Corporation*, pp. 117–28.

CATHREIN, A., 1883. Ueber Saussurit. *Zeitschrift für Krystallographie und Mineralogie*, **7**, 234–49.

CHACKO, T., KUMAR, G. R. R., MEEN, J. K. & RODGERS, J. J. W., 1992. Geochemistry of high-grade supracrustal rocks of the Kerala Khondalite Belt and adjacent massif charnockites, South India. *Precambrian Research*, **55**, 469–89.

CHAO, E. C. T., 1967a. Shock effects in certain rock-forming minerals. *Science*, **156**, 192–202.

CHAO, E. C. T., 1967b. Impact metamorphism. In P. H. Abelson (ed.), Researches in Geochemistry, vol. 2. *New York: Wiley*, pp. 204–33.

CHAO, E. C. T., 1968. Pressure and temperature histories of impact metamorphosed rocks, based on petrographic observations. In B. M. French & N. M. Short (eds.), Shock Metamorphism of Natural Materials. *Baltimore: Mono Book Corporation*, pp. 135–58.

CHELIUS, C., 1892. Das Granitmassiv des Melibocus and seine Ganggesteine. *Notizblatt des Vereins für Erdkunde und der Grossherzogischen Geologischen Landesanstalt zu Darmstadt*, Ser. 4, **13**, 1–13.

CHENG, Y. C., 1943. The migmatite area around Bettyhill, Sutherland. *Quarterly Journal of the Geological Society of London*, **99**, 107–54.

CHIDESTER, A. H., 1962. Petrology and geochemistry of selected talc-bearing ultramafic rocks and adjacent country rocks in north-central Vermont. *US Geological Survey Professional Paper*, **345**, 1–207.

CHILIGAR, G. V., BISSEL, H. J. & WOLF, K. H., 1967. Diagenesis in carbonate rocks. In G. Larsen & G. V. Chilingar (eds.), Diagenesis in Sediments. *Amsterdam: Elsevier*, pp. 179–322.

CHRISTIE, J. M., GRIGGS, D. T., HEUER, A. H. *et al.*, 1973. Electron petrography of Apollo 14 and 15 breccias and shock-produced analogs. *Proceedings of the 4th Lunar Science*

Conference (Supplement 4, Geochimica et Cosmochimica Acta), **1**, 365–82.

Church, W. R., 1968. Eclogites. In H. H. Hess & A. Poldervaart (eds.), Basalts, The Poldervaart Treatise on Rocks of Basaltic Composition, vol. 2. *New York: Wiley*, pp. 755–98.

Clark, B. H. & Peacor, D. R., 1992. Pyrometamorphism and partial melting of shales during combustion metamorphism; mineralogical, textural and chemical effects. *Contributions to Mineralogy and Petrology*, **112**, 558–68.

Cloos, E., 1946. Lineation: a critical review and annotated biography. *Geological Society of America Memoir*, **18**, 1–126.

Clough, C. T., 1888. The Geology of the Cheviot Hills. *Memoirs of the Geological Survey of England and Wales*, 1–60.

Cogné, J., 1960. Schistes cristallins et granites en Bretagne méridionale. Le domaine de l'anticlinal de Cornouaille. *Mémoire explicatif de la Carte géologique de France*.

Cogné, J. & Eller, J.-P. von, 1961. Défense et illustration des termes leptynite et granulite en pétrographie des roches métamorphiques. *Bulletin du Service de la Carte géologique d'Alsace-Lorraine*, **14**, 59–64.

Cohen, E., 1887. Geognostisch-petrographische Skizzen aus Süd-Afrika. *Neues Jahrbuch für Mineralogie, Geologie und Paläontologie*, **5**, 195–274.

Cole, G. A. J., 1902. On composite gneiss in Boylogh, West Donegal. *Proceedings of the Royal Irish Academy*, **24**, 203–30.

Cole, G. A. J., 1915. A composite gneiss near Borna (Co. Galway). *Quarterly Journal of the Geological Society of London*, **71**, 183–8.

Coleman, R. G., 1977. Ophiolites: Ancient oceanic lithosphere? *Berlin-Heidelberg-New York: Springer*.

Coleman, R. G. & Wang, X., 1995. Ultrahigh Pressure Metamorphism. *Cambridge University Press*.

Coleman, R. G., Lee, D. E., Beatty, L. B. & Brannock, W. W., 1965. Eclogites and eclogites; their differences and similarities. *Geological Society of America Bulletin*, **76**, 483–508.

Condie, K. C., 1981. Archean greenstone belts. Developments in Precambrian Geology, vol. 3. *Amsterdam: Elsevier*, **3**.

Cooke, H. C., 1927. Gold and Copper Deposits of Western Quebec. *Geological Survey of Canada*, Summary Report 1925, part C, 28–51.

Coombs, D. S., 1960. Lower grade mineral facies in New Zealand. *Report of the 21st International Geological Congress, Copenhagen*, **13**, 339–51.

Coombs, D. S., 1961. Some recent work on the lower grades of metamorphism. *The Australian Journal of Science*, **24**, 203–15.

Coombs, D. S., 1971. Present status of the zeolite facies. In Molecular Sieve Zeolites, Advances in Chemistry, Ser. 101. *Washington: American Chemical Society*, pp. 317–27.

Coombs, D. S., Ellis, A. J., Fyfe, W. S. & Taylor, A. M., 1959. The zeolite facies, with comments on the interpretation of hydrothermal syntheses. *Geochimica et Cosmochimica Acta*, **17**, 53–107.

Cooray, P. G., 1998. Usage of terms 'khondalite' and 'leptynite'. *Journal of the Geological Society of India*, **51**, 710.

Coquand, H., 1857. Traité des roches considérées au point de vue de leur origine, de leur composition, de leur gisement et leurs applications à la géologie et à l'industrie. *Paris: Baillière*.

Cordella, A., 1893. Laurium. In R. Lepsius (ed.), Geologie von Attika: Ein Beitrag zur Lehre vom Metamorphismus der Gesteine. *Berlin: Reimer*.

Cordier, P. L. A., 1842–8. Dictionnaire universel d'histoire naturelle (C. d'Orbigny, ed.). *Paris: Savy*.

Cordier, P. L. A., 1868. Description des roches composant l'écorce terrestre et des terrains cristallins constituant le sol primitif (C. d'Orbigny, ed.). *Paris: Savy, Dunod*.

Cosca, M. A., Essene, E. J., Geissman, J. W., Simmons, W. B. & Coates, D. A., 1989. Pyrometamorphic rocks associated with naturally burned coal beds, Powder River basin, Wyoming. *American Mineralogist*, **74**, 85–100.

Cossa, A., 1881. Ricerche chimiche e microscopiche su rocce e minerali d'Italia (1875–1880). *Regia Stazione Agraria Sperimentale. Torino: Bona*.

Cotta, B. von, 1846. Grundriss der Geognosie und Geologie, 2nd edn. 2 vols. *Dresden: Arnold*.

Cotta, B. von, 1855. Die Gesteinslehre. *Freiberg: Engelhardt*.

Cotta, B. von, 1862. Die Gesteinslehre, 2nd edn. *Freiberg: Engelhardt*.

Croft, S. K., 1980. Cratering flow fields: Implications for the excavation and transient expansion stages of crater formation. *Proceedings of the 11th Lunar Planetary Science Conference*, 2347–78.

CRONSTEDT, A., 1758. Försök til Mineralogie eller Mineral-rikets Uppstålning, 2nd edn. *Stockholm.* German translation by A. G. Werner, 1780, Versuch eines Mineralogie Systems. *Leipzig: Crusius* and *Freiberg: Vogel;* English translation by G. von Engestrøm, 1970, Towards a System of Mineralogy. *London: Dilly.*

CROSS, W., 1894. The laccolitic Mountain group of Colorado, Utah and Arizona. *US Geological Survey Annual Report*, **14**, 165–238.

CROSS, W., IDDINGS, J. P., PIRSSON, L. V. & WASHINGTON, H. S., 1906. The texture of igneous rocks. *Journal of Geology*, **14**, 692–707.

CROSS, W., IDDINGS, J. P., PIRSSON, L. V. & WASHINGTON, H. S., 1912. Modifications of the quantitative system of classification of igneous rocks. *Journal of Geology*, **20**, 550–61.

CUNNINGHAM-CRAIG, E. H., 1904. Metamorphism in the Loch Lomond district. *Quarterly Journal of the Geological Society of London*, **60**, 10–29.

DALY, R. A., 1917. Metamorphism and its phases. *Geological Society of America Bulletin*, **28**, 375–418.

D'AMICO, C., INNOCENTI, F. & SASSI, F. P., 1987. Magmatismo e metamorfismo. *Torino: UTET.*

DANA, E. S., 1892. The System of Mineralogy of J. D. Dana, 1837–1868, Descriptive Mineralogy, 6th edn. *New York: Wiley.*

DANA, J. D., 1850. A System of Mineralogy, 3rd edn. *New York: Wiley, Putman.*

DANA, J. D., 1886. On some general terms applied to metamorphism and to the porphyritic structure of rocks. *American Journal of Science*, **32**, 69–72.

DARWIN, C., 1846. Geological Observations in South America. *London: Smith, Elder & Co.*

DARWIN, C., 1860. Journal of Researches during the Voyage of H. M. S. Beagle. *London: Nelson.*

DAUBRÉE, A., 1857. Observations sur le métamorphisme et recherches expérimentales sur quelques-uns des agents qui ont pu le produire. *Annales des Mines*, sér. 5, **12**, 289–326.

DAUBRÉE, A., 1867. Classification adoptée pour la collection des roches du Muséum d'Histoire Naturelle de Paris. *Paris: Masson.*

DAUBRÉE, A., 1879. Études synthétiques de géologie expérimentale, part 1: Application de la méthode expérimentale à l'étude de divers phénomènes géologiques. *Paris: Dunod.*

DAVIS, G. H. & REYNOLDS, S. J., 1996. Structural Geology of Rocks and Regions, 2nd edn. *New York: Wiley.*

DEER, W. A., HOWIE, R. A. & ZUSSMAN, J., 1962. Rock-forming Minerals, vol. 1, Ortho- and Ring Silicates. *London: Longmans.*

DE LA BECHE, H. T., 1826. On the geology of southern Pembrokeshire. *Transactions of the Geological Society of London*, ser. 2, **2**, 1–20.

DELAMÉTHERIE, J. C., 1795. Théorie de la terre, vol. 2. *Paris: Maradan.*

DELAMÉTHERIE, J. C., 1806. Tableau des analyses chimiques des minéraux et d'une nouvelle classification de ces substances, fondée sur ces analyses. *Journal de Physique, de Chimie, d'Histoire naturelle et des Arts*, **62**, 319–405.

DELESSE, A., 1845. Sur un silicate d'alumine et de potasse hydraté d'une composition nouvelle. *Annales de Chimie et de Physique, Paris*, 3e sér., **15**, 248–55.

DELESSE, A., 1849. Sur la protogine des Alpes. *Annales de Chimie et de Physique, Paris*, 3e sér., **25**, 114–27.

DELESSE, A., 1852. Sur les variations des roches granitiques. *Bulletin de la Société géologique de France*, 2e sér., **9**, 464–82.

DELESSE, A., 1856. Sur la pierre ollaire. *Bulletin de la Société géologique de France*, 2e sér., **14**, 280–7.

DELESSE, A., 1857. Études sur le métamorphisme des roches. *Annales des Mines*, sér. 5a, **12**, 89–288, 417–516, 705–72; sér. 5a, **13**, 321–416.

DENCE, M. R., 1964. A comparative structural and petrographic study of probable Canadian meteorite craters. *Meteoritics*, **2**, 249–70.

DENCE, M. R., 1968. Shock zoning at Canadian craters, petrography and structural implications. In B. M. French & N. M. Short (eds.), Shock Metamorphism of Natural Materials. *Baltimore: Mono Book Corporation*, pp. 169–84.

DENCE, M. R., 1971. Impact melts. *Journal of Geophysical Research*, **76**, 5552–65.

DENCE, M. R., 2004. Structural evidence from shock metamorphism in simple and complex impact craters: linking observation to theory. *Meteoritics and Planetary Science*, **39**, 267–86.

DENCE, M. R. & ROBERTSON, P. B., 1989. Shock metamorphism. In D. R. Bowes (ed.),

Encyclopedia of Igneous and Metamorphic Petrology. *New York: Van Nostrand Reinhold*, pp. 526–30.

Dence, M. R., Grieve, R. A. F. & Robertson, P. B., 1977. Terrestrial impact structure: principal characteristics and energy considerations. In D. J. Roddy, R. O. Pepin & R. B. Lerrill (eds.), Impact and Explosion Cratering. *New York: Pergamon*, pp. 247–75.

Derby, O. A., 1901. On the manganese ore deposit of the Queluz (Lafayette) District, Minas Geraes, Brazil. *American Journal of Science*, **12** (162), 18–32.

Derby, O. A., 1908. On the original type of the manganese ore deposits of the Queluz District, Minas Geraes, Brazil. *American Journal of Science*, **25** (175), 213–16.

De Roever, W. P., 1972. Application of the facies principle to rocks metamorphosed in more than one metamorphic facies, with special reference to plurifacial metamorphism in southeastern Spain. *Koninklijke Nederlandse Akademie Wetenschappen Amsterdam*, **75** (B), 253–60.

De Roever, W. P. & Nijhuis, H. J., 1963. Plurifacial alpine metamorphism in the eastern Betic Cordilleras (SE Spain), with special reference to the genesis of the glaucophane. *Geologische Rundschau*, **53**, 324–36.

Desmons, J., Smulikowski, W. & Schmid, R., 1997. High-P/T rock terms: definitions proposed by SCMR. *Fourth International Eclogite Conference, Ascona.*

Desmons, J., Smulikowski, W. & Schmid, R., 2001. High-P/T rock terms: definitions proposed by SCMR. *Mineralogical Society of Poland, Special Papers* **19**, 36–8.

De Waard, D., 1950. Palingenetic structures in the augengneiss of the Sierra de Guadarama, Spain. *Comptes Rendus de la Société géologique de Finlande*, **23**, 51–66.

De Waard, D., 1973. Classification and nomenclature of felsic and mafic rocks of high-grade metamorphic terrains. *Neues Jahrbuch für Mineralogie*, Monatsheft, 381–92.

Dewey, J. F. & Pankhurst, R. J., 1970. The evolution of the Scottish Caledonides in relation to their isotopic age pattern. *Transactions of the Royal Society of Edinburgh*, **68**, 361–89.

Dietrich, R. V., 1959. Development of ptygmatic features within a passive host during partial anatexis. *Beiträge zur Mineralogie und Petrographie*, **6**, 357–66.

Dietrich, R. V., 1960. Nomenclature of migmatitic and associated rocks. *Geotimes*, **4**, 36–7, 50–1.

Dietrich, R. V., 1963. Banded gneisses of eight localities. *Norsk Geologisk Tidskrift*, **43**, 89–119.

Dietrich, R. V. & Mehnert, K. R., 1960. Proposal for the nomenclature of migmatites and associated rocks. *Report of the 21st International Geological Congress, Norden, Copenhagen*, part 26, sect. 14, 56–78.

Dietz, R. S., 1959. Shatter cones in cryptoexplosion structures. *Journal of Geology*, **67**, 496–505.

Dietz, R. S., 1960. Meteorite impact suggested by shatter cones in rock. *Science*, **131**, 1781–4.

Dietz, R. S., 1961. Astroblemes. *Scientific American*, **205**, 2–10.

Dietz, R. S., 1963. Cryptoexplosion structures: a discussion. *American Journal of Science*, **261**, 650–64.

Diller, J. S., 1898. The educational series of rock specimens, No. 11: Breccia. *US Geological Survey Bulletin*, **150**, 72–4.

Dobretsov, N. L., Bogatikov, O. A. & Rosen, O. M. (eds.), 1992. Classification and Nomenclature of Metamorphic Rocks, a Handbook (in Russian). *Novosibirsk: Nauka Press.*

Dodd, R. T. & Jarosewich, E., 1979. Incipient melting and shock classification of L-group chondrites. *Earth and Planetary Science Letters*, **44**, 335–40.

Doran, D. G. & Linde, R. K., 1966. Shock effects in solids. In F. Seitz & D. Turnbull (eds.), Solid State Physics. *New York: Academic Press*, pp. 229–90.

Dostal, J., Caby, R., Dupuy, C., Mével, C. & Owen, J. V., 1996. Inception and demise of a Neoproterozoic ocean basin: evidence from the Ougda complex, western Hoggar (Algeria). *Geologische Rundschau*, **85**, 619–31.

Dressler, B. O. & Reimold, W. U., 2001. Terrestrial impact melt rocks and glasses. *Earth-Science Reviews*, **56**, 205–84.

Dressler, B. O. & Reimold, W. U., 2004. Order or chaos? Origin and mode of emplacement of breccias in floors of large impact structures. *Earth-Science Reviews*, **67**, 1–54.

Dufrénoy, A. & Élie de Beaumont, L., 1841. Explication de la carte géologique de France, vol. 1. *Ministère des Travaux Publics.*

DUNN, J. A., 1942. Granite and magmatism and metamorphism. *Economic Geology*, **234**, 231–8.

DUNOYER DE SEGONZAC, G., 1970. The transformation of clay minerals during diagenesis and low-grade metamorphism: a review. *Sedimentology*, **15**, 281–346.

DUPARC, L. & HORNUNG, T., 1904. Sur une nouvelle théorie de l'ouralitisation. *Comptes rendus de l'Académie des Sciences, Paris*, **139**, 223–5.

DUPARC, L., MOLLY, E. & BORLOZ, A., 1927. Sur la birbirite, une roche nouvelle. *Comptes Rendus des Séances de la Société de Physique et d'Histoire Naturelle de Genève*, **44**, 137–9.

DURNEY, D. W., 1972. Solution-transfer, an important geological deformation mechanism. *Nature*, **235**, 315–17.

DUROCHER, J., 1846. Études sur le métamorphisme des roches. *Bulletin de la Société géologique de France*, 2e sér., **3**, 546–647.

DUROCHER, J., 1857. Essai de pétrologie comparée ou recherches sur la composition chimique et minéralogique des roches ignées, sur les phénomènes de leur émission et sur leur classification. *Annales des Mines, Paris*, sér. 5, **11**, 217–59, Appendix 676–80.

DUVALL, G. E., 1968. Shock waves in solids. In B. M. French & N. M. Short (eds.), Shock Metamorphism of Natural Materials. *Baltimore: Mono Book Corporation*, pp. 159–68.

DUVALL, G. E. & FOWLES, G. R., 1963. Shock waves. In R. S. Bradley (ed.), High Pressure Physics and Chemistry, vol. 2. *London and New York: Academic Press Inc.*, pp. 209–91.

EDGAR, A. D., 1974. Experimental Petrology. *Oxford: Clarendon*.

EDWARDS, A. B., 1954. Textures of Ore Minerals. *Australian Institute of Mining and Metallurgy*.

EICHWALD, E., 1846. Einige vergleichende Bemerkungen zur Geognosie Scandinaviens und der westlichen Provinzen Russlands. *Bulletin de la Société Impériale des Naturalistes de Moscou*, **19**, 1re partie, 3–156.

EINAUDI, M. T., MEINERT, L. D. & NEWBERRY, R. I., 1981. Skarn deposits. *Economic Geology*, anniversary volume, 317–91.

EISELE, H., 1907. Das Übergangsgebirge bei Baden-Baden, Ebersteinburg, Gaggenau und Sulzbach und seine Kontaktmetamorphose durch das Nordschwarzwälder Granitmassiv. *Zeitschrift der deutschen geologischen Gesellschaft*, **59**, 131–214.

ÉLIE DE BEAUMONT, J. B., 1847. Notes sur les émanations volcaniques et métallifères. *Bulletin de la Société géologique de France*, 2e sér., **4**, 1249–334.

EMMERLING, L. A., 1799. Lehrbuch der Mineralogie, 2nd edn, vol. 3. *Giesen: Heyes*.

ENGELBRECHT, C. A., 1802. Kurze Beschreibung des Weisssteins, einer im geognostischen System bis jetzt unbekannten gewesenen Gebirgsart. *Schriften der Linné'schen Gesellschaft zu Leipzig*, 26–34.

ENGELHARDT, W. VON, 1971. Detrital impact formations. *Journal of Geophysical Research*, **76**, 5566–74.

ENGELHARDT, W. VON & BERTSCH, W., 1969. Shock induced planar deformation structures in quartz from the Ries crater, Germany. *Contributions to Mineralogy and Petrology*, **20**, 203–34.

ENGELHARDT, W. VON & STÖFFLER, D., 1965. Spaltflächen in Quarz als Anzeichen für Einschläge grosser Meteoriten. *Naturwissenschaft*, **52**, 489–90.

ENGELHARDT, W. VON & STÖFFLER, D., 1968. Stages of shock metamorphism in crystalline rocks of the Ries basin, Germany. In B. M. French & N. M. Short. (eds.), Shock Metamorphism of Natural Materials. *Baltimore: Mono Book Corporation*, pp. 159–68.

ENGELHARDT, W. VON, ARNDT, J., STÖFFLER, D., MÜLLER, W. F., JEZIORKOWSKI, H. & GUBSER, R. A., 1967. Diaplektische Gläser in den Breccien des Ries von Nördlingen als Anzeichen für Stosswellenmetamorphose. *Contributions to Mineralogy and Petrology*, **15**, 93–102.

ENGELHARDT, W. VON, ARNDT, J., MÜLLER, W. F. & STÖFFLER, D., 1970. Shock metamorphism of lunar rocks and origin of the regolith at the Apollo 11 landing site. *Proceedings of the Apollo 11 Lunar Science Conference, Geochimica et Cosmochimica Acta*, Suppl. 1, 363–84.

ENGLAND, P. C. & RICHARDSON, S. W., 1977. The influence of erosion upon the mineral facies of rocks from different metamorphic environments. *Journal of the Geological Society*, **134**, 201–13.

ENGLUND, K. J. & ROEN, J. B., 1962. Origin of the Middlesboro Basins, Kentucky. *US Geological Survey Professional Paper*, **450-E**, 20–2.

ERDMANNSDOERFFER, O. H., 1924. Grundlagen der Petrographie. *Stuttgart: Enke.*

ERDMANNSDOERFFER, O. H., 1946. Über unausgereifte Magmatite (Aorite). *Nachrichten der Akademie der Wissenschaften Göttingen, Mathematisch-naturwissenschaftliche Klasse*, 96.

ERWIN, H. D., 1938. Some metamorphic terminology. *American Mineralogist*, **23**, 119–20.

ESCHWEGE, W. L. VON, 1822. Geognostisches Gemälde von Brasilien und wahrscheinliches Muttergestein der Diamanten. *Gr. H. S. priv. Landes-Industrie-Comptoirs, Weimar*, 28–30.

ESCHWEGE, W. L. VON, 1832. Beiträge zur Gebirgskunde Brasiliens. *Berlin: Reimer.*

ESKOLA, P., 1914. On the petrology of the Orijärvi region in southwestern Finland. *Bulletin de la Commission géologique de Finlande*, **40**, 1–277.

ESKOLA, P., 1915. Om sambandet mellan kemisk och mineralogisk sammansättning hos Orijärvitraktens metamorfa bergarter. *Bulletin de la Commission géologique de Finlande*, **44**, 1–145.

ESKOLA, P., 1920. The mineral facies of rocks. *Norsk Geologisk Tidskrift*, **6**, 143–94.

ESKOLA, P., 1929. Om mineralfacies. *Geologiska Föreningens i Stockholm Förhandlingar*, **51**, 157–72.

ESKOLA, P., 1932. On the principles of metamorphic differentiation. *Comptes Rendus de la Société géologique de Finlande*, **5**, 68–77.

ESKOLA, P., 1936. Wie ist die Anordnung der äusseren Erdsphären nach der Dichte zustandegekommen? *Geologische Rundschau*, **27**, 61–73.

ESKOLA, P., 1939. Die metamorphen Gesteine. In T. F. W. Barth, C. W. Correns & P. Eskola (eds.), Die Entstehung der Gesteine. *Heidelberg: Springer*, pp. 263–407.

ESKOLA, P., 1948. Über die Geologie Ostkareliens. *Geologische Rundschau*, **35**, 154–65.

ESKOLA, P., 1952. On the granulites of Lappland. *American Journal of Science*, Bowen vol., 133–72.

ESSENE, E. J. & FISHER, D. C., 1986. Lightning strike fusion: extreme reduction and metal-silicate liquid immiscibility. *Science*, **234**, 189–93.

ESTNER, 1795–7. Versuch der Mineralogie. 3 vols. *Wien.*

EVANS, B. W., 1990. Phase relations of epidote-blueschists. *Lithos*, **25**, 3–23.

EVANS, B. W. & BROWN, E. H., 1987. Reply on blueschists and eclogites. *Geology*, **15**, 773–4.

EVDOKIMOV, M. D., 1982. Fenites of Tur'insk alkali complex, Kola Peninsula (in Russian). *Leningrad: Leningrad University Press.*

FAIRBAIRN, H. W., 1935. Notes on the mechanics of rock foliation. *Journal of Geology*, **43**, 591–608.

FAIRBAIRN, H. W., 1949. Structural Petrology of Deformed Rocks. *Cambridge: Addison-Wesley.*

FAY, A. H., 1920. A glossary of the mining and mineral industry. *Bulletin of the US Bureau of Mines*, No. 95.

FENNER, C. N. F., 1937. A view of magmatic differentiation. *Journal of Geology*, **45**, 158–68.

FEOFILAKTOV, K., 1851. On the crystalline rocks of Kiev, Volhynia and Podolsk territories (in Russian). *Trudy kommisii vysochashe uchrezhdenoi pri Imperatorskom universitete Sv. Vladimira*, 1–32.

FERBER, J. J., 1776. Lettres sur la minéralogie et sur divers autres objets de l'histoire naturelle de l'Italie. Translated from the German by B. de Dietrich. *Strasbourg: Bauer et Treuttel.*

FERMOR, L. L., 1909. The manganese-ore deposits of India. Ch. 15, The gondite series. *Memoirs of the Geological Survey of India*, **37**, 1–1294.

FERMOR, L. L., 1927. Kata-metamorphism or hypo-metamorphism? *Geological Magazine*, **64**, 334–6.

FERSMAN, A. E., 1922. Geochemistry of Russia (in Russian). *Petrograd: Nauchnoe Kimichesko-Tekhnicheskoe izdatel'stvo (NKTI).*

FIEDLER, A., 1936. Über Verflößungserscheinungen von Amphibolit mit diatektischen Lösungen im östlichen Erzgebirge. *Tschermak's mineralogisch-petrographische Mitteilungen*, **47**, 470–516.

FISCHER, L. H., 1860. Neue Mineralien im Scharzwald. *Neues Jahrbuch für Mineralogie, Geognosie, Geologie und Petrefakten-Kunde*, 795–7.

FISCHER, L. H., 1861. Über den Kinzigit. *Neues Jahrbuch für Mineralogie, Geognosie, Geologie und Petrefakten-Kunde*, 641–54.

FIŠERA, M., 1968. Problémy geneze amfibolitů. Unpublished M.Sc. thesis, Charles University, Prague.

FIŠERA, M., 1973. Petrografie kutnohorského krystalinika mezi Plaňany a Kolinem. Unpublished Ph.D. thesis, Charles University, Prague.

FITCH, A. A., 1931. Barite and witherite from near El Portal, Mariposa County, California. *American Mineralogist*, **16**, 461–8.

FLAWN, P. T., 1951. Nomenclature of epidote rocks. *American Journal of Science*, **249**, 769–77.

FLEISCHER, M. & MANDARINO, J. A., 1991. Glossary of Mineral Species. *Tucson: The Mineralogical Record Inc.*

FLETT, J. S., 1907. In W. A. E. Ussher, The Geology of the Country Around Plymouth and Liskeard. *HMSO: Geological Survey of England and Wales,* Memoir 348.

FLETT, J. S., 1908. In H. Kynaston & H. B. Hill (eds.), The Geology of the Country near Oban and Dalmally, Explanation of Sheet 45. *Memoirs of the Geological Survey of Scotland.*

FLETT, J. S., 1910. In B. N. Peach *et al.* (eds.), The Geology of the Neighbourhood of Edinburgh, Sheet 32 with Part of 31. *Memoirs of the Geological Survey of Scotland.*

FLETT, J. S., 1912. Petrography of the older igneous rocks and hornfelses. In B. N. Peach, W. Gunn, C. T. Clough, L. W. Hinxman, C. B. Crampton & E. M. Anderson, (eds.) The Geology of Ben Wyvis, Carn Chuinneag, Inchbae and the Surrounding Countryside. *HMSO: Memoirs of the Geological Survey of Scotland.*

FOLK, R. L., 1965. Some aspects of recrystallisation in ancient limestones. In L. C. Pray & R. C. Murray (eds.), Dolomitization and Limestone Diagenesis. *Society of Economists, Paleontologists and Mineralogists,* Special Publication **13**, 14–48.

FOUCAULT, A. & RAOULT, J.-F., 1984. Dictionnaire de géologie, 2nd edn. *Paris: Masson.*

FOUCAULT, A. & RAOULT, J.-F., 2005. Dictionnaire de géologie, 6th edn. *Paris: Dunod.*

FOURNET, J., 1847. Résultats sommaires d'une exploration des Vosges. *Bulletin de la Société géologique de France*, 2e sér., **4**, 220–54.

FRANCHI, S., 1902. Ueber Feldspath-Uralitizirung der Natron-Thonerde-Pyroxene aus den eklogitischen Glimmerschiefern der Gebirge Biella (Graiische Alpen). *Neues Jahrbuch für Mineralogie, Geologie und Paläontologie*, **2**, 112–22.

FREDRIKSSON, K., DE CARLI, P. & AARAMÄE, A., 1963. Shock-induced veins in chondrites. In W. Priester (ed.), Space Research, III. *Amsterdam: North Holland Publishing Co.*, pp. 974–83.

FRENCH, B. M., 1968. Shock metamorphism as a geological process. In B. M. French & N. M. Short (eds.), Shock Metamorphism of Natural Materials. *Baltimore: Mono Book Corporation*, pp. 1–17.

FRENCH, B. M., 1998. Traces of Catastrophe: A Handbook of Shock-metamorphic Effects in Terrestrial Meteorite Impact Structures. LPI Contribution No. 954, *Houston: Lunar and Planetary Institute.*

FRENCH, B. M. & SHORT, N. M., 1968. Shock Metamorphism of Natural Materials. *Baltimore: Mono Book Corporation.*

FRENTZEL, A., 1911. Das Passauer Granitmassiv. *Geognostische Jahreshefte, München*, **24**, 105–92.

FRENZEL, G., IROUSCHEK-ZUMTHOR, A. & STÄHLE, V., 1989. Stosswellenmetamorphose, Aufschmelzung und Verdampfung bei Fulguritbildung an exponierten Berggipfeln. *Chemie der Erde*, **49**, 265–86.

FREY, M. (ed.), 1987a. Low Temperature Metamorphism. *Glasgow and London: Blackie.*

FREY, M., 1987b. Very low-grade metamorphism of clastic sedimentary rocks. In M. Frey (ed.), Low Temperature Metamorphism. *Glasgow and London: Blackie*, pp. 9–58.

FREY, M. & ROBINSON, D., 1999. Low-grade Metamorphism. *Oxford: Blackwell Scientific Publications.*

FRIEDMAN, G. M., 1956. The origin of spinel-emery deposits, with particular reference to those of the Cortlandt Complex, New York. *New York State Museum Bulletin*, 351.

FRIEDMAN, G. M., 1965. Terminology of crystallization textures and fabrics in sedimentary rocks. *Journal of Sedimentary Petrology*, **35**, 643–55.

FRITSCH, W., MEIXNER, H. & WIESENEDER, H., 1967. Zur quantitativen Klassifikation der kristallinen Schiefer. *Neues Jahrbuch für Mineralogie*, Monatsheft, **12**, 364–76.

FRITZ, J., GRESHAKE, A. & STÖFFLER, D., 2005. Ejection of Martian meteorites. *Meteoritics and Planetary Science*, **40**, 1393–1411.

FYFE, W. S., 1967. Metamorphism in mobile belts: the glaucophane schist problem. *Transactions Leicester Literary and Philosophical Society*, **61**, 36–54.

FYFE, W. S. & TURNER, F. J., 1966. Reappraisal of the metamorphic facies concept. *Contributions to Mineralogy and Petrology*, **12**, 354–64.

Fyfe, W. S., Turner, F. J. & Verhoogen, J., 1958. Metamorphic reactions and metamorphic facies. *Geological Society of America Memoir*, **73**, 1–259.

Fyodorov, E. S., 1903. Short presentation of the results of mineralogical and petrographical investigation of the White Sea shores. *Verhandlungen der Russisch-Kaiserlichen Mineralogischen Gesellschaft St Petersburg*, **40**, 211–20.

Gansser, A., 1964. Geology of the Himalayas. *London: Wiley Interscience.*

Gavrusevich, B. A., 1931. The problem of the genesis of Ukranian pelikanites (in Russian). *Travaux de l'Institut Minéralogique de l'Académie des Sciences de l'URSS*, **1**, 91–102.

Geijer, P., 1944. Omfattningen av termen leptit. *Geologiska Föreningens i Stockholm Förhandlingar*, **66** (3), 733–45.

Geijer, P. & Magnusson, N. H., 1944. De mellansvenska järnmalmernas geologi. *Sveriges Geologiska Undersökning*, ser. Ca., **35**, 1–654.

Geikie, A., 1879. Geology. In Encyclopedia Britannica, 9th edn. *London.*

Geikie, A., 1882. Textbook of Geology. *London: Macmillan.*

Geikie, A., 1903. Textbook of Geology. *London: Macmillan.*

Geoffroy, P. R. & Souza Santos, T. D., 1942. Nota sobre a geologia de Apiaì, S. Paulo. *Mineraçâo e Metalurgia*, **6**, (33), 109–10.

Gidon, M., 1987. Les structures tectoniques. *Éditions du Bureaude Recherches géologiques et minières*, Manuels et méthodes no. 15.

Gilbert, G. K., 1893. The Moon's face, a study of the origin of its features. *Philosophical Society of Washington, Bulletin*, **12**, 241–92.

Glass, B. P., 1990. Tektites and microtektites: key facts and inferences. In L. O. Nicolaysen & W. U. Reimold (eds.), Cryptoexplosions and catastrophes in the geological record, with a special focus on the Vredefort structure. *Tectonophysics*, **171**, 259–73.

Glass, B. P. & Burns, C. A., 1988. Mikrokrystites: a new term for impact-produced glassy spherules containing primary crystallites. *Lunar and Planetary Science 18th Conference, Proceedings, New York: Pergamon Press*, 455–8.

Glocker, E. F., 1845. Ueber den Saccharit. *Journal für praktische Chemie*, **34**, 494–501.

Goldschmidt, V. M., 1911. Die Kontaktmetamorphose im Kristianiagebiet. *Skrifter udgit av Videnskabs-Selskabet i Kristiania. I. Matematisk-naturvidenskabelig Klasse*, No. 10.

Goldschmidt, V. M., 1921. Die Injektionsmetamorphose im Stavanger-Gebiete. *Skrifter udgit av Videnskabsselskabet i Kristiania, I. Matematisk-naturvidenskabelig Klasse*, No. 11.

Goldsmith, R., 1959. Granofels, a new metamorphic rock name. *Journal of Geology*, **67**, 109–10.

Goodspeed, G. E., 1947. Xenoliths and skialiths. *Geological Society of America Bulletin*, **58**, 1251.

Goodspeed, G. E., 1949. Xenoliths and skialiths. *American Journal of Science*, **246**, 515–25.

Goodspeed, G. E. & Coombs, H. A., 1937. Replacement breccias of the Lower Keechelus. *American Journal of Science*, **234**, 12–23.

Gosselet, J., 1888. L'Ardenne. *Mémoire du Service de la Carte géologique de la France.*

Grabau, A. W., 1904. On the classification of sedimentary rocks. *American Geologist*, **33**, 228–47.

Grabau, A. W., 1920. A Textbook of Geology, Part 1. *Boston: Heath.*

Graham, R. A., 1993. Solids Under High-pressure Shock Compression: Mechanics, Physics and Chemistry. *New York, Berlin, Heidelberg: Springer.*

Grange, L. I., 1927. On the 'rodingite' of Nelson. *Transactions and Proceedings of the New Zealand Institute*, **58**, 160–6.

Grapes, R. H., 2003. Pyrometamorphic breakdown of cordierite-muscovite intergrowths. *Mineralogical Magazine*, **67**, 653–63.

Greenly, E., 1919. The Geology of Anglesey, 2 vols. *Memoirs of the Geological Survey of Great Britain.*

Grieve, R. A. F., 1987. Terrestrial impact structures. *Earth and Planetary Science, Annual Review*, **15**, 245–70.

Grieve, R. A. F., 1998. Extraterrestrial impacts on Earth: the evidence and the consequences. In M. M. Grady, R. Hutchinson, G. J. H. McCall & D. A. Rothery (eds.), Meteorites: Flux with Time and Impact Effects. *London: Geological Society, Special Publication* 140, pp. 105–31.

Grieve, R. A. F., Dence, M. R. & Robertson, P. B., 1977. Cratering processes as interpreted from the occurrence of impact melts. In D. J. Roddy, R. O. Pepin & R. B. Merrill (eds.),

Impact and Explosion Cratering. *New York: Pergamon Press*, 791–814.

Grieve, R. A. F., Sharpton, V. L. & Stöffler, D., 1990. Shocked minerals and the K/T controversy. *EOS, Transactions of the American Geophysical Union*, **71**, 1792.

Grout, F. F., 1932. Petrography and petrology. *New York, London: McGraw-Hill.*

Grubenmann, U., 1904. Die kristallinen Schiefer. I. Allgemeiner Teil. *Berlin: Borntraeger.*

Grubenmann, U., 1907. Die kristallinen Schiefer. II. Spezieller Teil. *Berlin: Borntraeger.*

Grubenmann, U. & Niggli, P., 1924. Die Gesteinsmetamorphose. Dritte, völlig umgearbeitete Auflage von die kristallinen Schiefer, eine Darstellung der Erscheinungen der Gesteinsmetamorphose und ihrer Produkte von U. Grubenmann, I. Allgemeiner Teil. *Berlin: Borntraeger.*

Gryaznov, O. N., 1992. Ore-bearing Metasomatic Formations of Folded Belts (in Russian). *Moscow: Nedra Press.*

Gueirard, S., 1957. Sur l'origine de la 'collobriérite' du massif des Maures (Var). *Comptes rendus des séances de l'Académie des Sciences, Paris*, **245**, 2339–41.

Guggenheim, S., Bain, D. C., Bergaya, F. *et al.*, 2002. Report of the Association internationale pour l'étude des Argiles (AIPEA) Nomenclature Committee for 2001: Order, disorder and crystallinity in phyllosilicates and the use of the 'crystallinity' index. *Clays and Clay Minerals*, **50**, 406–9.

Gümbel, C.W. von, 1868. Geognostische Beschreibung des ost-bayerischen Grenzgebirges oder des bayerischen und des oberpfälzer Waldgebirges. *Gotha: Perthes.*

Gümbel, C. W. von, 1870. Über den Riesvulkan und über vulkanische Erscheinungen im Rieskessel. *Sitzungsberichte der mathematisch-physikalischen Klasse der Bayerischen Akademie der Wissenschaften*, 153–200.

Gümbel, C. W. von, 1874. Die paläolithischen Eruptivgesteine des Fichtelgebirges. *München: Weiss.*

Gümbel, C. W. von, 1888. Geologie von Bayern, 1er Theil: Grundzüge der Geologie. *Kassel: Fischer.*

Gürich, G., 1905. Granit und Gneis. Ein Beitrag zur Lehre der Entstehung der Gesteine. *Himmel und Erde, Berlin*, **17**, 241–51.

Gutzweiler, E., 1912. Injektionsgneise aus dem Kanton Tessin. *Eclogae geologicae Helvetiae*, **12**, 5–64.

Hagge, R., 1871. Mikroskopische Untersuchungen über Gabbro und verwandte Gesteine. *Kiel: Verlag der Universitätsbuchhandlung.*

Haidinger, K., 1787. Systematische Eintheilung der Gebirgsarten. *Wien: Wappler.*

Haidinger, W., 1845. Handbuch der bestimmenden Mineralogie. *Wien: Braumüller-Seidel.*

Hall, A. L., 1910. The Geology of the country west and north-west of Lydenburg, including western Sekukuniland. *Report of the Geological Survey of Transvaal, 1909, Union of South Africa Mines Department*, 53–72.

Hameurt, J., 1967. Les terrains cristallins et cristallophylliens du versant occidental des Vosges. *Mémoire du Service de la Carte géologique d'Alsace et de la Lorraine*, **26**, 1–402.

Hammer, W., 1914. Das Gebiet der Bündnerschiefern im Tirolischen Oberinntal. *Jahrbuch der Kaiserlich-königlichen geologischen Reichsanstalt, Wien*, **64**, 443–567.

Hammer, W., 1925. Einige Ergebnisse der geologischen Landesaufnahme in den Westtiroler Zentralalpen. *Geologische Rundschau*, **16**, 147–60.

Hanmer, S. & Passchier, C., 1991. Shear-sense Indicators: A Review. *Geological Survey of Canada*, Paper 90–17.

Harker, A., 1885. Slaty cleavage and allied rock structures, with special reference to the mechanical theories of their origin. *British Association for the Advancement of Science*, Report 1885–6, 813–52.

Harker, A., 1889. Notes on the physics of metamorphism. *Geological Magazine*, Decades III, **6**, 15–20.

Harker, A., 1902. Pétrographie. Introduction à l'étude des roches au moyen du microscope. *Paris: Librairie Polytechnique Ch. Béranger.*

Harker, A., 1903. The overthrust Torridonian rocks of the Isle of Rhum and the associated gneisses. *Quarterly Journal of the Geological Society of London*, **59**, 189–216.

Harker, A., 1904. The Tertiary Igneous Rocks of Skye. *Glasgow: Memoirs of the Geological Survey of the United Kingdom.*

Harker, A., 1909. The Natural History of Igneous Rocks. *London: Methuen.*

Harker, A., 1918. The anniversary address of the President. *Quarterly Journal of the Geological Society of London*, **74**, 50–85.

Harker, A., 1932. Metamorphism: A Study of the Transformation of Rock Masses, 1st edn. *London: Methuen.*

Harker, A., 1939. Metamorphism, 2nd edn. *London: Methuen.*

Harker, A., 1950. Metamorphism, 3rd edn. *London: Methuen.*

Harker, A., 1954. Petrology for Students, 8th edn. *Cambridge University Press.*

Harley, S. L., 1989. The origins of granulites: a metamorphic perspective. *Geological Magazine*, **126**, 215–47.

Harper, C. T., 1967. On the interpretation of potassium-argon ages from Precambrian shields and Phanerozoic orogens. *Earth and Planetary Science Letters*, **3**, 128–32.

Harrassowitz, H., 1927. Anchimetamorphose, das Gebiet zwischen Oberflächen- und Tiefenumwandlung der Erdrinde. *Bericht der oberhessischen Gesellschaft für Natur- und Heilkunde zu Giessen, Naturwissenschaftliche Abteilung* (1928–9), **12**, 11–17.

Harte, B., 1977. Rock nomenclature with particular relation to deformation and recrystallization textures in olivine-bearing xenoliths. *Journal of Geology*, **85**, 279–88.

Harte, B. & Hudson, N. F. C., 1979. Pelite facies series and the temperatures and pressures of Dalradian metamorphism in E Scotland. In A. L. Harris, C. H. Holland & B. E. Leake (eds.), The Caledonides of the British Isles, Reviewed. *Edinburgh: Scottish Academic Press*, pp. 323–37.

Hartmann, W. K., 1973. Ancient lunar megaregolith and subsurface structure. *Icarus*, **18**, 634–6.

Hartmann, W. K., 2003. Megaregolith evolution and cratering cataclysm models. Lunar cataclysm as a misconception (28 years later). *Meteoritics and Planetary Science*, **38**, 579–93.

Hashimoto, M., 1966. On the prehnite-pumpellyite metagraywacke facies (in Japanese). *Journal of the Geological Society of Japan*, **72**, 253–65.

Hatch, F. H., 1888. Glossary of terms. In J. J. G. Teall (ed.), British Petrography. *London: Dulau & Co.*, pp. 423–51.

Hatch, F. H. & Rastall, R. H., 1913. The Petrology of the Sedimentary Rocks: A Description of the Sediments and their Metamorphic Derivatives. *London: Allen & Co.*

Hausmann, J. F. L., 1828. Handbuch der Mineralogie. *Göttingen: Vandenhoeck-Ruprecht.*

Haüy, R. J., 1822. Traité de minéralogie, 2nd edn, vol. 4. 4 vols. *Paris: Bachelier.*

Hawkins, J., 1822. On the nomenclature of the Cornish rocks. *Transactions of the Royal Geological Society of Cornwall*, **2**, 145–58.

Heiken, G., Vaniman, D. & French, B. M., 1991. Lunar Sourcebook: A User's Guide to the Moon. *New York: Cambridge University Press.*

Heim, A., 1891. Geologie der Hochalpen zwischen Reuss und Rhein. *Text zur geologischen Karte der Schweiz in 1:100 000, Blatt 14.*

Henkes, L. & Johannes, W., 1981. The petrology of a migmatite (Arvika, Värmland, western Sweden). *Neues Jahrbuch für Mineralogie*, Abhandlungen, **141**, 113–33.

Hentschel, H., 1943. Die kalksilikatischen Bestandmassen in den Gneisen des Eulengebirges (Schlesien). *Mineralogische und petrographische Mitteilungen*, **55**, H.1–3, 1–136.

Hentschel, H., 1951. Die Umbildung basischer Tuffe zu Schalsteinen. *Neues Jahrbuch für Mineralogie*, Abhandlungen, **82**, 199–230.

Herodotos, around 440–420 BC. Histories (or Quest), 9 Libri. Translated into French by A. Barguet, 1964, *Paris: La Pléiade, Gallimard.*

Hess, F. L., 1919. Tactite, the product of contact metamorphism. *American Journal of Science*, **48**, 377–8.

Hetzel, W. H., 1938. Boetoniet, een bijzonder gesteente van het eiland Boeton (Z. O.-Celebes). *De Ingenieur in Nederlandsch-Indië*, sect. 4, **5**, 150–5. Cf. *Mineralogical Abstracts*, 1941, **8**, 1, 32.

Heusser, C. & Claraz, G., 1859. Ueber die wahre Lagerstätte der Diamanten und anderen Edelsteine in der Provinz Minas Geraes in Brasilien. *Zeitschrift der deutschen geologischen Gesellschaft, Berlin*, **11**, 448–66.

Hieke, O., 1945. I giacimenti di contatto del Monte Costone (Adamello meridionale). *Memorie dell'Istituto di Geologia dell'Università di Padova*, **15**, 1–46.

Hietanen, A., 1967. On the facies series in various types of metamorphism. *Journal of Geology*, **75** (2), 187–214.

Higgins, M. W., 1971. Cataclastic rocks. *US Geological Survey Professional Paper*, **687**, 1–97.

HOBBS, B. E., MEANS, W. D. & WILLIAMS, P. F., 1976. An Outline of Structural Geology. *New York: Wiley*.

HOLLAND, T. H., 1907. General report of the Geological Survey of India for the year 1906. *Geological Survey of India*, **35**, part 1.

HOLLISTER, L. S., 1969. Contact metamorphism in the Kwoiek area of British Columbia: an end member of the metamorphic process. *Geological Society of America Bulletin*, **80**, 2465–94.

HOLMES, A., 1920. The Nomenclature of Petrology. *London: Thomas Murby*. Also 1971 (facsimile), *New York: Hafner Publishing Co*.

HOLMES, A., 1928. The Nomenclature of Petrology, 2nd edn. *London: Thomas Murby*.

HOLMES, A., 1978. Holmes, Principles of Physical Geology, 3rd edn. *London: Nelson*.

HOLMES, A. & REYNOLDS, D. L., 1947. A front of metasomatic metamorphism in the Dalradian of County Donegal. *Bulletin de la Commission géologique de Finlande*, **140**, 25–65.

HOLMQUIST, P. J., 1908. Utkast till ett bergartsschema för urbergsskiffrarna. *Geologiska Föreningens i Stockholm Förhandlingar*, **30**, 269–93.

HOLMQUIST, P. J., 1909. Slutord; gneisfragan. *Geologiska Föreningens i Stockholm Förhandlingar*, **31**, 108–12.

HOLMQUIST, P. J., 1910. Die Hochgebirgsbildungen am Torne Träsk in Lappland. *Geologiska Föreningens i Stockholm Förhandlingar*, **32**, 913–83.

HOLMQUIST, P. J., 1920. Om pegmatitpalingenes och ptygmatisk veckning. *Geologiska Föreningens i Stockholm Förhandlingar*, **42**, 191–213.

HOLMQUIST, P. J., 1921. Typen und Nomenklatur der Adergesteine. *Geologiska Föreningens i Stockholm Förhandlingar*, **43**, 612–31.

HONNOREZ, J., MÉVEL, C. & MONTIGNY, R., 1984. Geotectonic significance of gneissic amphibolite from the Vema fracture zone, equatorial Mid-Atlantic Ridge. *Journal of Geophysical Research*, **89**, 11379–400.

HOOSON, W., 1747. Miners Dictionary. *Wrexham: Payne*.

HORNE, J., 1886. The origin of the andalusite-schists of Aberdeenshire. *Mineralogical Magazine*, **6**, 98–100.

HÖRZ, F., 1968. Statistical measurements of deformation structures and refractive indices in experimentally shock loaded quartz. In B. M. French & N. M. Short (eds.), Shock Metamorphism of Natural Materials. *Baltimore: Mono Book Corporation*, pp. 243–53.

HÖRZ, F., CINTALA, M. J., SEE, T. H. & LE, L., 2005. Shock melting of ordinary chondrite powders and implications for asteroidal regoliths. *Meteoritics and Planetary Science*, **40**, 1329–46.

HUBER, H. M., 1943. Physiographie und Genesis der Gesteine im südlichen Gotthardmassiv. *Schweizerische mineralogische und petrographische Mitteilungen*, **23**, 72–360.

HULL, E., KINAHAN, G. H., NOLAN, J. *et al.*, 1891. North-West and Central Doneyol, Explanation of Sheets 3, 4, 5, 9, 10, 11, 15 and 16 (in part). *Memoir of the Geological Survey of Ireland*.

HUMBLE, W., 1860. Dictionary of Geology and Mineralogy, Comprising Such Terms in Natural History as are Connected with the Study of Geology, 3rd edn. *London: Griffin and Co*.

HUMBOLDT, A. VON, 1823. Essai géognostique. *Paris: Levrault*.

HUMBOLDT, A. VON, 1831. Essai de géologie et climatologie asiatique, vol. 1. 2 vols. *Paris: Gide*.

HUME, W. F., HARWOOD, H. F. & THEOBALD, L. S., with petrographical studies by A. I. Awad, 1935. Notes on some analyses of Egyptian igneous and metamorphic rocks. *Geological Magazine*, **72**, 8–32.

HUMMEL, D., 1875. Om Sveriges lagrade urberg, jemförda med sydvestra Europas. *Kungliga svenska Vetenskaps-akademiens Handlingar*, **3**, part 2, 1–68.

HUNT, T. S., 1852. *Geological Survey of Canada, Annual Report*.

HUTCHINGS, W. M., 1890. Notes on the probable origin of some slates. *Geological Magazine*, **27**, 264–73 and 316–22.

IRVING, A., 1889. Chemical and Physical Studies in the Metamorphism of Rocks. *London: Longmans*.

IRVING, J. D., 1911. In H. F. Bain (ed.), Types of Ore Deposits. *San Francisco: Mining and Scientific Press*.

ISSEL, A., 1880. Osservazioni intorno a certe roccie anfiboliche della Liguria, a proposito d'una nota del Prof. Bonney concernente alcune serpentine della Liguria e della Toscana. *Bolletino del Regio Comitato Geologico d'Italia*, **11**, 183–92.

IVANOV, B. A. & ARTEMIEVA, N. A., 2002. Numerical modelling of the formation of large impact craters. In Catastrophic events and mass extinctions: Impact and beyond. *Geological Society of America, Special Paper*, **356**, 619–30.

Jacquet, E. & Michel-Lévy, A., 1886. Sur une roche anomale de la vallée d'Aspe (Basses-Pyrénées). *Comptes rendus de l'Académie des Sciences, Paris*, **102**, 523–5.

James, O. B., 1969. Shock and thermal metamorphism of basalt by nuclear explosions, Nevada Test site. *Science*, **166**, 1615–20.

Jameson, R., 1817. A Treatise on the External, Chemical, and Physical Characters of Minerals, 3rd edn. *Edinburgh: Constable*.

Jaupart, C. & Provost, A., 1985. Heat focussing, granite genesis and inverted metamorphic gradients in continental collision zones. *Earth and Planetary Science Letters*, **73**, 385–97.

Johannsen, A., 1931. A Descriptive Petrography of the Igneous Rocks, vol. 1. *Chicago University Press*.

Johannsen, A., 1932. A Descriptive Petrography of the Igneous Rocks, vol. 2. *Chicago University Press*.

Johannsen, A., 1939. A Descriptive Petrography of the Igneous Rocks, 2nd edn, vol. 1. *Chicago University Press*.

Johnston-Lavis, H. J., 1914. Saturation of minerals and genesis of igneous rocks. *Geological Magazine*, Decade 6, **1**, 381–3.

Jongmans, D. & Cosgrove, J. W., 1993. Observations structurales dans la région de Bastogne. *Annales de la Société géologique de Belgique*, **116**, 129–36.

Joplin, G. A., 1935. The exogenous contact zone at Ben Bullen, New South Wales. *Geological Magazine*, **72**, 385–400.

Joplin, G. A., 1968. A Petrography of Australian Metamorphic Rocks. *Sydney: Angus and Robertson*.

Judd, J. W., 1886. On the gabbros, dolerites and basalts of Tertiary age in Scotland and Ireland. *Quarterly Journal of the Geological Society of London*, **42**, 49–97.

Judd, J. W., 1889. On statical and dynamical metamorphism. *Geological Magazine*, **6**, 243–9.

Jung, J., 1958. Précis de pétrographie. *Paris: Masson*.

Jung, J., 1963. Précis de pétrographie, 2nd edn. *Paris: Masson*.

Jung, J. & Chenevoy, M., 1951. Sur la présence dans les Vosges d'un gisement de durbachite de Sainte-Croix-aux-Mines et sur l'origine de cette formation. *Comptes rendus de l'Académie des Sciences, Paris*, **232**, 868–9.

Jung, J. & Roques, M., 1936. Les zones d'isométamorphisme dans le terrain cristallophyllien du Massif Central français. *Revue des Sciences naturelles d'Auvergne*, n. sér., **2**, 38–84. Also in *Travaux du Laboratoire de Géologie de l'Université de Clermont-Ferrand*.

Jung, J. & Roques, M., 1952. Introduction à l'étude zonéographique des formations cristallophylliennes. *Bulletin du Service de la Carte géologique de la France*, **50** (235), 1–62.

Jung, J., Roques, M. & Richard, J., 1938. Les schistes cristallins du massif Central. *Bulletin du Service de la Carte géologique de France*, **39** (197), 120–48.

Jurine, L., 1806. Lettre de M. le Professeur Jurine, de Genève, à M. Gillet-Laumont, membre du Conseil des Mines, correspondant de l'Institut. *Journal des Mines, Paris*, **19**, 367–78.

Justi, J. H. G. von, 1757. Grundriss des gesamten-Mineralreiches worinnen alle Fossilien in einem ihren wesentlichen Beschaffenheite gemässen, Zusammenhänge vorgestellt und beschrieben werden. *Göttingen: Witwe Vandenhöck*.

Kalkowsky, E., 1886. Elemente der Lithologie. *Heidelberg: Winter*.

Karamata, S., 1968. Zonality in contact metamorphic rocks around the ultramafic mass of Brezovica (Serbia, Yugoslavia). *Report of the 23rd International Geological Congress, Prague*, sect. 1, 197–207.

Karamata, S., 1985. Metamorphism in the contact aureole of Brezovica (Serbia, Yugoslavia) as a model of metamorphism beneath obducted hot ultramafic bodies. *Bulletin de l'Académie Serbe des Sciences et des Arts, Classe des Sciences naturelles*, **90** (26), 51–68.

Karsten, D. L. G., 1800. Mineralogische Tabellen mit Rücksicht auf die neusten Entdeckungen. *Berlin: Rottman*.

Karsten, G., 1789. Museum Leskeanum. Regnum minerale, vol. 2. 2 vols. *Lipsiae (Leipzig): Muller*.

Kayser, E., 1870. Ueber die Contactmetamorphose der körnigen Diabase im Harz. *Zeitschrift der deutschen geologischen Gesellschaft*, **22**, 103–72.

KEIL, K., STÖFFLER, D., LOVE, S. G. & SCOTT, R. D., 1997. Constraints on the role of impact heating and melting of asteroids. *Meteoritics and Planetary Science*, **32**, 349–63.

KENNAN, P. S., 1986. The coticule package: a common association of some very distinctive lithologies. *Aardkund Mededelingen*, **3**, 139–48.

KERRICK, D. M., 1991. Overview in contact metamorphism. In D. M. Kerrick (ed.), Contact Metamorphism. *Mineralogical Society of America, Reviews in Mineralogy*, **26**, 1–12.

KERRICK, D. M., LASAGA, A. C. & RAEBURN, S. P., 1991. Kinetics of heterogeneous reactions. In D. M. Kerrick (ed.), Contact Metamorphism. *Mineralogical Society of America, Reviews in Mineralogy*, **26**, 583–671.

KESSLERN, P., 1921. Über Hochverwitterung und ihre Beziehungen zur Metharmose (Umbildung) der Gesteine. *Geologische Rundschau*, **12**, 237–70.

KIEFFER, S. W., 1971. Shock metamorphism of the Coconino sandstone at Meteor crater, Arizona. *Journal of Geophysical Research*, **76**, 5449–73.

KIEFFER, S. W., 1975. From regolith to rock by shock. *The Moon*, **13**, 301–20.

KIEFFER, S. W. & SIMONDS, C. S., 1980. The role of volatiles and lithology in the impact cratering process. *Review of Geophysics and Space Physics*, **18**, 143–81.

KIEFFER, S. W., SCHAAL, R., GIBBONS, R. *et al.*, 1976. Shocked basalt from Lonar impact crater, India, and experimental analogues. *Proceedings of the 7th Lunar Planetary Science Conference*, 1391–412.

KINAHAN, G. H., 1873. On the nomenclature of schistose rocks. *Geological Magazine*, **10**, 354–5.

KINAHAN, G. H., 1878. Manual of the Geology of Ireland. *London: Paul.*

KING, W. & FOOTE, R. B., 1865. The geology of sheet 79. *Memoir of the Geological Survey of India*, **4**, art. 2.

KIRWAN, R., 1794. Elements of Mineralogy, 2nd edn, vol. 1. *London: Nicholls.*

KISCH, H. J., 1983. Mineralogy and petrology of burial diagenesis (burial metamorphism) and incipient metamorphism in clastic rocks. In G. Larsen & G. V. Chilinar (eds.), Diagenesis in Sediments and Sedimentary Rocks, 2: Developments in Sedimentology 25B. *Amsterdam: Elsevier.*

KISCH, H. J., 1987. Correlation between indicators of very low-grade metamorphism. In M. Frey (ed.), Low Temperature Metamorphism. *Glasgow and London: Blackie*, pp. 227–300.

KISCH, H. J., 1991. Illite crystallinity: recommendations on sample preparation, X-ray diffraction settings, and interlaboratory samples. *Journal of Metamorphic Geology*, **9**, 665–70.

KISCH, H. J., ÁRKAI, P. & COVADONGA, B., 2004. Calibration of illite Kübler index (illite 'crystallinity'). *Schweizerische mineralogische und petrographische Mitteilungen*, **84**, 323–31.

KLÁPOVÁ, H., 1977. Metabazity strážeckého moldanubika. Unpublished M.Sc. thesis, Charles University, Prague.

KLAPROTH, M. H., 1795–7. Beytrage zur chemischen Kenntnis der mineral Körper, 2 vols. *Berlin.*

KLEMM, G., 1926. Petrographische Mitteilungen aus dem Odenwalde. *Notizblatt des Vereins für Erdkunde und der Hessichen geologischen Landesanstalt zu Darmstadt*, **9**, 104–17.

KNILL, J., 1960. A classification of cleavages, with special reference to the Craignish district of the Scottish Highlands. *Report of the 21st International Geological Congress, Norden-Copenhagen*, part 18, 317–25.

KNOPF, E. B., 1931. Metamorphism and phyllonitization. *American Journal of Science*, **21**, 1–27.

KNOPF, E. B., 1933. Petrotectonics. *American Journal of Science*, **25**, 433–70.

KNOPF, E. B. & INGERSON, E., 1938. Structural petrology. *Geological Society of America Memoir*, **6**, 1–270.

KOLODNY, Y. & GROSS, S., 1974. Thermal metamorphism by combustion of organic matter: isotopic and petrological evidence. *Journal of Geology*, **82**, 489–506.

KOONS, P. O., RUBIE, D. C. & FRUEH-GREEN, G., 1987. The effects of disequilibrium and deformation on the mineralogical evolution of quartz diorite during metamorphism in the eclogite facies. *Journal of Petrology*, **28**, 679–700.

KORNPROBST, J., 2002. Metamorphic Rocks and Their Geodynamic Significance. *Dordrecht: Kluwer Academic Publishers.*

KORZHINSKII, D. S., 1946. Metasomatic zonation near fractures and veins (in Russian). *Zapiski Vsesoyuznogo Mineralogicheskogo Obshchestva*, **75** (4), 321–32.

KORZHINSKII, D. S., 1953. Outline of metasomatic processes. In Basic Problems in Magmatogenic

Ore Deposits Theory (in Russian). *Moscow: Nauka Press*, pp. 334–456. German edn, 1965. Abriss der metasomatischen Prozesse. *Berlin: Akademie Verlag*.

Korzhinskii, D. S., 1957. Physico-chemical Basis for the Analysis of Mineral Parageneses (in Russian). *Moscow: Nauka Press* (2nd edn in 1973). English edn, 1959. Physico-chemical Basis for the Analysis of the Paragenesis of Minerals. *New York: Consultants Bureau*. French edn, Bases physico-chimiques de l'analyse des paragenèses de minéraux. *Bureau de Recherches Géologiques et Minières*, service d'information géologique, trad. 2294.

Korzhinskii, D. S., 1970. Theory of Metasomatic Zoning (translated from Russian by J. Argell). *Oxford: Clarendon Press*.

Kouznetsov, G., 1924. Geologico-petrographical outline and genesis of the Karachaevskoye silver-lead-zinc deposit (in Russian). *Matériaux pour la géologie appliquée*, **2**, 1–65.

Kramm, U., 1976. The coticule rocks (spessartine quartzites) of the Venn-Stavelot massif, Ardennes, a volcanoclastic metasediment? *Contributions to Mineralogy and Petrology*, **56**, 135–55.

Kretz, R., 1983. Symbols for rock-forming minerals. *American Mineralogist*, **68**, 277–9.

Kretz, R., 1994. Metamorphic Crystallization. *Chichester: Wiley & Sons*.

Kübler, B., 1967. La cristallinité de l'illite et les zones tout à fait supérieures du métamorphisme. In Étages tectoniques, colloque de Neuchâtel. *Neuchâtel: A la Baconnière*, pp. 105–21.

Kübler, B., 1968. Évaluation quantitative du métamorphisme par la cristallinité de l'illite. *Bulletin du Centre de Recherches de Pau, SNPA*, **2**, 385–97.

Kübler, B., 1984. Les indicateurs des transformations physiques et chimiques dans la diagenèse, température et calorimétrie. In M. Lagache (ed.), Thermobarométrie et barométrie géologiques. *Société française de Minéralogie et Cristallographie*, pp. 487–596.

Lacroix, A., 1889. Contributions à l'étude des gneiss à pyroxène et des roches à wernérite. *Bulletin de la Société de Minéralogie*, **12**, 83–364.

Lacroix, A., 1892. Sur l'axinite des Pyrénées, ses formes et les conditions de son gisement. *Comptes rendus de l'Académie des Sciences, Paris*, **115**, 739–41.

Lacroix, A., 1893a. Les enclaves des roches volcaniques. *Paris: Masson*.

Lacroix, A., 1893b. Minéralogie de la France et des colonies. *Paris: Béranger*. (New printing: Minéralogie de la France et de ses anciens territoires d'outre-mer. *Paris: Blanchard*)

Lacroix, A., 1897. Minéralogie de la France et de ses colonies, vol. 2. *Paris: Baudry*.

Lacroix, A., 1899. Le granite des Pyrénées et ses phénomènes de contact. *Bulletin du Service de la Carte géologique de la France*, **10** (64), 241–308.

Lacroix, A., 1910. Minéralogie de la France, vol. 4. *Paris: Baudry*.

Lacroix, A., 1914. Sur un nouveau type pétrographique (manjakite). *Bulletin de la Société française de Minéralogie*, **37**, 68–75.

Lacroix, A., 1916. Sur quelques roches volcaniques mélanocrates des possessions françaises de l'océan Indien et du Pacifique. *Comptes rendus de l'Académie des Sciences, Paris*, **163**, 177–83.

Lacroix, A., 1917. Sur un nouveau type ferrifère de schistes cristallins (collobriérite). *Bulletin de la Société française de Minéralogie*, **40**, 62–9.

Lacroix, A., 1920. Les roches éruptives du Crétacé pyrénéen et la nomenclature des roches éruptives modifiées. *Comptes rendus de l'Académie des Sciences, Paris*, **170**, 685–90.

Lacroix, A., 1933. Communication orale. *Compte rendu sommaire des séances de la Société géologique de France*, fasc. 6, 62.

Lacroix, A., 1936. Le volcan actif de l'île de la Réunion et ses produits. *Paris: Gauthier-Villars*.

Lacroix, A., 1939. La lamboanite, schiste cristallin à faciès gneissique dépourvu de quartz et la pegmatite à cordiérite qui l'accompagne à Ankaditany (S Madagascar). *Bulletin de la Société française de Minéralogie*, **62**, 289–98. Cf. *Mineralogical Abstracts*, 1947–9, **10** (2), 89.

Lambert, P., 1981. Breccia dikes: geological constraints on the formation of complex craters. In P. H. Schultz & R. B. Merrill (eds.), Multi-ring Basins. *Proceedings of the 12th Lunar and Planetary Science Conference. New York: Pergamon Press*, pp. 56–78.

Lambert, R. St. J., 1965. The metamorphic facies concept. *Mineralogical Magazine*, **34**, 283–91.

Landes, K. K., 1967. Eometamorphism, and oil and gas in time and space. *American Association of Petroleum Geologists*, **51**, 828–41.

Lane, A. C., 1903. Porphyritic appearance of rocks. *Geological Society of America Bulletin*, **14**, 385–406.

Lapparent, A. de, 1882. Traité de géologie. *Paris: Savy.*

Lapparent, A. de, 1893. Traité de géologie, 3rd edn, vol. 2. 2 vols. *Paris: Savy.*

Lapparent, J. de, 1937. L'émeri de Samos. *Mineralogische und petrographische Mitteilungen*, **49**, 1–30.

Lapworth, C., 1885. The Highland controversy in British geology: its causes, course, and consequences. *Nature*, **32**, 558–9.

Lasaulx, A. von, 1875. Elemente der Petrographie. *Bonn: Strauss.*

Lasaulx, A. von, 1878. Ueber den Saccharit. *Neues Jahrbuch für Mineralogie, Geologie und Paläontologie*, 623–9.

Leake, B. E. *et al.*, 1997. Nomenclature of amphiboles: Report of the Subcommittee on Amphiboles of the International Mineralogical Association, Commission on new Minerals and Mineral Names. *American Mineralogist*, **82**, 1019–37.

Le Fort, P., 1975. Himalayas: the collided range. Present knowledge of the continental arc. *American Journal of Science*, **275**, 1–44.

Lehmann, E., 1933. Das Nebengestein des Erzlagers Theodor bei Aumenau. *Neues Jahrbuch für Mineralogie, Geologie und Paläontologie*, Abt. A., **67**, 69–117.

Lehmann, J., 1884. Untersuchungen über die Entstehung der altkrystallinischen Schiefergebirge, mit besonderer Bezugnahme auf das Sächsische Granulitgebirge. *Bonn: Hochgürtel.*

Leith, C. K., 1905. Rock cleavage. *US Geological Survey Bulletin*, **239**, 1–216.

Leith, C. K., 1923. Structural geology. *New York: Holt.*

Le Maitre, R. W. (ed.), 1989. A Classification of Igneous Rocks and Glossary of Terms. Recommendations of the International Union of Geological Sciences, Subcommission on the Systematics of Igneous Rocks. *Oxford: Blackwell Scientific Publications.*

Le Maitre, R. W. (ed.), 2002. A Classification of Igneous Rocks and Glossary of Terms. *Cambridge University Press.*

Lemberg, J., 1872. Ueber die Kontaktbildungen bei Predazzo. *Zeitschrift der deutschen geologischen Gesellschaft*, **24**, 187–264.

Lemeyrie, A., 1878. Éléments de minéralogie et de lithologie, ouvrage complémentaire des Éléments de géologie, 3rd edn. *Paris: Masson.*

Lempe, J. F., 1785. Beschreibung des Bergbaues aus dem Sächsischen Zinnwalde. *Magazin der Bergbaukunde, Waltherische Hofbuchhandlung, Dresden*, 1.

Leonhard, K. C. von, 1823. Charakteristik der Felsarten, vol. 1. *Heidelberg: Engelmann.*

Leonhard, K. C. von, 1824. Charakteristik der Felsarten, vol. 2. *Heidelberg: Engelmann.*

Leonhard, K. C. von, 1832. Die Basaltgebilde in ihre Beziehungen zu normalen und abnormen Felsmassen. 2 vols. *Stuttgart: Engelmann.*

Lieber, O., 1860. Der Itacolumit, seine Begleiter und die Metallführung desselben. In B. von Cotta & H. Müller (eds.), Gangstudien oder Beiträge zur Kenntniss der Erzgänge, vol. 3. *Freiberg: Engelhardt*, pp. 359–71.

Lightfoot, B., 1929. The geology of the central part of the Wankie coalfield. *Bulletin of the Geological Survey of S-Rhodesia*, **15**, 1–61.

Limur, Cte de, 1884. Catalogue raisonné des minéraux du Morbihan. *Vannes: Galles.*

Lindacker, J. T., 1792. Beschreibung einer harten, im Bruche dichtfaserigen Steinart, die ich Faserkiesel nenne. *Sammlung physikalischer Aufsätze* (published by J. Mayer), **2**, 277–80.

Lindgren, W., 1912. The nature of replacement. *Economic Geology*, **7**, 521–35.

Lindgren, W., 1919. Mineral Deposits, 2nd edn. *New York: McGraw-Hill.*

Lindgren, W., 1925. Metasomatism. *Geological Society of America Bulletin*, **36**, 247–62.

Lindgren, W., 1928. Mineral Deposits, 3rd edn. *New York: McGraw-Hill.*

Lindgren, W., 1933. Mineral Deposits, 4th edn. *New York: McGraw-Hill.*

Liou, J. G., Maruyama, S. & Cho, M., 1985. Phase equilibria and mineral parageneses of metabasites in low-grade metamorphism. *Mineralogical Magazine*, **49**, 321–33.

Liou, J. G., Maruyama, S. & Cho, M., 1987. Very low-grade metamorphism of volcanic and volcanoclastic rocks: Mineral assemblages and mineral facies. In M. Frey (ed.), Low Temperature Metamorphism. *Glasgow and London: Blackie*, pp. 59–113.

List, K., 1852. Chemisch-mineralogische Untersuchungen der Taunusschiefer. *Annalen der Chemie und Pharmacie*, **5**, 181–205.

Lister, G. S. & Snoke, A. W., 1984. S-C mylonites. *Journal of Structural Geology*, **6**, 617–38.

Loewinson-Lessing, F., 1888. Zur Bildungsweise und Classification der klastischen Gesteine. *Mineralogische und petrographische Mitteilungen*, **9**, 528–39.

Loewinson-Lessing, F., 1893–4. Petrographisches Lexicon. Repertorium der petrographischen Termini und Benennungen. *Jurjew (Dorpat, now Tartu), Estonia: Mattiesen* (with suppl. 1898).

Loewinson-Lessing, F., 1899. Studien über die Eruptivgesteine. *Comptes rendus du 7e Congrès géologique international, St-Pétersbourg*, 193–464.

Loewinson-Lessing, F., 1905. Eine petrographische Excursion auf dem Tagil (in Russian). *Zapiski imperatorskago Sankt-Peterburgskago mineralogicheskago obshchestva*, **43**, 543–86. Also *Izvestiya Sankt-Peterburgskogo politekhnicheskogo instituta*, **3**, 1–40.

Loewinson-Lessing, F., 1911. On the chemical characteristics of feldspar amphibolites (in Russian). *Izvestiya Sankt-Peterburgskogo politekhnicheskogo instituta*, **15**, 559–73.

Loewinson-Lessing, F., 1925. Petrography. *Leningrad: ONTI.*

Loewinson-Lessing, F., 1933. Petrography, 3rd edn. *Leningrad: ONTI.*

Loewinson-Lessing, F., 1949. The problem of the origin of magmatic rocks, new edn (first publ. in 1934). In Selected Works. *Moscow and Leningrad: Academy of Science of the USSR*, pp. 420–55.

Loewinson-Lessing, F. & Struwie, E. A., 1937. Petrography (in Russian), 2nd edn. *Leningrad: ONTI.*

Loewinson-Lessing, F. & Struwie, E. A., 1963. Petrography (in Russian), new edn. *Moscow: Gosgeoltechizdat.*

Logvinenko, N. V., 1968. Postdiagenetic Changes in Sedimentary Rocks (in Russian). *Leningrad: Nauka Press.*

Lohest, M., 1909. De l'origine des veines et des géodes des terrains primaires de Belgique, 3e note. *Annales de la Société géologique de Belgique*, **36**, B.275–82.

Lohest, M., Stainier, X. & Fourmarier, P., 1908. Compte rendu de la session extraordinaire de la Société géologique de Belgique tenue à Eupen et à Bastogne les 29, 30 et 31 août et 1er, 2 et 3 septembre 1908. *Annales de la Société géologique de Belgique*, **35** (25), B.365–80.

Lorenz, W., 1980. Petrographische Nomenklatur metamorpher Gesteine. 1. Grundlage, Konzeption, Systematik. *Zeitschrift für geologische Wissenschaften, Berlin*, **12**, 1479–509.

Lorenz, W., 1981. Petrographische Nomenklatur metamorpher Gesteine. 2. Klassifikatorische und nomenklatorische Spezialfragen. *Zeitschrift für geologische Wissenschaften, Berlin*, **9**, 137–56.

Lorenz, W., 1996. Gneisses and mica schists. *Chemie der Erde*, **56**, 79–84.

Lorenz, W., 1998. Granulites and non-granulites: a contribution to the classification and nomenclature of metamorphic rocks (2). *Chemie der Erde*, **58**, 98–118.

Loretz, H., 1881. Beitrag zur geologischen Kenntniss der cambrisch-phyllitischen Schieferreihe in Thüringen. *Jahrbuch der Königlich-Preussischen geologischen Landesanstalt und Bergakademie zu Berlin*, **2**, 175–257.

Lossen, K. A., 1869. Metamorphische Schichten aus der paläozoischen Schichtenfolge des Ostharzes. *Zeitschrift der deutschen geologischen Gesellschaft*, **21**, 281–340.

Lossen, K. A., 1872. Ueber den Spilosit und Desmosit Zincken's, ein Beitrag zur Kenntnis der Contactmetamorphose. *Zeitschrift der deutschen geologischen Gesellschaft*, **24**, 701–86.

Lossen, K. A., 1875. Die Porphyroide des Harz sind abnormale Schichtglieder des herzynischen Schiefergebirges. *Zeitschrift der deutschen geologischen Gesellschaft*, **27**, 967–71.

Lossen, K. A., 1884. Ueber die Anforderungen der Geologie an die petrographische Systematik. *Jahrbuch der königlichen preussischen geologischen Landesanstalt*, 486–513.

Lovering, T. S., 1941. The origin of the tungsten ores of Boulder County, Colorado. *Economic Geology*, **36**, 229–79.

Lovering, T. S., 1949. Rock alteration as a guide to ore, East Tintic District, Utah. *Economic Geology Monograph*, **1**, 1–64.

Luchitzky, V., 1922. Textbook of Petrology. *Moscow.*

Lyell, C., 1830. Principles of Geology. *London: Murray.*

Lyell, C., 1833. Principles of Geology being an Attempt to Explain the Former Changes of the Earth's Surface, by Reference to Causes Now in Operation. *London: Murray.*

Lyell, C., 1835. Principles of Geology. *London: Murray.*

Lyell, C., 1851. A Manual of Elementary Geology. *London: Murray.*

Lyell, C., 1854. Principles of Geology, or the Modern Changes of the Earth and its Inhabitants Considered as Illustration of Geology, 9th edn, vol. 3. *New York: Appleton.*

Macculloch, J., 1819. A Description of the Western Islands of Scotland, Including the Isle of Man, comprising an Account of their Geological Structure; with Remarks on their Agriculture, Scenery and Antiquities, vol. 3. 3 vols. *London: Constable.*

Macculloch, J., 1821. Geological Classification of Rocks. *London: Longman, Hurst, Rees, Orme and Brown.*

MacGregor, A. G., 1931. Scottish pyroxene-granulite hornfelses and Odenwald beerbachites. *Geological Magazine*, **68**, 506–21.

Magnan, H., 1877. Matériaux pour l'étude stratigraphique des Pyrénées et des Corbières, notes rédigées en 1874. *Mémoire de la Société géologique de France*, 2e sér., **1**, 1–111.

Magnusson, N. H., 1936. The evolution of the lower Archæan rocks in Central Sweden and their iron, manganese, and sulphide ores. *Quarterly Journal of the Geological Society of London*, **92**, 332–59.

Malaroda, R., 1993. Permian anatexis and anatectonics in the southern Argentera (Maritime Alps). *Memorie della Società Geologica Italiana*, **49**, 257–71.

Mallet, F. R., 1875. On the geology and mineral resources of the Darjiling district and western Duars. *Geological Survey of India, Memoirs*, **11**, 1–50.

Mandarino, J. A. & Back, M. E., 2004. Fleischer's Glossary of Mineral Species, 9th edn. *Tucson: The Mineralogical Record Inc.*

Marignac, C., Semiani, A., Fourcade, S. *et al.*, 1996. Metallogenesis of the late Pan-African gold-bearing East Ouzzel shear zone (Haggar, Algeria). *Journal of Metamorphic Geology*, **14**, 783–801.

Marr, J. E., 1916. The Geology of the Lake District. *Cambridge University Press.*

Masaitis, V. L., 1994. Impactites from Popigai crater. In B. O. Dressler, R. A. F. Grieve & V. L. Sharpton (eds.), Large Meteorite Impacts and Planetary Evolution, *Geological Society of America, Special Paper*, **293**, 153–62.

Masaitis, V. L., Mikhailov, M. V. & Selivanovskaya, T. V., 1975. Popigai Meteorite Crater (in Russian). *Moscow: Nauka Press.*

Mashkovtsev, S. F., 1937. Metaquartzites (kazakhites) (in Russian). *17th International Geological Congress, Abstracts, ONTI Publications, Moscow.*

Mason, R., 1978. Petrology of the Metamorphic Rocks. *London: Allen & Unwin.*

Matthes, S. & Krämer, A., 1955. Die Amphibolite und Hornblendegneise im mittleren kristallinen Vor-Spessart und ihre petrogenetische Stellung. *Neues Jahrbuch für Mineralogie*, **88**, 225–72.

Matthes, S. & Okrusch, M., 1965. Spessart. *Berlin: Sammlung geologischer Führer, Borntraeger*, vol. 44.

Matthes, S., Okrusch, M., Röhr, C., Schüssler, U., Richter, P. & Gehlen, K. von, 1995. Talc-chlorite-amphibole felses of the KTB pilot hole, Oberpfalz, Bavaria; protolith characteristics and phase relationships. *Mineralogy and Petrology*, **52**, 25–59.

Mazzuoli, L. & Issel, A., 1881. Relazione degli studi fatti per un rilievo delle masse ofiolitiche nella Riviera di Levante (Liguria). *Bolletino del Regio Comitato Geologico d'Italia*, **12**, 313–49.

McIntyre, D. B., 1962. Impact metamorphism at Clearwater Lake, Quebec. *Journal of Geophysical Research*, **67**, 1647–53.

McIntyre, D. B. & Reynolds, D. L., 1947. Chilled and 'baked' edge as criteria of relative age. *Geological Magazine*, **84**, 61–4.

McLintock, W. F. P., 1932. On the metamorphism produced by the combustion of hydrocarbons in the Tertiary sediments of south-west Persia. *Mineralogical Magazine*, **23**, 207–26.

Medlicott, H. B., 1864. On the geological structure and relations of the southern portion of the Himalayan ranges between the rivers Ganges and Ravee. *Geological Survey of India, Memoirs*, **3** (2), 1–212.

Mehnert, K. R., 1968. Migmatites and the Origin of Granitic Rocks. *Amsterdam: Elsevier.*

Mehnert, K. R., 1972. Granulites. Results of a discussion, 2. *Neues Jahrbuch für Mineralogie*, Monatsheft, 139–50.

MELOSH, H. J., 1989. Impact Cratering: A Geological Process. *Oxford University Press.*

MEMPEL, G., 1935–6. Über den Begriff 'Adinol'. *Zentralblatt für Mineralogie, Geologie und Paläontologie*, Abt. A, 13–18.

MERRILL, G. P., 1891. Stones for Building and Decoration. *New York: Wiley.*

MERRILL, G. P., 1897. A Treatise on Rocks, Rock-weathering and Soils. *New York: Macmillan.*

MERRIMAN, R. J. & FREY, M., 1999. Patterns of very low-grade metamorphism in metapelitic rocks. In M. Frey & D. Robinson (eds.), Low-grade Metamorphism. *Oxford: Blackwell Scientific Publications*, pp. 61–107.

MERRIMAN, R. J. & PEACOR, D. R., 1999. Very low-grade metapelites: mineralogy, microfabrics and measuring reaction progress. In M. Frey & D. Robinson (eds.), Low-grade Metamorphism. *Oxford: Blackwell Scientific Publications*, pp. 10–60.

MEYER, J., 1983. Mineralogie und Petrologie des Allalingabbros. Unpublished Ph.D. thesis, Basel University.

MICHEL-LÉVY, A., 1874. Note sur une classe de roches éruptives intermédiaires entre les granites porphyroïdes et les porphyres granitoïdes; groupe des granulites. *Bulletin de la Société géologique de France*, **3**, 2, 177–89.

MICHEL-LÉVY, A., 1888. Sur l'origine des terrains cristallins primitifs. *Bulletin de la Société géologique de France*, 3e sér., **16**, 102–13.

MILCH, L., 1894. Beiträge zur Lehre von der Regionalmetamorphose. *Neues Jahrbuch für Mineralogie, Geologie und Paläontologie*, Beilage **9**, 101–28.

MILCH, L., 1917. Über Adinolen und Adinolschiefer des Harzes. *Zeitschrift der deutschen geologischen Gesellschaft*, **69**, 349–486.

MILCH, L., 1922. Die Umwandlung der Gesteine. In W. Salomon (ed.), Grundzüge der Geologie, 7th edn, vol. 1, Allgemeine Geologie, T.1 Innere Dynamik. *Stuttgart: Schweizerbart*, pp. 267–327.

MILTON, D. J. & DE CARLI, P. S., 1963. Maskelynite: formation by explosive shock. *Science*, **140**, 670–1.

MIYASHIRO, A., 1961. Evolution of metamorphic belts. *Journal of Petrology*, **2**, 277–311.

MIYASHIRO, A., 1972. Pressure and temperature conditions and tectonic significance of regional and ocean-floor metamorphism. *Tectonophysics*, **13**, 141–79.

MIYASHIRO, A., 1973a. Metamorphism and Metamorphic Belts. *London: Allen & Unwin.*

MIYASHIRO, A., 1973b. Paired and unpaired metamorphic belts. *Tectonophysics*, **17**, 241–54.

MIYASHIRO, A., 1994. Metamorphic Petrology. *London: University College Press.*

MIYASHIRO, A. & BANNO, S., 1958. Nature of glaucophanitic metamorphism. *American Journal of Science*, **256**, 97–110.

MIYASHIRO, A., SHIDO, F. & EWING, M., 1971. Metamorphism in the Mid-Atlantic Ridge near 24° and 30° N. *Transactions of the Royal Society of London*, **A268**, 589–603.

MÖHL, H., 1873. Die südwestlichsten Ausläufer des Vogelsgebirges. *Bericht des Offenbacher Vereins für Naturkunde*, **14**, 51–101.

MOORHOUSE, W. W., 1959. The Study of Rocks in Thin Section. *New York: Harper & Brothers.*

MORCHE, W., 1979. Petrographische und geochemische Untersuchungen am Durbachit des Scharzwaldes. Unpublished thesis, Universität Freiburg-im-Bresgau.

MORIMOTO, N., FABRIÈS, J., FERGUSON, A. K. *et al.*, 1988. Nomenclature of pyroxenes. *Mineralogical Magazine*, **52**, 535–50.

MORLOT, A. VON, 1847. Ueber Dolomit und seine künstliche Darstellung aus Kalkstein. Naturwissenschaftliche Abhandlungen, gesammelt und durch Subscription, herausgegeben von W. Haidinger. *Wien: Braunmuller und Seidel*, pp. 308–15.

MÜGGE, O., 1893. Untersuchungen über die 'Lenneporphyre' in Westfalen und den angrenzenden Gebieten. *Neues Jahrbuch für Mineralogie, Geologie und Paläontologie*, Beilageband **8**, 535–721.

MÜGGE, O., 1930. Bewegungen von Porphyroblasten in Phylliten und ihre Messung. *Neues Jahrbuch für Mineralogie, Geologie und Paläontologie*, Beilageband **61**, Abteilung A, 469–510.

MÜLLER, G., 1967. Diagenesis in argillaceous sediments. In G. Larsen & G. V. Chilingar (eds.), Diagenesis in Sediments. *Amsterdam: Elsevier*, pp. 127–77.

MULLIS, J., 1987. Fluid inclusion studies during very low-grade metamorphism. In M. Frey (ed.), Low Temperature Metamorphism. *Glasgow and London: Blackie*, pp. 162–99.

Murchison, R. I., 1839. Silurian System. *London: Murray*.

Nakovnik, N. I., 1954. Greisens: Altered Wall Rocks and Their Significance for Prospecting (in Russian). *Moscow: Gosgeoltekhizdat*, pp. 53–81.

Nakovnik, N. I., 1965. Secondary Quartzites of the USSR and Connected Mineral Deposits (in Russian). *Moscow: Nedra Press*, pp. 33–44.

Napione, C. A., 1797. Elementi di mineralogia. *Torino: Reale Stamperia*.

Nasir, S., 1993. Classification and nomenclature of metamorphic rocks. *Chemie der Erde, Jena*, **53**, 71–8.

Naumann, C. F., 1826. Lehrbuch der Mineralogie. *Leipzig: Engelmann*.

Naumann, C. F., 1849–53. Lehrbuch der Geognosie, 1st edn. *Leipzig: Engelmann*.

Naumann, C. F., 1858. Lehrbuch der Geognosie, 2nd edn, vol. 1. *Leipzig: Engelmann*.

Necker de Saussure, L. A., 1821. Voyage en Écosse et aux îles Hébrides, vol. 1. 3 vols. *Genève and Paris: Pachoud*.

Neuendorff, K. K. E., Mehl, J. P. Jr & Jackson, J. A. (eds.), 2005. Glossary of Geology, 5th edn. *Virginia: American Geological Institute*.

Neukum, G., Ivanov, B. & Hartmann, W. K., 2001. Cratering records in the inner solar system in relation to the lunar reference system. *Space Science Review*, **96**, 55–86.

Niggli, P., 1948. Gesteine und Minerallagerstätten. *Basel: Birkhäuser*.

Niggli, P., 1954. Rocks and Mineral Deposits. *San Francisco: Freeman*.

Noë-Nygaard, A., 1955. Comparaison entre les roches grenues appartenant à deux orogénies précambriennes voisines au Groenland. *Sciences de la Terre, Nancy*, **3**, 61–75.

Nöggerath, J. J., 1822. Das Gebirge in Rheinland-Westphalien, 1. *Bonn: Weber*.

Norton, W. H., 1917. Studies for students: a classification of breccias. *Journal of Geology*, **25**, 160–94.

Novarese, V., 1895. Nomenclatura e sistematica delle roccie verdi nelle Alpi Occidentali. *Bolletino del Regio Comitato Geologico d'Italia*, **26**, 3rd ser., (6), 164–81.

Nutall, T., 1822. Observations on the serpentine rocks of Hoboken, in New Jersey, and on the minerals which they contain. *American Journal of Science*, **4**, 16–23.

Nyquist, L. E., Bogard, D. D., Shih, C.-Y. *et al.*, 2001. Ages and geologic histories of Martian meteorites. *Space Science Review*, **96**, 105–64.

Och, D. J., Leitch, E. C., Caprarelli, G. & Watanabe, T., 2003. Blueschist and eclogite in tectonic mélange, Port Macquarie, New South Wales, Australia. *Mineralogical Magazine*, **67**, 609–24.

Oehlert, D. P., 1862. Notes géologiques sur le département de la Mayenne. *Angers: Germain et G. Grassin*.

Oen, Ing Sen, 1962. Hornblendic rocks and their polymetamorphic derivatives in area NW of Ivigtut, south Greenland. *Meddelelser om Grønland*, **6**, 169–84.

O'Keefe, J. & Ahrens, T. J., 1978. Impact flows and crater scaling on the Moon. *Physics of the Earth and Planetary Interiors*, **16**, 341–51.

Okrusch, M., Seidel, E., Kreuzer, H. & Harre, W., 1978. Jurassic age of metamorphism at the base of Brezovica peridotite (Yugoslavia). *Earth and Planetary Science Letters*, **39**, 291–7.

Omalius d'Halloy, J. J., 1828. Mémoires pour servir à la description géologique des Pays-Bas, de la France et de quelques contrées voisines. *Namur: Gérard*, pp. 3.

Omalius d'Halloy, J. J., 1831. Éléments de géologie. *Paris: Levrault*.

Omalius d'Halloy, J. J., 1868. Précis élémentaire de géologie, 8th edn. *Bruxelles: Murquardt, Paris: Savy*.

Omel'yanenko, B. I., 1978. Wall Rock Hydrothermal Alteration (in Russian). *Moscow: Nedra Press*.

Orphal, D. L., Borden, W. F., Larson, S. A. & Schultz, P. H., 1980. Impact melt generation and transport. *Proceedings of the 11th Lunar Planetary Science Conference*, 2309–23.

Ostertag, R., 1983. Shock experiments on feldspar crystals. *Proceedings of the 14th Lunar Planetary Science Conference, Journal of Geophysical Research*, **88** Suppl., B364–76.

Park, A. F., 1989a. Astrobleme. In D. R. Bowes (ed.), Encyclopedia of Igneous and Metamorphic Petrology. *New York: Van Nostrand Reinhold*, pp. 40–2.

Park, A. F., 1989b. Tektite. In D. R. Bowes (ed.), Encyclopedia of Igneous and

Metamorphic Petrology. *New York: Van Nostrand Reinhold*, p. 554.

PARK, R. G., 1983. Foundations of Structural Geology. *Glasgow & London: Blackie*.

PASSCHIER, C. W. & SIMPSON, C., 1986. Porphyroclast systems as kinematic indicators. *Journal of Structural Geology*, **8**, 831–43.

PASSCHIER, C. W. & TROUW, R. A. J., 1996. Microtectonics. *Berlin-Heidelberg-New York: Springer*.

PATERSON, M. S. & WEISS, L. E., 1961. Symmetry concepts in the structural analysis of deformed rocks. *Geological Society of America Bulletin*, **72**, 841–82.

PATTEN, H., 1888. Die Serpentin- und Amphibolgesteine nördlich von Marienbad in Böhmen. *Mineralogische und petrographische Mitteilungen*, **9**, 89–143.

PATTISON, D. R. M. & TRACY, R. J., 1991. Phase equilibria and thermobarometry of metapelites. In D. M. Kerrick (ed.), Contact metamorphism. *Mineralogical Society of America, Reviews in Mineralogy*, **26**, pp. 105–206.

PEACH, B. N., HORNE, J., GUNN, W., CLOUGH, C. T. & HINXMAN, L. W., 1907. The Geological Structure of the North-west Highlands of Scotland. *Memoirs of the Geological Survey of Great Britain*.

PECK, F. B., 1901. Preliminary notes on the occurrence of serpentine and talc at Easton, Pennsylvania. *Annals of the New York Academy of Sciences*, **13**, 419–30.

PECORA, W. T., 1960. Coesite craters and space geology. *Geotimes*, **5** (2), 16–19 and 32.

PERTSEV, N. N., 1977. High-temperature Metamorphism and Metasomatism of Carbonate Rocks (in Russian). *Moscow: Nauka Press*.

PERTSEV, N. N., 1991. Magnesian skarns (in Russian). In Skarns, their Genesis and Metallogeny. *Athens: Theophrastus Press*, pp. 299–324.

PEŠKOVÁ, J., 1973. Amfibolitové horniny moldanubika z oblasti Dolní Rožinky. Unpublished M.Sc. thesis, Charles University, Prague.

PETTERSEN, K., 1883. Sagvandit, eine neue enstatitführende Gebirgsart. *Neues Jahrbuch für Mineralogie, Geologie und Paläontologie*, **2**, 247.

PETTIJOHN, F. J., 1957. Sedimentary Rocks, 2nd edn. *New York: Harper*.

PETTIJOHN, F. J., 1975. Sedimentary Rocks. *New York: Harper and Row*.

PETZHOLDT, A., 1843. Beiträge zur Geognosie von Tyrol. Skizzen auf einer Reise durch Sachsen, Bayern, Salzkammergut, Salzburg, Tyrol, Österreich. *Leipzig: Weber*.

PHILLIPS, W., 1818. Geology of England and Wales. *Phillips*.

PHILLIPS, W., 1819. Mineralogy. *Phillips*.

PINKERTON, J., 1811a. Petralogy: A Treatise on Rocks, vol. 1. *London: White, Cochrane & Co*.

PINKERTON, J., 1811b. Petralogy: A Treatise on Rocks, vol. 2. *London: White, Cochrane*.

PITCHER, W. S. & READ, H. H., 1963. Contact metamorphism in relation to manner of emplacement of the granites of Donegal, Ireland. *Journal of Geology*, **71**, 261–96.

PLAYFAIR, J., 1822. Collected Works of John Playfair, Esq., with a Memoir of the Author. 4 vols. *Edinburgh: Constable*.

PLAZA, G. R., 1945. Patoita. *Boletin de la Sociedad Geologica de Peru*, **18**, 89–99.

PLINY, C. THE ELDER (CAIUS PLINIUS SECUNDUS MAJOR), 80 (posthumous). Naturalis historia. Translated into French by E. Littré, 1877. 2 vols. *Paris: Firmin-Didot*, vol. 2. English translation by D. E. Eichholz, 1962. *London: Heinemann*.

PLYUSHCHEV, E. V., 1981. Altered Rocks and their Prospecting Significance. *Moscow: Nedra Press*.

PLYUSNINA, L. P., LIKHOIDOV, G. G. & ZARAISKII, G. P., 1993. Physico-chemical conditions of rodingite formation from experimental data (in Russian). *Petrology*, **1** (5), 557–68.

POHL, J., STÖFFLER, D., GALL, J. & ERNSTSON, K., 1977. The Ries impact crater. In D. J. Roddy, R. O. Pepin & R. B. Merrill (eds.), Impact and Explosion Cratering. *New York: Pergamon Press*, pp. 343–404.

POLENOV, B., 1897. Die massigen Gesteine vom nördlichen Theile des Witim-Plateau (Ostsibirien) (in Russian). *Travaux de la Société Impériale des Naturalistes de St-Pétersbourg*, section de Géologie et de Minéralogie, **27**, livre 5, 90–488.

POLKANOV, A., 1935. Geological-petrographical Description of the NW Part of the Kola Peninsula (in Russian), part 1. *Academy of Science of the USSR, Moscow & Leningrad*.

PONOMARENKO, A. I., 1975. Alkremite, a new variety of aluminous ultramafic rock in xenoliths from the Udachnaya kimberlite pipe (in

Russian). *Translations of Doklady Akademii Nauk SSSR*, **225**, 155–7.

Powell, C. McA., 1979. A morphological classification of rock cleavage. *Tectonophysics*, **58**, 21–34.

Price, N. J. & Cosgrove, J. W., 1990. Boudinage and Pinch-and-swell Structures. *Cambridge University Press.*

Pustovalov, L. V., 1940. Petrography of Sedimentary Rocks, part 1. *Moscow: Gostoptekhizdat.*

Quaide, W. L. & Bunch, T. E., 1970. Impact metamorphism of lunar surface materials. *Proceedings of the Apollo 11 Lunar Science Conference, Geochimica et Cosmochimica Acta*, suppl. **1**, 711–29.

Quensel, P., 1916. Zur Kenntnis der Mylonitbildung, erläutert an Material aus dem Kebnekaisergebiet. *Uppsala Universitet Geologiska Institut Bulletin*, **15**, 91–116.

Quirke, T. T. & Lacy, W. C., 1941. Deep-zone dome and basin structures. *Journal of Geology*, **49**, 589–609.

Raikes, S. A. & Ahrens, T. J., 1979. Post-shock temperatures in minerals. *Journal of the Royal Astronomical Society*, **58**, 717–47.

Ramberg, H., 1951. Remarks on the average chemical composition of granulite facies and amphibolite to epidote-amphibolite facies gneisses in West Greenland. *Meddedelser fra Dansk Geologisk Forening*, **12**, 27–34.

Ramberg, H., 1955. Natural and experimental boudinage and pinch-and-swell structures. *Journal of Geology*, **63**, 512–26.

Ramond, L., 1801. Voyage au Mont Perdu et dans la partie adjacente des hautes Pyrénées. *Paris: Belin.*

Ramsay, J., 1967. Folding and Fracturing in Rocks. *New York: McGraw-Hill.*

Ramsay, J. & Huber, M. I., 1983. The Techniques of Modern Structural Geology. 1. Strain Analysis. *New York: Academic Press.*

Ramsay, J. & Huber, M. I., 1987. The Techniques of Modern Structural Geology. 2. Folds and Fractures. *New York: Academic Press.*

Ramsay, J. & Lisle, R. J., 2000. Techniques of Modern Structural Geology. 3. Applications of Continuum Mechanisms in Structural Geology. *New York: Academic Press.*

Rankine, W. J. M., 1870. On the thermodynamic theory of waves of finite disturbance. *Transactions of the Royal Society of London*, **160**, 277–88.

Ransome, F. M., 1912. Copper deposits near Superior, Arizona. *US Geological Survey Bulletin*, **540**, 139–66.

Rast, N., 1965. Nucleation and growth of metamorphic minerals. In W. S. Pitcher & G. W. Flinn (eds.), Controls of Metamorphism. *Edinburgh: Oliver & Boyd*, pp. 73–102.

Raumer, K. von, 1819. Das Gebirge Nieder-Schlesiens der Grafschaft Glatz und eines Theils von Böhmen und der Ober-Lausitz, geognostisch dargestellt. *Berlin: Reimer.*

Read, H. H., 1931. The Geology of Central Sutherland. *Memoir of the Geological Survey of Scotland.*

Read, H. H., 1944. Meditations on granite, part 2. *Proceedings of Geologists' Association*, **55**, 45–93.

Read, H. H., 1949. A contemplation of time in plutonism. *Quarterly Journal of the Geological Society of London*, **105**, 101–56.

Read, H. H., 1951. Metamorphism and granitisation. *Transactions of the Geological Society of South Africa*, **54**.

Read, H. H., 1952. Metamorphism and migmatisation in the Ythan Valley, Aberdeenshire. *Transactions of the Edinburgh Geological Society*, **15**, 265–79.

Reimold, W. U., 1995. Pseudotachylite-generation by friction melting or shock brecciation. A review and discussion. *Earth-Science Reviews*, **39**, 247–64.

Reinhard, M., 1909. Die Kristallinen Schiefer der Făgăraşer Gebirges in den Rumänischen Karpaten. *Anuarul Institutuliec Geologie al Romaniei*, **3**, 224–59.

Reinhard, M., 1935. Ueber Gesteinsmetamorphose in den Alpen. *Jaarboek van de Mijnbouwkundige Vereeniging te Delft*, 39–45.

Reinsch, D., 1977. High pressure rocks from Val Chiusella (Sesia-Lanzo zone, Italian Alps). *Neues Jahrbuch für Mineralogie*, Abhandlungen, **130**, 89–102.

Renard, A., 1878. Sur la structure et la composition minéralogique du coticule et sur les rapports avec le phyllade oligistifère. *Mémoires couronnés de l'Académie Royale de Belgique*, **41**, 1–42.

Reverdatto, V. V., 1973. The facies of contact metamorphism. Translated by D. A. Brown.

Australian National University, Canberra, Dept. of Geology, No. **233**, 1–263.

Reverdatto, V. V., Sharapov, V. N. & Melamed, V. G., 1970. The controls and selected peculiarities of the origin of contact metamorphic zonation. *Contributions to Mineralogy and Petrology*, **29**, 310–37.

Reverdatto, V. V., Sharapov, V. N., Lavrent'ev, Yu. G. & Pokachalova, O. S., 1974. Investigations in isochemical contact metamorphism. *Contributions to Mineralogy and Petrology*, **48**, 287–9.

Reynolds, D. L., 1936. Demonstrations in petrogenesis from Kiloran Bay, Colonsay. *Mineralogical Magazine*, **24**, 367–407.

Reynolds, D. L., 1946. The sequence of geochemical changes leading to granitization. *Quarterly Journal of the Geological Society of London*, **102**, 390–446.

Reynolds, S. H., 1928. Breccias. *Geological Magazine*, **65**, 97–107.

Rice, H. H., McQueen, R. G. & Walsh, J. M., 1958. Compression of solids by strong shock waves. In F. Seitz & D. Turnbull (eds.), Solid State Physics 6. *New York: Academic Press Inc.*, pp. 1–63.

Richard, J., 1938. Étude de la série cristallophyllienne renversée de la vallée de la Sioule aux confins de l'Auvergne et du Bourbonnais. *Revue des Sciences naturelles d'Auvergne*, **4**, fasc. 1, 1–37.

Richardson, S. W. & England, P. C., 1979. Metamorphic constraints of crustal eclogite production in overthrust orogenic zones. *Earth and Planetary Science Letters*, **42**, 183–90.

Richtofen, F. von, 1868. The natural system of volcanic rocks. *Memoirs of the Californian Academy of Sciences*, **1**, part 2, 1–94.

Rickard, M. J., 1961. A note on cleavages in crenulated rocks. *Geological Magazine*, **98**, 324–32.

Rietmeijer, F. J. M., 2004. Dynamic pyrometamorphism during atmospheric entry of large (~10 micron) pyrrhotite fragments from cluster IDPs. *Meteoritics and Planetary Science*, **39**, 1869–87.

Rinne, F., 1908. Praktische Gesteinskunde, 3rd edn. *Hannover: Jänecke*.

Rinne, F., 1920. Die geothermischen Metamorphosen und die Dislokationen der deutschen Kalisalzlagerstätten. *Fortschritte der Mineralogie, Kristallographie und Petrographie*, **6**, 101–36.

Rinne, F., 1928. Über die Auslösung tektonischer Spannungen in Tonschiefer und Diabas an Hand von Beobachtungen bei Goslar am Harz. *Helsinki: Fennia*, **50** (3), 1–11.

Rinne, F., 1928–40. Gesteinskunde, 12th edn. *Leipzig: Jänecke Verlagsbuchhandlung*.

Rinne, F. & Boeke, H. E., 1908. Ueber Thermometamorphose und Sammelkrystallisation. *Tschermak's mineralogische und petrographische Mitteilungen*, **27**, 393–8.

Robertson, P. B., Dence, M. R. & Vos, M. A., 1968. Deformation in rock-forming minerals from Canadian craters. In B. M. French & N. M. Short (eds.), Shock Metamorphism of Natural Materials. *Baltimore: Mono Book Corporation*, pp. 433–52.

Robertson, S., 1999. BGS Rock Classification Scheme. 2. Classification of Metamorphic Rocks. *British Geological Survey*, Research Report No. 99–02, 1–24.

Robinson, D., 1987. Transition from diagenesis to metamorphism in extensional and collision settings. *Geology*, **15**, 866–9.

Robinson, D. & Bevins, R. E., 1989. Diastathermal (extensional) metamorphism at very low grades and possible high grade analogues. *Earth and Planetary Science Letters*, **92**, 81–8.

Robinson, D. & Bevins, R. E., 1999. Patterns of regional low-grade metamorphism in metabasites. In M. Frey & D. Robinson (eds.), Low-grade Metamorphism. *Oxford: Blackwell Scientific Publications*, pp. 143–68.

Roddy, D. J., Pepin, R. O. & Merrill, R. B. (eds.), 1977. Impact and Explosion Cratering: Planetary and Terrestrial Implications. *New York: Pergamon Press*.

Rogers, A. F., 1918. An American occurrence of periclase and its bearing on the origin and history of calcite-brucite rocks. *American Journal of Science*, **196**, 581–6.

Rolle, F., 1879. Mikropetrographische Beiträge aus den Rhätischen Alpen. *Wiesbaden: Bergmann*.

Romé de l'Isle, J. B., 1783. Cristallographie ou description des formes propres à tous les corps du règne minéral, 2nd edn. 4 vols. *Paris: Imprimerie de Monsieur*.

Roques, M., 1936. Sur les relations des amphibolites et des péridotites à Sarrazac

(Dordogne). *Comptes rendus de l'Académie des Sciences, Paris*, **202**, 332–4. Also *Mineralogical Abstracts*, **6**, 299.

Roques, M., 1941. Les schistes cristallins de la partie sud-ouest du Massif Central Français. *Mémoires du Service de la Carte Géologique de France*, **24**.

Rose, G., 1831. Ueber die Notwendigkeit, Augit und Hornblende in einer Gattung zu vereinigen. *Annalen der Physik und Chemie, Leipzig*, **22**, 321–43.

Rose, G., 1837. Mineralogisch-geognostische Reise nach dem Ural, dem Altai und dem Kaspischen Meere. *Berlin: Sanderschen Buchhandlung*.

Rose, G., 1842. Mineralogisch-geognostische Reise nach dem Ural, dem Altai und dem Kaspischen Meere, vol. 2. *Berlin: Reiner*.

Rosen, O. M., Fettes, D. & Desmons, J., 2005. Chemical and mineral compositions of metacarbonate rocks under regional metamorphism conditions and guidelines on rock classification. *Russian Geology and Geophysics*, **46** (4), 357–66.

Rosenbusch, H., 1873. Mikroskopische Physiographie der petrographisch wichtigen Mineralien. *Stuttgart: Schweizerbart*.

Rosenbusch, H., 1877. Die Steiger Schiefer und ihre Contactzone an den Granititen von Barr-Andlau und Hohwald. *Abhandlungen zur geologischen Spezialkarte von Elsass-Lothringen*, **1**, H.2, 79–393.

Rosenbusch, H., 1887. Mikroskopische Physiographie der massigen Gesteine, vol. 2. *Stuttgart: Schweizerbart*.

Rosenbusch, H., 1896. Mikroskopische Physiographie der Mineralien und Gesteine: Ein Hilfsbuch bei mikroskopischen Gesteinsstudien. *Stuttgart: Schweizerbart*.

Rosenbusch, H., 1898. Elemente der Gesteinslehre. *Stuttgart: Schweizerbart*.

Rosenbusch, H., 1901. Elemente der Gesteinslehre, 2nd edn. *Stuttgart: Schweizerbart*.

Rosenbusch, H., 1907. Mikroskopische Physiographie der Mineralien und Gesteine. 2. Massigen Gesteine, 4th edn. *Stuttgart: Schweizerbart*.

Rosenbusch, H., 1910. Elemente der Gesteinslehre, 3rd edn. *Stuttgart: Schweizerbart*.

Rosenbusch, H., 1923. Elemente der Gesteinslehre, 4th edn by A. Osann, *Stuttgart: Schweizerbart*.

Roth, J., 1851. Bemerkungen über die Verhältnisse von Predazzo. *Zeitschrift der deutschen geologischen Gesellschaft*, **3**, 140–8.

Roth, J., 1887. Über den Zobtenit. *Sitzungsberichte der Königlich-Preussischen Akademie der Wissenschaften zu Berlin*, No. **32**, 611–30.

Rumpf, J., 1873. Ueber krystallisirte Magnesite aus den nord-östlichen Alpen. *Mineralogische Mittheilungen*, **4**, 263–72.

Rundquist, D. V. & Pavlova, I. G., 1974. An attempt at distinguishing families of hydrothermal-metasomatic rocks (in Russian). *Zapiski Vsesoyuznogo Mineralogicheskogo Obshchestva*, **103** (3), 289–304.

Rundquist, D. V., Denisenko, V. K. & Pavlova, I. G., 1971. Greisen Deposits: Ontogenesis, Phylogenesis (in Russian). *Moscow: Nedra Press*.

Ruppel, C. & Hodges, K. V., 1994. Pressure–temperature–time paths from two-dimensional thermal models: prograde, retrograde and inverted metamorphism. *Tectonics*, **13**, 17–44.

Rusinov, V. L., 1972. Geological and Physicochemical Regularities of Propylitisation (in Russian). *Moscow: Nauka Press*.

Rusinov, V. L., 1989. Metasomatic Processes in Volcanic Rocks (in Russian). *Moscow: Nauka Publishing*.

Rutter, E. H., 1983. Pressure solution in nature, theory and experiment. *Journal of the Geological Society of London*, **140**, 725–40.

Ryka, W. & Maliszewska, A., 1982. Slownik petro-graficzny. *Wydawnictwa geologiczene, Warszawa*. (Russian edn, 1989, *Moskva Hedra*.)

Saggerson, E. P., 1989. Regional metamorphism. In D. R. Bowes (ed.), The Encyclopedia of Igneous and Metamorphic Petrology. *New York: Van Nostrand Reinhold*.

Sainte-Claire Deville, H., 1861. Sur un nouveau mode de reproduction du fer oligiste et de quelques oxydes métalliques de la nature. *Comptes rendus de l'Académie des Sciences, Paris*, **52**, 1264–7.

Salomon, W., 1890. Geologische und petrographische Studien am Monte Aviolo im italienischen Antheil der Adamellogruppe. *Zeitschrift der deutschen geologischen Gesellschaft*, **42**, 450–556.

Salomon, W., 1891. Über einige Einschlüsse metamorpher Gesteine im Tonalit. *Neues*

Jahrbuch für Mineralogie, Geologie und Paläontologie, Beilageband **7**, 471–87.

Salomon, W., 1898. Ueber Alter, Lagerungsform und Entstehungsart der periadriatischen granitisch-körnigen Massen. *Tschermak's mineralogische und petrographische Mitteilungen*, **17**, 109–283.

Salomon, W., 1908. Die Adamellogruppe, ein alpines Zentralmassiv, und seine Bedeutung für die Gebirgsbildung und unser Kenntnis von dem Mechanismus der Intrusionen. *Verhandlungen der Kaiserlich-königlichen geologischen Reichsanstalt, Wien*, **21**, 1–603.

Salomon, W., 1915. Die definitionen von Grauwacke, Arkose und Ton. *Geologische Rundschau*, **6**, 398–404.

Sander, B., 1911. Über Zusammenhänge zwischen Teilbewegung und Gefüge in Gesteinen. *Tschermak's mineralogische und petrographische Mitteilungen*, **30**, 281–314.

Sander, B., 1912. Über einige Gesteinsgruppen des Tauernbestandes. *Jahrbuch der Kaiserlich-königlichen geologischen Reichsanstalt, Wien*, **62**, 249–57.

Sander, B., 1930. Gefügekunde der Gesteine mit besonderer Berücksichtigung der Tektonite. *Vienna: Springer*.

Santallier, D., Briand, B., Ménot, R.P. & Piboule, M., 1988. Les complexes leptyno-amphiboliques (C.L.A.): revue critique et suggestions pour un meilleur emploi de ce terme. *Bulletin de la Société géologique de France*, 8e sér., **4**, 3–12.

Sargent, H.C., 1917. Reports of societies and academies. *Nature*, **99**, 59.

Sauer, A., 1890. Der Granitit von Durbach im nördlichen Scharzwald und seine Grenzfazies von Glimmersyenite (Durbachit). *Mitteilungen der Grossherzoglich Badischen Geologischen Landesanstalt, Heidelberg*, **2**, 231–76.

Sauer, A., 1901. Petrographische Studien an den Lavabomben aus dem Ries. *Jahreshefte des Vereins für vaterländische Naturkunde in Württemberg*, **57**, 88.

Saussure, H.-B. de, 1779. Voyages dans les Alpes, précédés d'un essai sur l'histoire naturelle des environs de Genève, vol. 1. 8 vols. *Neuchâtel: Louis Fauché*.

Saussure, H.-B. de, 1786. Voyages dans les Alpes, précédés d'un essai sur l'histoire naturelle des environs de Genève, vol. 2. *Fauché, Père et Fils, Imprimeurs-Libraires du Roi*, 595–615. Also *Genève: Barde*.

Saussure, H.-B. de, 1796. Voyages dans les Alpes, précédés d'un essai sur l'histoire naturelle des environs de Genève. *Neuchâtel: Louis Fauché-Borel, Imprimeur du Roi*.

Saussure, T. De, 1806. *Annales de Chimie et de Physique*, 3e sér., 16.

Schaal, R.B. & Hörz, F., 1977. Shock metamorphism of lunar and terrestrial basalts. *Proceedings of the 8th Lunar Planetary Science Conference*, 1697–729.

Schaal, R.B. & Hörz, F., 1980. Experimental shock metamorphism of lunar soil. *Proceedings of the 11th Lunar Planetary Science Conference*, 1679–95.

Schaal, R.B., Hörz, F., Thompson, T.D. & Bauer, J.F., 1979. Shock metamorphism of granulated lunar basalt. *Proceedings of the 10th Lunar Planetary Science Conference*, 2547–71.

Scharbert, H.G., 1963. Zur Nomenklatur der Gesteine in Granulitfazies. *Tschermak's mineralogisch-petrographische Mitteilungen*, **8**, 591–8.

Scharizer, R., 1879. Notizen über einige österreichische Mineralvorkommnisse. *Verhandlungen der Kaiserlich-königlichen geologischen Reichsanstalt, Wien*, 243–7.

Scheumann, K.H., 1936a. Zur Nomenklatur migmatitischer und verwandter Gesteine. *Tschermak's mineralogisch-petrographische Mitteilungen*, **48**, 297–302.

Scheumann, K.H., 1936b. Metatexis und Metablastesis. *Mineralogische und petrographische Mitteilungen, Leipzig*, **48**, 402–12.

Scheumann, K.H., 1936c. Ueber eine Gruppe bisher wenig beachteter Orthogneise des Granulitgebirges und deren Einschlichtung. *Tschermak's mineralogische und petrographische Mitteilungen*, **47**, 403–69.

Scheumann, K.H., 1954. Bemerkungen zur Genese der Gesteins- und Mineralfazies der Granulite. *Geologie, Berlin*, **2**, 99–154.

Scheumann, K.H., 1955. Ueber Akyrosome. *Neues Jahrbuch für Mineralogie, Geologie und Paläontologie*, Monatsheft, 11–24.

Scheumann, K.H., 1961. 'Granulit', eine petrographische Definition. *Neues Jahrbuch für Mineralogie*, Monatsheft, 75–80.

Schiel, J., 1857. Ueber die Zusammensetzung einiger amerikanischer Felsarten. *Annalen der Chemie und Pharmacie*, **27**, 119–20.

Schmid, R., 1968. Schwierigkeiten der Nomenklatur und Klassifikation massiger Katamorphite, erläutert am Beispiel der Zone Ivrea-Verbano (Norditalien). *Schweizerische mineralogische und petrographische Mitteilungen*, **48**, 81–90.

Schmid, R. & Sassi, F. P., 1986. On the way to a recommended nomenclature and classification of metamorphic rocks. *Rendiconti della Società Italiana di Mineralogia e Petrologia*, **41/2**, 201–4.

Schmidt, R. G., 1985. High-alumina hydrothermal system in volcanic rocks and their significance to mineral prospecting in the Carolina Slate Belt. *US Geological Survey Bulletin*, **1562**, 1–59.

Schmitt, R. T., 2000. Shock experiments with the H6 chondrite Kernouvé: pressure calibration of microscopic shock effects. *Meteoritics and Planetary Science*, **35**, 545–60.

Schrauf, A., 1882. Beiträge zur Kenntnis des Associationskreises der Magnesiasilikate. *Zeitschrift für Krystallographie und Mineralogie, Leipzig*, **6**, 321–88.

Schreyer, W., 1973. Whiteschist, a high-pressure rock and its geologic significance. *Journal of Geology*, **81**, 735–9.

Schwartz, G. M., 1924. The contrast in the effect of granite and gabbro intrusions on the Ely Greenstone. *Journal of Geology*, **32**, 89–138.

Schwartz, G. M., 1939. Hydrothermal alteration of igneous rocks. *Geological Society of America Bulletin*, **50**, 181–238.

Scott, J. S. & Drever, H. I., 1953. Frictional fusion along a Himalayan thrust. *Proceedings of the Royal Society of Edinburgh*, **65B**, 121–42.

Sederholm, J. J., 1891. Studien über Archäis, die Eruptivgesteine aus dem südwestlichen Finnland. *Tschermak's mineralogische und petrographische Mitteilungen*, **12**, 97–142.

Sederholm, J. J., 1893. Om berggrunden i södra Finland. *Fennia, Helsingfors*, **8**, 1–165.

Sederholm, J. J., 1897–9. Über eine archäische Sedimentformation im südwestlichen Finnland und ihre Bedeutung für die Erklärung der Entstehungsweise des Grundgebirges. *Bulletin de la Commission géologique de Finlande*, **6** (1), 1–254.

Sederholm, J. J., 1907. Om granit och gneiss. *Bulletin de la Commission géologique de Finlande*, **4**, No. 23.

Sederholm, J. J., 1916. On synantetic minerals and related phenomena. *Bulletin de la Commission géologique de Finlande*, **48**, 1–148.

Sederholm, J. J., 1923. On migmatites and associated Precambrian rocks of southwestern Finland. *Bulletin de la Commission géologique de Finlande*, **58**, 1–153.

Sederholm, J. J., 1926. On migmatites and associated Precambrian rocks of southwestern Finland, part 2. *Bulletin de la Commission géologique de Finlande*, **77**, 1–143.

Sederholm, J. J., 1931. On the sub-Bothnian unconformity and on Archaean rocks formed by secular weathering. *Bulletin de la Commission géologique de Finlande*, **95**, 1–81.

Sederholm, J. J., 1934. On migmatites and associated Pre-Cambrian rocks of southwest Finland, part 3. The Aland islands. *Bulletin de la Commission géologique de Finlande*, **107**, 1–68.

Sedgewick, A., 1835. Remarks on the structure of large mineral masses, and especially on the chemical changes produced in the aggregation of stratified rocks during different periods after deposition. *Transactions of the Geological Society of London*, 2nd ser., **3**, 461–86.

Seki, Y., 1961. Pumpellyite in low-grade regional metamorphism. *Journal of Petrology*, **2**, 407–23.

Semenenko, N. P., 1964. The genesis and the classification of skarns. *Kristallinikum*, **2**, 61–9.

Senft, F., 1857. Classification und Beschreibung der Felsarten. *Breslau: Korn.*

Serres, M. de, 1863. Traité des roches simples et composées. *Paris: Lacroix.*

Shabynin, L. I., 1973. Formation of the Magnesian Skarns (in Russian). *Moscow: Nauka Press.*

Shand, S. J., 1910. On borolanite and its associates in Assynt. *Transactions of the Edinburgh Geological Society*, **9**, 376–424.

Shand, S. J., 1916. The pseudotachylite of Parijs (Orange Free State) and its relation to 'trapshotten gneiss' and 'flinty crush rock'. *Quarterly Journal of the Geological Society of London*, **77**, 198–221.

Shand, S. J., 1931. The Study of Rocks. *London: Murby.*

Shand, S. J., 1945. Coronas and coronites. *Geological Society of America Bulletin*, **56**, 247–65.

Shand, S. J., 1951. The Study of Rocks, 3rd edn. *London: Murby.*

Shcherban', L. P., 1975. Conditions of the Formation of Low-temperature Wall-rock Metasomatites, on the Example of the Altai-Sayan Region (in Russian). *Novosibirsk: Nauka Press.*

SHCHERBAN', L. P., 1996. Ore-bearing Near-vein Metasomatites (in Russian). *Kiev: Lebed' Publishing*, pp. 10–30.

SHELLEY, D., 1993. Igneous and Metamorphic Rocks under the Microscope. *London: Chapman & Hall.*

SHI, Y. & WANG, C.-Y., 1987. Two dimensional modeling of the P–T–t paths of regional metamorphism in simple overthrust terrains. *Geology*, **15**, 1048–51.

SHLYGIN, A. E. & GUKOVA, V. D., 1981. The experience from small-scale mapping of pneumato-hydrothermal fluidogenic formations, on the example of Kazakhstan (in Russian). *Izvestiya (Proceedings) of the Academy of Science of Kazakhstan, Geological Series*, **6**, 71–8.

SHOEMAKER, E. M., HAIT, M. H., SWANN, G. A. *et al.*, 1970. Lunar regolith at Tranquillity Base. *Science*, **167**, 452–5.

SHROCK, R. R., 1948. Sequence in Layered Rocks. *New York: McGraw-Hill.*

SIBSON, R. H., 1977. Fault rocks and fault mechanisms. *Quarterly Journal of the Geological Society of London*, **133**, 191–213.

ŠICHTÁŘOVÁ, I., 1977. Amfibolitové horniny východnĭ části moravského moldanubika. M.Sc. thesis, Charles University, Prague.

SIMLER, T., 1862. Ueber die Petrogenese im Allgemeinen und Bunsen'sche Gesetz der syntektischen Gesteinsbildung, angewendet auf die Verrucane des Kantons Glarus, im Besonderen. *Bern: Huber*.

SMITH, R. E., 1968. Redistribution of major elements in the alteration of some basic lavas during burial metamorphism. *Journal of Petrology*, **9**, 191–219.

SMOLIN, P. P., 1959. On the foundations for rational classification of metamorphosed carbonate rocks (in Russian). *Izvestiya (Proceedings) of the Academy of Science of the USSR, Geological Series*, **12**, 36–53.

SMULIKOWSKI, K., 1947. Petrological studies in the granitic areas of North-Volhynia (with an English summary). *Archives de Minéralogie de la Société des Sciences et des Lettres de Varsovie*, **16**, 258–321.

SMULIKOWSKI, K., 1950. On anatectic differentiation in granitic areas. *Report of the 18th International Geological Congress*, part 2, 131–8.

SMULIKOWSKI, K., 1958. Problems of genetic classification of granitoids. *Studia Geologica Polonica*, **1**, 59–115.

SMULIKOWSKI, K., 1964a. An attempt at eclogite classification. *Bulletin de l'Académie Polo-naise des Sciences, sér. Sciences géologiques et géographiques*, **12**, 27–33.

SMULIKOWSKI, K., 1964b. Le problème des éclogites. *Geologia Sudetica*, **1**, 13–52.

SMULIKOWSKI, K., 1972. Classification of eclogites and allied rocks. *Krystalinikum*, **9**, 107–30.

SMULIKOWSKI, K., 1989. Eclogite. In D. R. Bowes (ed.), The Encyclopedia of Igneous and Metamorphic Petrology. *New York: Van Nostrand Reinhold*, pp. 137–42.

SMYTHE, H. L., 1891. Structural geology of Steep Rock Lake, Ontario. *American Journal of Science*, 3rd ser., **42**, 317–31.

SNOKE, A. W., TULLIS, J. & TODD, V. R., 1998. Fault-related Rocks: A Photographic Atlas. *Princeton University Press.*

SOBOLEV, N. V., ZYUZIN, N. I. & KUZNETSOVA, I. K., 1966. A continuous series of pyrope-grossular garnets in grospydite (in Russian). *Doklady Akademii Nauk SSSR*, **167** (1), 126–9; (4), 902–5.

SOKOL, E., VOLKOVA, N. & LEPEZIN, C., 1998. Mineralogy of pyrometamorphic rocks associated with naturally burned coal-bearing spoil heaps of the Chelyabinsk coal basin, Russia. *European Journal of Mineralogy*, **10**, 1003–14.

SORBY, H. C., 1879. President's anniversary address. *Quarterly Journal of the Geological Society of London*, **39**, 1–95.

SPEAR, F. S., 1993. Metamorphic phase equilibria and pressure–temperature–time paths. *Mineralogical Society of America.*

SPENCER, L. J., 1932. Meteorite craters. *Nature*, **129**, 781–4.

SPENCER, L. J., 1933a. Origin of tektites. *Nature*, **131**, 117.

SPENCER, L. J., 1933b. Meteorite craters as topographical features on the earth's surface. *Geographical Journal London*, **81**, 227–48.

SPENCER, L. J., 1933c. Meteoritic iron and silica glass from the meteorite craters of Henbury (central Australia) and Wabar (Arabia). *Mineralogical Magazine*, **23**, 387–404.

SPOONER, E. T. C. & FYFE, W. S., 1973. Subsea-floor metamorphism, heat and mass transfer. *Contributions to Mineralogy and Petrology*, **42**, 287–304.

Spray, J. G., 1998. Localised shock- and friction-induced melting in response to hypervelocity impact. In M. M. Grady, R. Hutchinson, G. J. H. McCall & D. A. Rothery (eds.), Meteorites: Flux with Time and Impact Effects. *Geological Society of London, Special Publication*, **140**, 195–204.

Spry, A., 1969. Metamorphic Textures. *Oxford: Pergamon Press.*

Spry, A. & Solomon, M., 1964. Columnar buchites at Apsley, Tasmania. *Quarterly Journal of the Geological Society of London*, **120**, 519–45.

Spurr, J. E., 1923. The Ore Magmas, vol. 2. *New York: McGraw-Hill*, pp. 431–915.

Spurr, J. E., Garrey, G. H. & Fenner, C. N., 1912. Study of a contact metamorphic ore deposit, the Dolores Mine at Matehuala, S. L. P., Mexico. *Economic Geology*, **7**, 444–92.

Stache, G., 1876. Die Eruptivgesteine des oberen Addagebietes (Veltlin) zwischen Bormio und Boladore. *Verhandlungen der Kaiserlich-königlichen geologischen Reich-sanstalt, Wien*, pp. 357–8.

Stache, G. & John, C., 1877. Geologische und petrographische Beiträge zur Kenntnis der älteren Eruptiv- und Massengesteine der Mittel- und Ost-Alpen. Nr. 1. *Jahrbuch der Kaiserlich-königlichen geologischen Reich-sanstalt, Wien*, **27**, 143–242.

Staub, R., 1915. Petrographische Untersuchungen im westlichen Berninagebirge. *Naturforschende Gesellschaft Zürich Vierteljahrsschrift*, **60**, 55–336.

Steinmann, G., 1908. Die Entstehung des Nephrits in Ligurien und die Schwellungsmetamorphose. *Sitzungsberichte der Niederrheinischen Gesellschaft für Natur- und Heilkunde zu Bonn*, 1–13.

Steinmann, G., 1925. Gibt es fossile Tiefseeablagerungen von erdgeschichtlicher Bedeutung? *Geologische Rundschau*, **16**, 435–68.

Stelzner, A., 1871. Untersuchungen im Gebiete des sächsischen Granulitgebirges. *Neues Jahrbuch für Mineralogie, Geologie und Paläontologie*, 244–9.

Stelzner, A., 1885. Beiträge zur Geologie und Paläontologie der Argentischen Republik. 1. Geologischer Teil. *Cassel-Berlin: Fischer.*

Stemprok, M., 1987. Greisenization (a review). *Geologische Rundschau*, **76**, 169–75.

Stewart, D. B., 1975. Apollonian metamorphic rocks. *Lunar and Planetary Science*, **6**, 774–6.

Stifft, C. E., 1807. Beiträge zu einer Beschreibung der Gangformationen in den Fürstentümern Dillenburg und Siegen. *Efemeriden der Berg- und Hüttenkunde, Nürnberg*, **3**, 377–99.

Stillwell, F. L., 1918. The metamorphic rocks of Adelie Land. *Scientific Report on the Australian Antarctic Expedition* 1911–1914, ser. A, **3**, 1–230.

Stöffler, D., 1966. Zones of impact metamorphism in the crystalline rocks of the Nördlinger Ries crater. *Contributions to Mineralogy and Petrology*, **12**, 15–24.

Stöffler, D., 1971. Progressive metamorphism and classification of shocked and brecciated crystalline rocks at impact craters. *Journal of Geophysical Research*, **76**, 5541–51.

Stöffler, D., 1972. Deformation and transformation of rock-forming minerals by natural and experimental shock processes. 1. Behavior of minerals under shock compression. *Fortschritte der Mineralogie*, **49**, 50–113.

Stöffler, D., 1974. Deformation and transformation of rock-forming minerals by natural and experimental shock processes. 2. Physical properties of shocked minerals. *Fortschritte der Mineralogie*, **51**, 256–89.

Stöffler, D., 1977. Research drilling Nördlingen 1973: polymict breccias, crater basement and cratering model of the Ries impact structure. *Geologica Bavarica*, **75**, 443–58.

Stöffler, D., 1984. Glasses formed by hypervelocity impact. *Journal of Non-Crystalline Solids*, **67**, 465–502.

Stöffler, D. & Grieve, R. A. F., 1994. Classification and nomenclature of impact metamorphic rocks: a proposal to the IUGS Subcommission on the Systematics of Metamorphic Rocks. *Lunar and Planetary Science* 25. *Houston: The Lunar and Planetary Science Institute*, 1347–8.

Stöffler, D. & Grieve, R. A. F., 1996. IUGS classification and nomenclature of impact metamorphic rocks: towards a final proposal. *International Symposium on the Role of Impact Processes in the Geological and Biological Evolution of Planet Earth, Postojna, Slovenia*, 27.9.–2.10.1996, Abstract.

Stöffler, D. & Langenhorst, F., 1994. Shock metamorphism of quartz in nature and experiment. I. Basic observation and theory. *Meteoritics*, **29**, 155–81.

Stöffler, D. & Ryder, G., 2001. Stratigraphy and isotope ages of lunar geologic units: chronological standard for the inner solar system. *Space Science Review*, **96**, 9–54.

Stöffler, D., Gault, D. E., Wedekind, J. & Polkowski, G., 1975. Experimental hypervelocity impact into quartz sand: Distribution and shock metamorphism of ejecta. *Journal of Geophysics*, **80** (29), 4062–77.

Stöffler, D., Knöll, H.-D. & Maerz, U., 1979. Terrestrial and lunar impact breccias and the classification of lunar highland rocks. *Proceedings of the 10th Lunar Planetary Science Conference*, 639–75.

Stöffler, D., Knöll, H.-D., Marvin, U. B., Simonds, C. H. & Warren, P. H., 1980. Recommended classification and nomenclature of lunar highland rocks. In J. J. Papike & R. B. Merrill (eds.), Proceedings of the Conference on Lunar Highland Crust. *New York: Pergamon Press*, pp. 51–70.

Stöffler, D., Ostertag, R., Jammes, C. *et al.*, 1986. Shock metamorphism and petrography of the Shergotty achondrite. *Geochimica et Cosmochimica Acta*, **50**, 889–903.

Stöffler, D., Bischoff, L., Oskierski, W. & Wiest, B., 1988. Structural deformation, breccia formation, and shock metamorphism in the basement of complex terrestrial impact craters: implications for the cratering process. In A. Boden & K. G. Eriksson (eds.), Deep Drilling in Crystalline Bedrock, vol. 1. *Berlin-Heidelberg-New York: Springer*, pp. 277–97.

Stöffler, D., Keil, K. & Scott, E. R. D., 1991. Shock metamorphism of ordinary chondrites. *Geochimica et Cosmochimica Acta*, **55**, 3845–87.

Stokes, G. G., 1848. On a difficulty of the theory of sound. *Philosophical Magazine*, ser. 3, **33**, 349–56.

Stoppani, A., 1873. Corso di geologia, vol. 3, Geologia Endografica. *Milano: Bernardoni & Brigola.*

Strakhov, N. M., 1958. Schéma de la diagenèse des dépôts marins. *Eclogae geologicae Helvetiae*, **51**, 761–7.

Strakhov, N. M., 1967. Principles of Lithogenesis, vol. 1. *Edinburgh and London: Oliver & Boyd.*

Streckeisen, A., 1967. Classification and nomenclature of igneous rocks. Final report of an inquiry. *Neues Jahrbuch für Mineralogie, Geologie und Paläontologie*, Abhandlungen, **107**, 144–214.

Strøm, H., 1762. Physisk og oeconomisk Beskrivelse over Fogderiet Søndmør, beliggende i Bergens Stift, i Norge.

Suess, F. E., 1900. Die Herkunft of Moldavite und verwandter Gläser. *Jahrbuch der Kaiserlich-königlichen geologischen Reichsanstalt, Wien*, **50**, 193–382.

Suess, F. E., 1931. A suggested interpretation of the Scottish Caledonide structure. *Geological Magazine*, **83**, 71–81.

Suess, F. E., 1937. Bausteine zu einem System der Tektogenese. 1, Periplutonische und anorogene Regionalmetamorphose in ihrer tektogenetischen Bedeutung. *Fortschritte der Geologie und Paläontologie*, **13** (42), 1–86.

Suk, M., 1983. Petrology of Metamorphic Rocks. *Amsterdam: Elsevier.*

Svenonius, F., 1894. Fjellproblemet i ofre Norrland. *Geologiska Förningensi Stockholm Fördandlinger*, **16**, 244–5.

Swapp, S. M. & Hollister, L. S., 1991. Inverted metamorphism within the Tibetan Slab of Buthan: evidence for a tectonically transported heat-source. *Canadian Mineralogist*, **29**, 1019–41.

Tacconi, E., 1911. La massa calcare ed i calcefiri di Candoglia in valle del Toce. *Atti della Società Italiana di Scienze Naturali*, **50**, 55–94.

Tatarsky, V. B., 1949. On the dedolomitized rocks extent (in Russian). *Doklady Akademii Nauk SSSR*, **69**, 839–51.

Taylor, S. R., 1982. Planetary Science: A Lunar Perspective. *Houston: Lunar and Planetary Institute.*

Teall, J. J. H., 1887. On the origin of certain banded gneisses. *Geological Magazine*, **34**, 484–93.

Teall, J. J. H., 1888. British Petrography. *London: Dulau.*

Teall, J. J. H., 1902. Anniversary address of the President: the evolution of petrological ideas. *Quarterly Journal of the Geological Society of London*, **58**, lxiii–lxxviii.

Teall, J. J. H., 1903. On dedolomitisation. *Geological Magazine*, **10** (5), 513–14.

Teall, J. J. H., 1918. Dynamic metamorphism, a review, mainly personal. *Proceedings of Geologists' Association*, **29**, 1–15.

TEICHMÜLLER, M., 1987. Organic material and very low-grade metamorphism. In M. Frey (ed.), Low Temperature Metamorphism. *Glasgow and London: Blackie*, pp. 114–61.

TERMIER, P., 1903. Les schistes cristallins des Alpes occidentales. *Compte rendu de la 8e Session du Congrès géologique international, Vienne*, **2**, 1–20.

TERMIER, P., 1912. Sur la genèse des terrains cristallophylliens. *Compte rendu de la 11e Session du Congrès géologique international, Stockholm 1910*, **1**, 587–95.

TESTER, A. C. & ATWATER, G. I., 1934. The occurrence of authigenic feldspars in sediments. *Journal of Sedimentary Petrology*, **4**, 23–31.

THEOPHRASTOS (TYRTAMOS said Theophrastos, god-like talker), *c.* 325 BC. History of Stones. Translated into German by C. Schmeider, 1807. *Freiberg: Craz und Gerlachschen Buchhandlung*.

THOMAS, H. H., 1922. On certain xenolithic Tertiary minor intrusions in the island of Mull (Argyllshire). *Quarterly Journal of the Geological Society of London*, **78**, 229–60.

THOMPSON, J. B., 1957. The graphical analysis of mineral assemblages in pelitic schists. *American Mineralogist*, **42**, 842–58.

THOMSON, R., 1935. Sudburite, a metamorphic rock near Sudbury, Ontario. *Journal of Geology*, **43**, 427–35.

TIEJE, A. J., 1921. Suggestions as to the description and naming of sedimentary rocks. *Journal of Geology*, **29**, 650–66.

TILLEY, C. E., 1920. The metamorphism of the Pre-Cambrian dolomites of the southern Eyre Peninsula, South Australia. *Geological Magazine*, **57**, 449–500.

TILLEY, C. E., 1924a. Facies of metamorphic rocks. *Geological Magazine*, **31**, 167–71.

TILLEY, C. E., 1924b. Contact metamorphism in the Comrie area of the Perthshire Highlands. *Quarterly Journal of the Geological Society of London*, **80**, 22–71.

TILLEY, C. E., 1925. Metamorphic zones in the southern Highlands of Scotland. *Quarterly Journal of the Geological Society of London*, **81**, 100–12.

TILLEY, C. E., 1935. Metasomatism associated with the greenschist-hornfelses of Kendijack and Botallack, Cornwall. *Mineralogical Magazine*, **24**, 181–202.

TILLEY, C. E., 1936. Enderbite, a new member of the charnockite series. *Geological Magazine*, **73**, 312–16.

TOMKEIEFF, S. I., 1940. The dolerite plugs of Tieveragh and Tievebulliagh near Cushendall, Co. Antrim, with a note on buchite. *Geological Magazine*, **77**, 54–64.

TOMKEIEFF, S. I., 1954. Coals and Bitumens and Related Fossil Carbonaceous Substances: Nomenclature and Classification. *London: Pergamon*.

TOMKEIEFF, S. I., 1983. Dictionary of Petrology (E. K. Walton, B. A. O. Randall, M. H. Battey & O. Tomkeieff, eds.). *Chichester-New York-Brisbane-Toronto-Singapore: Wiley*.

TONIKA, J., 1969. Survey classification and terminology of metamorphosed amphibole-bearing rocks (in Czech). *Čas. Mineral. Geol. Prague*, **14**, 373–7.

TÖRNEBOHM, A. E., 1877. Om Sveriges viktigare diabas- och gabbro-arter. *Kungliga Svenska Vetenskapsakademiens, Handlingar, Stockholm*, **13**, part 14, 1–55.

TÖRNEBOHM, A. E., 1880. Några ord om granit och gneiss. *Geologiska Föreningens i Stockholm Förhandlingar*, **5**, 233–48.

TÖRNEBOHM, A. E., 1881a. Geologische Übersichtskarte der Statthalterschaft Vermland im Massstab 1:400 000 nebst Beschreibung. *Stockholm*. Also in E. Cohen 1882, *Neues Jahrbuch für Mineralogie, Geologie und Paläontologie*, **1**, 200–201.

TÖRNEBOHM, A. E., 1881b. Geologisk Öfversigtskarte öfver Mellersta Sveriges Berglag Blad 2–5 och 7. Also in E. Cohen 1882, *Neues Jahrbuch für Mineralogie, Geologie und Paläontologie*, **1**, 395–400.

TOURET, J. & NIJLAND, T. G., 2002. Metamorphism today: new science, old problems. In D. R. Oldroyd (ed.), The Earth Inside and Out: Some Major Contributions to Geology in the Twentieth Century. *Geological Society of London, Special Publication*, **192**, 113–41.

TRACY, R. J. & FROST, B. R., 1991. Phase equilibria and thermobarometry of calcareous, ultramafic and mafic rocks, and iron formations. In D. M. Kerrick (ed.), Contact Metamorphism. *Mineralogical Society of America, Reviews in Mineralogy*, **26**, 207–89.

TRÖGER, W. E., 1938. Spezielle Petrographie der Eruptivgesteine. Eruptivgesteinsnamen.

Fortschritte der Mineralogie, Kristallographie und Petrographie, **23**, 41–90.

TROPPER, P., RECHEIS, A. & KONZETT, J., 2004. Pyrometamorphic formation of phosphorus-rich olivines in partially molten metapelitic gneisses from a prehistoric sacrificial burning site (Ötz Valley, Tyrol, Austria). *European Journal of Mineralogy*, **16**, 631–40.

TSCHERMAK, G., 1872. Die Meteoriten von Shergotty und Gopalpur. *Sitzungsberichte der Kaiserlichen Akademie der Wissenschaften, Mathematisch-Naturwissenschaftliche Klasse, Wien*, **65** (1), 122–45.

TULLIS, J., SNOKE, A. W. & TODD, V. R., 1982. Significance and petrogenesis of mylonitic rocks. *Geology*, **10**, 227–30.

TURNER, F. J., 1938. Progressive regional metamorphism in southern New Zealand. *Geological Magazine*, **75**, 160–74.

TURNER, F. J., 1948. Mineralogical and structural evolution of the metamorphic rocks. *Geological Society of America Memoir*, **30**, 1–342.

TURNER, F. J., 1968. Metamorphic Petrology, Mineralogical and Field Aspects. *New York: McGraw-Hill.*

TURNER, F. J., 1981. Metamorphic Petrology, 2nd edn. *New York: McGraw-Hill.*

TURNER, F. J. & VERHOOGEN, J., 1951. Igneous and Metamorphic Petrology. *New York: McGraw-Hill.*

TURNER, F. J. & VERHOOGEN, J., 1960. Igneous and Metamorphic Petrology, 2nd edn. *New York: McGraw-Hill.*

TURNER, F. J. & WEISS, L. E., 1963. Structural Analysis of Metamorphic Tectonites. *New York: McGraw-Hill.*

TURNER, H. W., 1899. The granitic rocks of the Sierra Nevada. *Journal of Geology*, **7**, 141–62.

TURNER, H. W., 1900. The nomenclature of feldspathic granulites. *Journal of Geology*, **8**, 105–11.

TYRRELL, G. W., 1921. Some points in petrographic nomenclature. *Geological Magazine*, **58**, 494–502.

TYRRELL, G. W., 1926. The Principles of Petrology, an Introduction to the Science of Rocks. *London: Methuen.*

TYRRELL, G. W., 1929. The Principles of Petrology, an Introduction to the Science of Rocks, 2nd edn. *London: Methuen.*

USSHER, W. A. E., REID, C., FLETT, J. S. & MACALISTER, D. A., 1913. The Geology of the Country around Newton Abbot. *Memoir of the Geological Survey of England and Wales.* Explanation of sheet 339.

VAN HISE, C. R., 1904. A Treatise on Metamorphism. *US Geological Survey Monograph*, **47**.

VANNAY, J.-C. & GRASEMANN, B., 2001. Himalayan inverted metamorphism and syn-convergence extension as a consequence of a general shear extrusion. *Geological Magazine*, **138**, 253–76.

VAN'T HOFF, J. H., 1905. Zur Bildung der ozeanischen Salzablagerungen. *Braunschweig: Wieweg und Sohn.*

VASARI, G., 1568. Le vite dei più excellenti pittori, sculptori e architetti, 2nd edn, part 1–3. Reprinted in Rome by *Newton & Compton.*

VÄYRYNEN, H., 1938. Notes on the geology of Karelia and Onega region in the summer of 1937. *Bulletin de la Commission géologique de Finlande, Comptes rendus*, **123**, 65–80.

VENDL, M., 1929. Die Geologie der Umgebung von Sopron. 1. Die kristallinen Schiefer. *Bányamérnöki és Erdömérnöki Föiskola Bányászati és Kohászati Osztály Közleményei*, **1**, 225–91.

VERNON, R. H., 2004. A Practical Guide to Rock Microstructure. *Cambridge University Press.*

VIOLA, C., 1892. Nota preliminare sulla regione dei gabbri e delle serpentine nell'alta valle del Sinni in Basilicata. *Bollettino del Regio Comitato Geologico d'Italia*, ser. 3, **4**, fasc. 2, 105–25.

VIRLET D'AOUST, T., 1844. Sur les filons en général et le rôle qu'ils paraissent avoir joué dans l'opération du métamorphisme. Note sur les roches d'imbibition, etc. *Bulletin de la Société géologique de France*, 2e sér., **1**, 825–55.

VIRLET D'AOUST, T., 1847. Observations sur le métamorphisme normal et la probabilité de la non-existence de véritables roches primitives à la surface du globe. *Bulletin de la Société géologique de France*, 2e sér., **4**, 498–505.

VOGEL, D. E., 1967. Petrology of an eclogite and pyrigarnite-bearing polymetamorphic rock complex at Cabo Ortegal, NW Spain. *Leidse geologische Mededelingen*, **40**, 121–213.

VOGT, J. H. L., 1905. Über anchi-eutektische und anchi-monomineralische Eruptivgesteine. *Norsk geologisk Tidsskrift*, **1**, (2), 1–33.

VOGT, T., 1927. Sulitjelmafeltets geologi og petrografi. *Norges Geologisk Undersökkelse*, **121**, 1–560.

VUAGNAT, M., 1952. Sur une structure nouvelle observée dans les roches vertes du Montgenèvre (Hautes-Alpes). *Archives des Sciences, Genève*, **5**, 191–3.

VUAGNAT, M. & JAFFÉ, F., 1953. Observations sur quelques ophisphérites de la région des Gets (Haute-Savoie). *Archives des Sciences, Genève*, **6**, 413.

VUAGNAT, M. & PUSZTASZERI, L., 1964. Ophisphérites et rodingites dans diverses serpentinites des Alpes. *Schweizerische mineralogische und petrographische Mitteilungen*, **44**, 12–15.

WAGER, R., 1938. Studien im Gneisgebirge des Schwarzwaldes. 9. Über die Kinzigitgneise von Schenkenzell und die Syenite vom Typ Erzenbach. *Sitzungsberichte der Heidelberger Akademie der Wissenschaften, Mathematisch-naturwissenschaftliche Klasse*, Abhandlung **4**, 1–51.

WALKER, T. L., 1902. The geology of the Kalahandi State, Central Provinces. *Memoirs of the Geological Survey of India*, **33**, part 3, 1–22.

WALKER, T. L., 1930. Dalmatianite, the spotted greenstone from Amulet mine, Noranda, Quebec. *University of Toronto Studies, Geological Series*, **29**, 9–12.

WALTHER, J., 1894. Einleitung in die Geologie als historische Wissenschaft: Beobachtungen über die Bildung der Gesteine und ihre organischen Einschlüsse. *Jena: Fischer*.

WARNER, J. L. & SIMONDS, C. H., 1989. Lunar petrology. In D. R. Bowes (ed.), Encyclopedia of Igneous and Metamorphic Petrology. *New York: Van Nostrand Reinhold*, pp. 284–99.

WARNER, J. L., PHINNEY, W. C., BICKEL, C. E. & SIMONDS, C. H., 1977. Feldspathic granulitic impactites and pre-final bombardment lunar evolution. *Proceedings of the 8th Lunar Science Conference*, 2051–66.

WARR, L. N. & RICE, A. H. N., 1994. Interlaboratory standardization and calibration of clay mineral crystallinity and crystallite size data. *Journal of Metamorphic Geology*, **12**, 141–52.

WASHINGTON, H. S., 1922a. The jade of the Tuxtla statuette. *Proceedings of the United States Natural Museum*, **60**, art. 14, No. 2409, 1–12.

WASHINGTON, H. S., 1922b. The jades of Middle America. *Proceedings of the National Academy of Sciences of the USA*, **8**, 319–26.

WATERS, A. C. & CAMPBELL, C. D., 1935. Mylonites from the San Andreas Fault Zone. *American Journal of Science*, **29**, 473–503.

WATERS, A. C. & KRAUSKOPF, K., 1941. Protoclastic border of the Colville batholith. *Geological Society of America Bulletin*, **52**, 1355–416.

WEBSTER, R., 1975. Gems: Their Sources, Descriptions and Identification. *London: Newnes-Butterworths*.

WEINSCHENK, E., 1902a. Vergleichende Studien über den Contactmetamorphismus. *Zeitschrift der deutschen geologischen Gesellschaft*, **54**, 441–79.

WEINSCHENK, E., 1902b. Allgemeine Gesteinskunde als Grundlage der Geologie, vol. 6. *Freiburg-im-Bresgau: Herdersche*.

WEISS, S. C., 1803. Über die Gebirgsart des sächsischen Erzgebirges, welche unter dem Namen Weissstein neuerlich bekannt gemacht worden ist. *Neue Schriften der Gesellschaft naturforschender Freunde zu Berlin, Neue Schriften*, **4**, 342–66.

WERNER, A. G., 1780. Cronstedt's Versuch einer Mineralogie. *Freiberg, Leipzig: Vogel*.

WERNER, A. G., 1786. Kurze Klassifikation und Beschreibung der verschiedenen Gebirgsarten. *Abhandlungen der Böhmischen Gesellschaft der Wissenschaften*, 272–97. Also published as a pamphlet in 1787, *Dresden*.

WERNER, A. G., 1791–2. Ausführliches und systematisches Verzeichnis des Mineralien-Kabinetts des Pabst von Ohain. 2 vols. *Freyberg*.

WHITE, S., 1976. The effects of strain on the microstructures, fabrics, and deformation mechanisms in quartzites. *Transactions of the Royal Society of London*, ser. A, **283**, 69–86.

WILLEMS, H. W. V., 1934. Astridiet, een chroomrijk gesteente van Nieuw-Guinea. *De Ingenieur in Nederlandisch-Indië*, Bandoeng, **1** (7), 120–1.

WILLIAMS, G. H., 1890. The greenstone schist areas of the Menominee and Marquette regions of Michigan. *US Geological Survey Bulletin*, **62**, 1–217.

WILLIAMS, H. & SMYTH, W. R., 1973. Metamorphic aureoles beneath ophiolite suites and Alpine peridotites: tectonic implications with West Newfoundland examples. *American Journal of Science*, **173**, 594–621.

WILLIAMS, H., TURNER, F. J. & GILBERT, C. M., 1954. Petrography. An Introduction to the Study of Rocks in Thin Sections. *San Francisco: Freeman*.

Williams, H., Turner, F. J. & Gilbert, C. M., 1982. Petrography. An Introduction to the Study of Rocks in Thin Sections, 2nd edn. *San Francisco: Freeman.*

Wilson, G., 1953. Mullion and Redding structures in the Moine Series of Scotland. *Proceedings of the Geologists' Association*, **64**, 118–51.

Wilson, G., 1961. The tectonic significance of small scale structures, and their importance to the geologist in the field. *Annales de la Société géologique de Belgique*, **84**, 423–548.

Winchell, A. N., 1900. Mineralogical and petrographic study of the gabbroid rocks of Minnesota, and more particularly, of the plagioclasytes. *American Geologist*, **26**, ch. 4–7, 261–306, and ch. 8–10, 348–88.

Winkler, H. G. F., 1967. Petrogenesis of Metamorphic Rocks, 2nd edn. *New York: Springer.*

Winkler, H. G. F., 1970. Abolition of metamorphic facies, introduction of metamorphic stage, and of a classification based on isograds in common rocks. *Neues Jahrbuch für Mineralogie*, Monatsheft, **5**, 189–248.

Winkler, H. G. F., 1974. Petrogenesis of Metamorphic Rocks, 3rd edn. *New York, Heidelberg: Springer.*

Winkler, H. G. F. & Sen, S. K., 1973. Nomenclature of granulites and other high grade metamorphic rocks. *Neues Jahrbuch für Mineralogie*, Monatsheft, 393–402.

Wiseman, J. D. H., 1934. The central and southwest Highland epidiorites. A study in progressive metamorphism. *Quarterly Journal of the Geological Society of London*, **90**, 354–417.

Woodward, J., 1728. A Catalogue of English Fossils, vols. 2–3, *London: Fayram, Senex, Osborn and Longman.*

Yachevskii, L., 1909. Chrysotile deposits in the Bis-Tag ridge in the Minusinsk district of the Yenisei territory (in Russian). *Geologicheskaya isledovaniya v 'Zolotonosnykh Oblastyakh' Sibiri, Sankt-Petersburg*, **8**, 31–50.

Yardley, B. W. D., 1989. An introduction to metamorphic petrology. *Longman Earth Science Series, Singapore.*

Yoder, H. S., 1952. The MgO-Al_2O_3-SiO_2-H_2O system and the related metamorphic facies. *American Journal of Science*, **250a**, 569–627.

Yund, R. A. & Tullis, J., 1991. Composition changes of minerals associated with dynamic recrystallisation. *Contributions to Mineralogy and Petrology*, **108**, 346–55.

Zanzucchi, G. (coord.), 1994. Appennino ligure-emiliano. Guide Geologiche Regionali. *Società Geologica Italiana, BE-MA Editrice.*

Zaraiskii, G. P., 1989. Zonation and Conditions of Metasomatic Rock Formation (in Russian). *Moscow: Nauka Press.*

Zealley, A. E. V., 1919. On certain felsitic rocks, hitherto called 'banded ironstone', in the ancient schists around Gatooma, Rhodesia. *Transactions of the Geological Society of South Africa*, **21**, 43–52.

Zellmer, H., 1997. Über den Zusammenhang zwischen Vulkanismus und Kieselschieferbildung im Harz. *Zeitschrift der deutschen Gesellschaft für Geowissenschaften*, **148**, 457–77.

Zharikov, V. A., 1959. Geology and metasomatic processes at the skarn base-metal deposits of the western Karamazar (in Russian). *Transactions of the IGEM AN SSSR*, 14.

Zharikov, V. A., 1968. Skarn deposits. In Genesis of Endogenic Ore Deposits (in Russian). *Moscow: Nedra Publications*, pp. 220–302.

Zharikov, V. A., 1970. Skarns. *International Geological Review*, **12**, 541–59.

Zharikov, V. A., 1976. Grounds of Physicochemical Petrology (in Russian). *Moscow: MGU Press.*

Zharikov, V. A., 1991. Types of skarns, formation and ore mineralization. In Skarns, their Genesis and Metallogeny. *Athens: Theophrastus Press*, pp. 455–66.

Zharikov, V. A. & Omel'yanenko, B. I., 1978. Classification of metasomatites (in Russian). In Metasomatism and Ore Deposition, *Moscow: Nauka Press*, pp. 9–28.

Zharikov, V. A., Omel'yanenko, B. I. & Pertsev, N. N., 1992. Classification, systematics and nomenclature of metasomatic rocks. In Classification and Nomenclature of Metamorphic Rocks. *Novosibirsk: Nauka Press*, pp. 123–9.

Zharikov, V. A., Rusinov, V. L., Marakushev, A. A. *et al.*, 1998. Metasomatism and Metasomatic Rocks (in Russian). *Moscow: Nauchnyi Mir.*

Zhdanov, V. V., Rundqvist, D. V. & Baskov, E. A., 1978. The experience of regional metasomatites and their rows systematics (in Russian).

Zapiski Vsesoyuznogo Mineralogicheskogo Obshchestva, **116**, 408–16.

Zincken, J. K. L., 1841. Ueber die Granitränder der Gruppe des Rambergs und der Rosstrappe. *Karsten's und v. Dechen's Archiv*, **15**.

Zincken, J. K. L., 1845. Ueber die Granitränder der Gruppe des Rambergs und der Rosstrappe. *Karsten's und v. Dechen's Archiv*, **19**, 583–6.

Zirkel, F., 1866a. Lehrbuch der Petrographie, vol. 1. *Bonn: Marcus.*

Zirkel, F., 1866b. Lehrbuch der Petrographie, vol. 2. *Bonn: Marcus.*

Zirkel, F., 1873. Die mikroskopische Beschaffenheit der Mineralien und Gesteine. *Leipzig: Engelmann.*

Zirkel, F., 1879. Limurit aus der Vallée de Lesponne. *Neues Jahrbuch für Mineralogie, Geologie und Paläontologie*, part 3, 379–82.

Zirkel, F., 1893. Lehrbuch der Petrographie, 2nd edn, vol. 1. *Leipzig: Engelmann.*

Zirkel, F., 1894a. Lehrbuch der Petrographie, 2nd edn, vol. 2. *Leipzig: Engelmann.*

Zirkel, F., 1894b. Lehrbuch der Petrographie, 2nd edn, vol. 3. *Leipzig: Engelmann.*

Appendix A

A.1 Members of the Subcommission: past and present

Árkai, P. (Hungary)
Brodie, K. (UK)
Bryhni, I. (Norway)
Callegari, E. (Italy)
Coleman, R. G. (USA)
Coutinho, J. M. V. (Brazil)
Davis, E. (Greece)
Desmons, J. (France)
Dudek, A. (Czech Republic)
Fettes, D. J. (UK)
Frey, M. (Switzerland)
Gorbatschev, R. (Sweden)
Harte, B. (UK)
Hashimoto, M. (Japan)
Hatch, N. (USA)
Hollister, L. S. (USA)
Karamata, S. (Serbia)
Kräutner, H.-G. (Germany)
Kretz, R. (Canada)
Liou, J. G. (USA)
Masch, L. (Germany)
Meyer-Marsilius, H.-J. (Switzerland)
Peinado, M. (Spain)
Pertsev, N. (Russia)
Rosen, O. M. (Russia)
Sassi, F. P. (Italy)
Schmid, R. (Switzerland)
Sen, S. K. (India)
Shen, Qi-Han (China)
Siivola, J. (Finland)
Smulikowski, W. (Poland)
Stöffler, D. (Germany)
Teruggi, M. E. (Argentina)
Thompson, P. (Canada)
Wimmenauer, W. (Germany)
Xu, Shu-Tong (China)

A.2 Leaders of Study Groups

Types of metamorphism: Smulikowski, W. (Poland)
Structural terms: Brodie, K. (UK)
High-pressure rocks: Desmons, J. (France)
Low-grade rocks: Árkai, P. (Hungary)
Migmatites: Wimmenauer, W. (Germany)
Metacarbonate rocks: Rosen, O. (Russia)
Granulites, amphibolites: Dudek, A. (Czech Republic) and S. Sen (India)
Metasomatic rocks: Pertsev, N. (Russia)
Contact metamorphic rocks: Callegari, E. (Italy)
Impactites: Stöffler, D. (Germany)
Mineral abbreviations: Siivola, J. (Finland)

A.3 Members of the Working Group

Aires-Barros, L. (Portugal)
Amstutz, G. C. (Germany)
Andreasson, P. (Sweden)
Ashworth, J. (UK)
Baltatzis, E. (Greece)
Banno, S. (Japan)
Barbey, P. (France)
Bard, J.-P. (France)
Barnicoat, A. (UK)
Barr, D. (UK)
Berza, T. (Romania)
Best, M. G. (USA)
Blenkinsop, T. G. (Zimbabwe)
Brew, D. A. (USA)
Brime, C. (Spain)
Brothers, R. N. (New Zealand)
Coombs, D. S. (New Zealand)
Cortesogno, L. (Italy)
Crawford, M. L. (USA)
Dimitrescu, R. (Romania)
Dobretsov, N. L. (Russia)
Don, J. (Poland)
Droop, G. (UK)
Efremova, S. V. (Russia)
Essene, E. (USA)
Evans, B. W. (USA)
Fediuk, F. (Czech Republic)
Frost, R. (USA)
Fry, N. (UK)
Fuck, R. (Brazil)
Garcia de Figuerola, L. C. (Spain)
Ghent, E. D. (Canada)
Giusca, D. (Romania)
Gorkovets, (Russia)
Greenwood, H. J. (Canada)
Guidotti, C. V. (USA)
Hänny, R. (Switzerland)
He, (China)
Heitzmann, P. (Switzerland)
Helmers, H. (Netherlands)
Hinterlechner-Ravnik, A. (Slovenia)
Hoinkes, G. (Austria)
Hudson, N. (UK)

Hyndman, D. W. (USA)
Iancu, V. (Romania)
Janardhan, A. S. (India)
Jung, D. (Germany)
Kano, T. (Japan)
Katagos, (Greece)
Kilmurray, J. O. (Argentina)
Komatsu, M. (Japan)
Kozlowski, (Poland)
Kraft, (Germany)
Kruhl, J. H. (Germany)
Kurat, G. (Austria)
Laajoki, K. (Finland)
Lal, R. K. (India)
Lehtinen, M. (Finland)
Lindh, A. (Sweden)
Liu, X. (China)
Logvinenko, (Russia)
Long, B. (Ireland)
Lorenz, W. (Germany)
Majer, V. (Croatia)
Majerowicz, A. (Poland)
Marakushev, A. (Russia)
Matthes, S. (Germany)
Michot, P. (Belgium)
Miyashiro, A. (USA)
Mukhopadhyay, M. (India)
Munoz, M. (Spain)
Oliver, (Canada)
Oliver, R. L. (Australia)
Plyusnina, I. (Russia)
Powell, R. (Australia)
Purtscheller, F. (Austria)
Radulescu, D. (Romania)
Restrepo, J. (Colombia)
Robinson, D. (UK)
Sabine, P. (UK)
Schmid, S. (Switzerland)
Shaw, D. (Canada)
Skjernaa, L. (Denmark)
Smith, D. (France)
Suk, M. (Czech Republic)
Tagiri, M. (Japan)
Thompson, A. Jr (USA)
Toselli, A. J. (Argentina)
Touret, J. L. (Netherlands)
Triboulet, C. (France)
Trow, R. (Brazil)
Trzcienski, W. Jr (Canada)
Tyler, I. (Australia)
Vernon, R. (Australia)
Visonà, D. (Italy)
Vrana, S. (Czech Republic)
Wenk, H. (Switzerland)
Winter, J. (Australia)
Witt, W. K. (Australia)
You, (China)